Reverse Micelles

Biological and Technological Relevance of Amphiphilic Structures in Apolar Media

Reverse Micelles

Biological and Technological Relevance of Amphiphilic Structures in Apolar Media

Edited by

P. L. Luisi

and

B. E. Straub

Institute for Polymers
Zurich, Switzerland

Springer Science+Business Media, LLC

Library of Congress Cataloging in Publication Data

European Science Foundation. Workshop (4th: 1982: Rigi-Kaltbad, Switzerland)
  Reverse micelles.

  "Proceedings of the Fourth European Science Foundation Workshop, held
September 21–October 2, 1982, in Rigi-Kaltbad, Switzerland"—T.p. verso.
  Includes bibliographical references and indexes.
  1. Micelles—Congresses. I. Luisi, P. L. II. Straub, B. E. III. Title.
QD549.E9   1982   574.87′3   83-26958

Proceedings of the Fourth European Science Foundation Workshop,
held September 21–October 2, 1982, in Rigi-Kaltbad, Switzerland

ISBN 978-1-4757-6426-0   ISBN 978-1-4757-6424-6 (eBook)
DOI 10.1007/978-1-4757-6424-6

© 1984 Springer Science+Business Media New York
Originally published by Plenum Press, New York in 1984

PREFACE

There are several reasons for our fascination with micelles. Above all, they represent a form of self-organization of matter. Somehow, at certain critical conditions, a chaotic mixture of surfactant molecules assembles into spatially ordered macromolecular structures. Order is created from disorder, at least at a first, superficial level of analysis. Immediately, questions as to how and why arise. The physical chemists in particular have been asking questions about the thermodynamics and kinetics of these disorder/order transitions.

Micelles can also be seen in a different light. Both normal and reverse micelles contain two distinct regions which are quite different from each other: the shell and the interior, constituting geometrically closed forms. Thus, the biologist can see in them a similarity to cells or simple organelles and the chemist can view them as mobile, tailored-to-size microreactors, where guest molecules can be induced to react. Technologists and engineers see micelle chemistry as part of surfactant chemistry with all its implications for the oil lubrication, cosmetic and food industries.

How do the different researchers interact with each other? The meeting in Rigi-Kaltbad and the corresponding Proceedings arose from the desire to catalyze such interactions. The theme of the meeting was restricted to reverse micelles, first of all in order to reduce the complexity of the endeavor and because no specific meeting has yet been devoted to this subject alone. But also, the discussion of reverse micelles appeared to be particularly timely. In fact, in the last two to three years it has been reported that reverse micelles may be present in membranes and perform there important biological functions. Along a different line of research, it has been reported that enzyme-containing reverse micelles may offer novel tools for biotechnology. They have also been utilized to prepare colloidal forms of metal catalysts and to form water soluble polymers.

Most of these aspects are covered in this Proceedings volume. Indeed, the reader will find a most comprehensive picture of the state of the art, the variety of the disciplines, and the variety of the scientists' backgrounds. Also the variety of "languages" used, will be apparent. In this regard, no extensive changes in the terminology used in the reports have been made by the editors. It is our hope that this publication is a first step towards the foundation of a more uniform system of concepts, techniques and terminology.

This meeting would not have been possible without the generous support of the European Science Foundation (ESF), the Swiss National Science Foundation, the Swiss Chemical Industry, the Swiss Biochemical Society and the Swiss Federal Institute of Technology (ETH) in Zürich. Much gratitude goes to Mrs. R. Graber for her valuable and skillful assistance in the organization of the meeting and of this volume. The editors thank the authors for their contributions and for their interest. We are indebted to Dr. V. Rizzo, Dr. H. Jäckle, Dr. A. Pande and Prof. H. Eicke for their indispensable collaboration and for their many helpful suggestions. Finally, we express our heartfelt thanks to Mrs. R. Schulthess, who typed the camera-ready copy of this volume, for her painstaking efforts and cheefulness throughout this seemingly endless project.

P.L. Luisi and B.E. Straub
Institut für Polymere
ETH-Zentrum
CH-8092 Zürich/Switzerland

CONTENTS

METHODOLOGY

BIOLOGICAL RELEVANCE

APPLICATIONS

MICELLES AND REVERSED MICELLES: A HISTORICAL OVERVIEW

Per Stenius

Institute for Surface Chemistry
Box 5607
114 86 Stockholm, Sweden

This is a review of the historical development of
models for micelle formation in aqueous solution, lyotropic liquid crystals, reversed micelles and microemulsions. The current model of micelle formation was suggested about fifty years ago, while the quantitative
development of thermodynamic and kinetic models is
relatively recent. A very important development was
the mapping of complete phase equilibria of surfactant/
water/amphiphilic or non-polar solvent systems that was
started about thirty years ago. At about the same time,
the structure of lyotropic liquid crystals formed in
such systems was clarified. Investigations of surfactant
association in organic solvents in the presence of water
were begun already in the 1930's but so far these
studies have not yielded a comprehensive description of
the structure and properties of the different association structures that may occur. Recent thermodynamic
and spectroscopic studies indicate that it is not possible to assume that well-defined spherical aggregates
are always formed when large amounts of water are incorporated into the solutions. A very important research topic appears to be the structure and dynamics
of the surfactant layer between oil and water domains
in isotropic solutions of surfactant (and co-surfactant),
oil and water.

The literature on micelles and reversed micelles is immense
and a reasonably complete overview would be far beyond the scope

1

of this introductory lecture. In the following are described some experimental results and comprehensive theories that have been of importance to the development of our knowledge about surfactant systems. More complete reviews of the development and present state of our knowledge about the physical chemistry of surfactant systems are given in References 1-16.

Micelles

Unambiguous evidence for the reversible formation of colloidal aggregates by fatty acid soaps was presented by McBain and his co-workers in a series of papers published from 1910 onward (see, for example, References 17-20). Through his studies of the colligative properties and molar conductivities of soap solutions, McBain concluded that part of the soap formed large aggregates, micelles, in aqueous solution. Although the details of his original model have been revised, McBain caught the essential feature of micelle formation in his idea that colloidal particles could be formed by reversible association of ions with equal charges. In 1910-1920 this was a rather bold suggestion, because at that time it was generally thought that colloidal systems were never thermodynamically stable and that their properties depended on the method of preparation.

As early as 1913 Reychler[21] showed that sodium cetylsulfate behaved very similarly to the fatty acid soaps and suggested that the molecules in micelles are arranged in such a way that a hydrophobic core is surrounded by hydrophilic end-groups.

One reason for the opposition to McBain's ideas about reversible association certainly was that the nature of the cooperative force that would hold together a large micellar aggregate of limited size in reversible equilibrium with monomers was incompletely understood. However, it is worth pointing out that already in 1919 W.D. Harkins had concluded that the immiscibility of water and octane was primarily due to the strong interaction between the water molecules. His statement that "while the molecules of water attract each other more than those of octane, on the other hand, the octane molecules attract those of octane no more than they do those of water" does not convey the correct appreciation of the entropy effects involved in hydrophobic bonding; however, it does give a description of the essential driving force behind micelle formation. This was correctly interpreted by Hartley in 1936[1]; our understanding of the details of hydrophobic bonding has, of course, developed extensively since then. A good review is given in Reference 29.

McBain originally suggested[18,20] that two types of micelles
were formed: a large, nearly neutral aggregate consisting of a bi-
molecular layer of soap molecules and a highly charged micelle formed by the association of a small number of soap ions of like charge.
This model was suggested in order to interpret the molar conductivity of the soap. Typical conductivity curves are shown in Figure 1.
It was thought that the rapid decrease in conductivity was due to
the formation of the neutral aggregate, while the subsequent increase was a consequence of an increasing concentration of small,
highly charged aggregates.

It was pointed out by Hartley[1] that this model is thermodynamically unlikely and that it is much more reasonable to assume that
the decrease is due to the adherence of counter ions, which would
partially neutralize the micellar charge; the subsequent increase
would be due to interactions between charged micelles that might
lead to changes in the micellar shape.

In 1927, Grindley and Bury[23,24] showed by the simple application
of the law of mass action that cooperative association would lead to
a sharply defined concentration region in which the concentration of
micelles would begin to rise steeply. This introduced the concept of
"critical micelle concentration", cmc. The concept was suggested almost simultaneously by Ekwall[25] who did not present the quantitative
treatment put forward by Grindley and Bury.

The model for micellar structure that is still generally accepted was described in 1936 by Hartley : a spherical aggregate
(Figure 2) which has a very fluid character and therefore a form
that may be modified locally by the thermal motion of the molecules.
The diameter of the sphere would be approximately twice the molecular
length of the surfactant molecule. The surfactant ions are aggregated with their chains jumbled together so as to minimize the

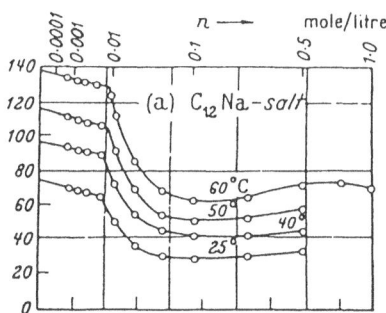

Figure 1. *Molar conductivity of sodium dodecyl sulfate at different
temperatures*[21].

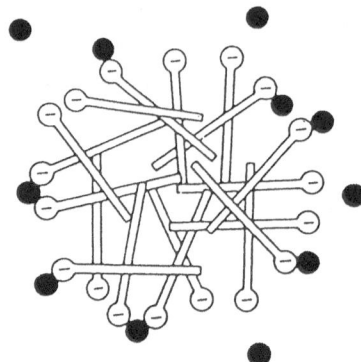

Figure 2. The Hartley micelle, as it was originally depicted.

hydrocarbon/water contact. The driving force for micelle formation
is the strong interaction between water molecules, while the growth
of the aggregates is essentially limited by repulsive interactions
between the head groups.

A very important contribution by Hartley was the realization
that a rapid and reversible formation of micellar aggregates must
imply that the associating molecules are essentially fluid in
character[26]. This was further substantiated by studies of partial
molar volumes of the hydrocarbon chains in micelles which were found
to be very similar to those in liquid hydrocarbons[27].

Little has been added to the essential features of this model
since then. A truly amazing number of experiments have been performed that confirm this structure (or just confirm the existence
of a cmc[30]) and add details about micellar interactions[31-33], pre-
cursors to micellar formation[34-36] and changes in size and shape
as the concentration of micelles increases[37-46]. During the last
two decades however, substantial progress has been made in the
formulation of quantitative theories for micellar equilibria   as
well as in an understanding of the kinetics[55,5.6] of the processes
leading to micelle formation. In combination with spectroscopic[8,9]
and kinetic experiments[8,9,57] these theories have verified and substantiated the highly dynamic character of micellar equilibria. By
now we have a fairly comprehensive picture of the essential mechanisms involved in the formation of micelles in aqueous solution.
The properties of concentrated micellar solutions however, are not
completely understood, especially in the case of non-ionic systems.

Lyotropic Liquid Crystals in Two-Component Systems

In parallel with the early studies of micellar systems,exten-
sive mappings were made by McBain and his co-workers of the phase
equilibria of surfactant/water systems and many phase diagrams wer·

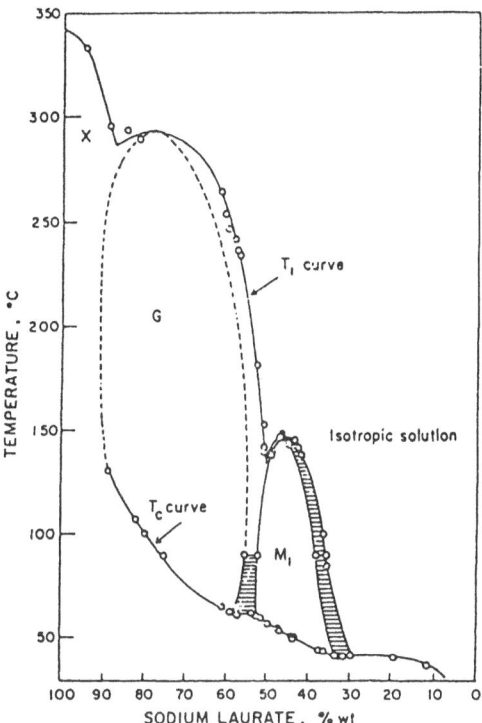

Figure 3. Phase diagram for the sodium laurate-water system.
M = "middle" soap (hexagonal phase); G = "neat"soap(lamellar phase).

published [58,59] . A typical diagram is shown in Figure 3. These studies
showed that liquid crystalline phases were formed in such systems.
Important contributions to the knowledge about such phases were also
made by Winsor [2,37] as well as by Luzzati and his associates in
Strasbourg, whose pioneering X-ray work in the 1960's led to the
clarification of the structure of several of these phases [60-64]
Again however, much of the basic knowledge about the regions where
such phases exist and the realization that the old descriptions
"neat" and "middle" phases really represent thermodynamically stable
single phases goes back to a much earlier date.The first reasonable
complete phase diagrams were published by McBain et al. in 1926 [65]

It is not intended that the structure of lyotropic liquid crystals will be discussed here. It is however worth while to point out
that the association behavior of surfactant systems is so complicated
that it cannot be understood without a detailed determination of
phase equilibria. Failure to realize this has caused much confusion,
particularly in the studies of reversed systems. The detailed investigations of two-component systems have made it possible to
make inferences by analogy about the structures of many of the
phases occurring in three- or four-component systems[12].

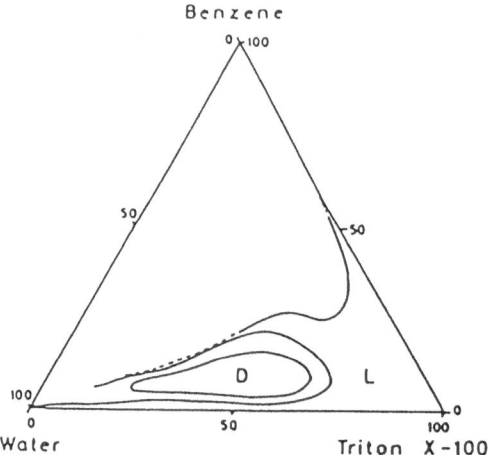

Figure 4. Phase diagram for the Triton X-100/water/hexane system.
Triton X-100 is a nonylphenol polyethylenoxide ether with an average
of 9-10 ethylene oxide monomer units in the hydrophilic chain.D =
liquid crystal, L = solution.

Solubilization

     Solubilization, i.e. the ability of surfactants in solution to
increase the solubility of hydrophobic compounds in water, has been
utilized for as long a time as soap itself. Indeed, in their classic
book on solubilization, Laing et al. [3] cite reports dating from 1846
on solubilization phenomena.

     Solubilization had of course already been studied by many authors
when the picture of micelle structure was first developed and it was
evident from the beginning that solubilizates are incorporated into
the micellar interior.

     In 1936 McBain [66] showed that the solutions formed through the
action of solubilizing agents are thermodynamically stable and exhibit definite,reproducible saturation values.

     One of the very first phase diagrams illustrating solubilization
that also shows a solution which would by now be characterized as a
microemulsion was published by Marsden and McBain in 1948 (Figure
4) [67]. The diagram is incomplete and does not refer to the ionic systems which were usually employed in fundamental studies of surfactant
systems at that time, but it very clearly indicates the complexity
of multi-component surfactant systems.

     Solubilization has been studied very extensively by many authors

although only a few groups (Lumb[68], Winsor[2,37] ,Ekwall[12]) have done
so by a really systematic mapping of multi-component phase equlibria.
In order to demonstrate the structural properties of inverted systems
and to understand their relevance in technical applications and biological systems, it is of interest to review some of the different
types of solubilization behavior[12].   These are illustrated in
Figures 5 to 8.

First we have typical solubilization in micellar systems. One
should make a clear distinction between (i) non-polar solvents and
polarized solvents that are solubilized in the micellar interior or
are located in the vicinity of the polar head groups (Figure 5), (ii)
amphiphilic compounds that are built into the micelle as co-surfactants (Figure 6) and (iii) compounds that associate only with the
polar end-groups. These different locations have been confirmed by
spectroscopic studies, with nmr methods playing an important role
[71-74].   The different locations also produce quite different solubilization capacities. As Figures 5 and 6 show, solubilization in
the micellar interior at low micellar concentration is low, while
solubilization of amphiphilic compounds is quite substantial.

However, it should be observed that it is certainly not always
possible to describe increased solubility in surfactant solutions

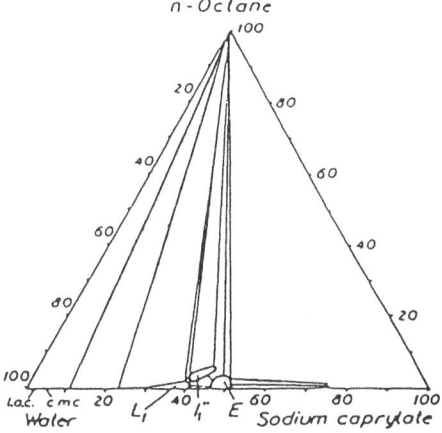

Figure 5. Phase diagram for the sodium caprylate n-octane in
water system, 20ºC.

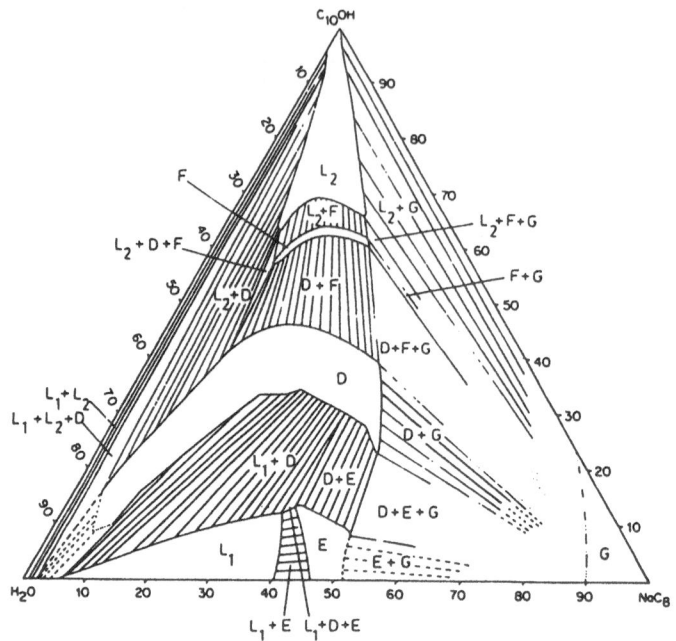

Figure 6. Phase diagram for the sodium caprylate/n-deconal/water system, 20°C

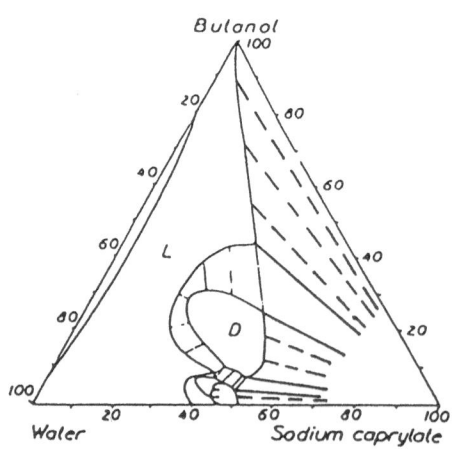

Figure 7. Phase diagram for the sodium caprylate/butan-1-ol/water system, 20°C [12].

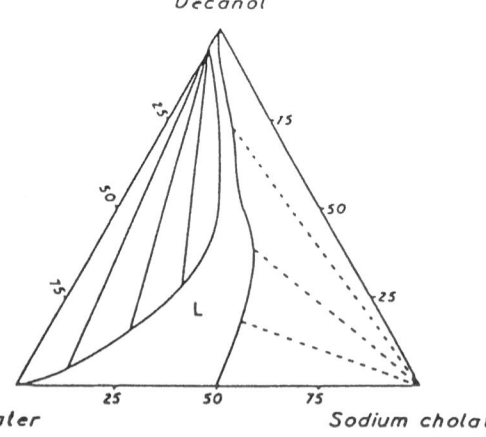

Figure 8. Phase diagram for the sodium cholate/decan-1-ol/water system, 20°C [70].

as solubilization in micelles. Figure 7 shows that the addition of
sodium octanoate results in complete miscibility of butanol with
the aqueous solution; it is certainly not conceivable that this is
solely due to incorporation of butanol into micelles. Another, more
drastic example is shown in Figure 8; here the solubility of decanol
increases to infinity when sodium cholate is added to water. Recent
studies of self-diffusion in systems of the type illustrated in
Figure 7 (sodium dodecyl sulfate/water/various solubilizates) indicate that as more and more solubilizate is added, there is a transition from well-defined closed micellar aggregates to smaller aggregates[75].

Surfactants in Organic Solvents

I have described here the progressive development of our
knowledge about surfactant systems, starting from the properties of
aqueous surfactant solutions. The systems composed by surfactant/
water/apolar solvent, in which the solvent is the dominant component, can be seen as a logical extension of the work on solubilization. Thus, it is obvious that the determination of phase equilibria
must also form the basis of a proper understanding of such multi-
component systems. This, of course, may become a cumbersome task,
particularly because many of the most interesting systems involve
at least four components: water, oil, surfactant and co-surfactant.
In recent years, nevertheless, several such systems have been investigated in detail[12,69,76-89] and equilibria involving isotropic
solutions, liquid crystals and solids have been unambiguously defined according to the conditions set out by Gibbs' phase rule. Once
we know the phase equilibria, we can start discussing the possible
structure of the various phases and to formulate molecular models
for the interpretation of these equilibria and of the aggregation
processes in the isotropic solutions. It is only after we have a
clear picture of the validity of such models that we can introduce
unambiguous descriptions such as micellar solutions, reversed micellar solutions or microemulsions.

For the obvious reason that there is such a multitude of different possibilities, research on surfactants in apolar media has
been much more diverse than the work on aqueous solutions. It is
difficult to form a comprehensive picture of the development of the
concepts of aggregate structures in such systems.

Early studies of surfactant association and solubility in a
large variety of solvents were - in many cases quite successfully -
interpreted with the theory of regular solutions developed by
Hildebrand and Scott. This subject was discussed in some detail in
a well-known paper by Little and Singleterry[90].

However, the fact that surfactants do associate into small
aggregates in polar solvents, was established much earlier, for
example, with studies of alkylammonium salts by Batson and Kraus[91]
in 1934 and alkylamine soaps by Hoerr and Ralston[92] in 1942. In 1955
Singleterry published a list of micelle-forming surfactants in non-
polar solvents[93]; the conclusion was reached quite early that when
both cations and anions are large organic molecules with high solubility in the solvent, the associated ion pair is too soluble to
promote micelle formation. However, if one of the ions is a small
inorganic ion, the salt would be sufficiently insoluble to form
micellar aggregates, which most early investigations showed to be
quite small (aggregation numbers from 4 to 10). Fowkes[97] also referred to Lewis' acid-base theory for an explanation of these interactions and gave several examples of the use of this theory for a
qualitative understanding of the effect of polar solvents. The next
step to suggest that such aggregates incorporate water and form
roughly spherical aggregates was very obvious, and thus already in
1954 Winsor published[2] a scheme clearly showing all the different
types of aggregates formed in surfactant systems (Figure 9). The
role played by the water in the stabilization of the reversed micellar aggregates has been much discussed, in particular in connection with studies of Aerosol OT (AOT, sodium bis(2-ehtylhexyl)sulfo-
succinate)[94,95]. As was emphasized by Ekwall and his co-workers[12,99],
the phase equilibria in many cases give direct indications that a
minimum amount of water is required to obtain large solubilities of

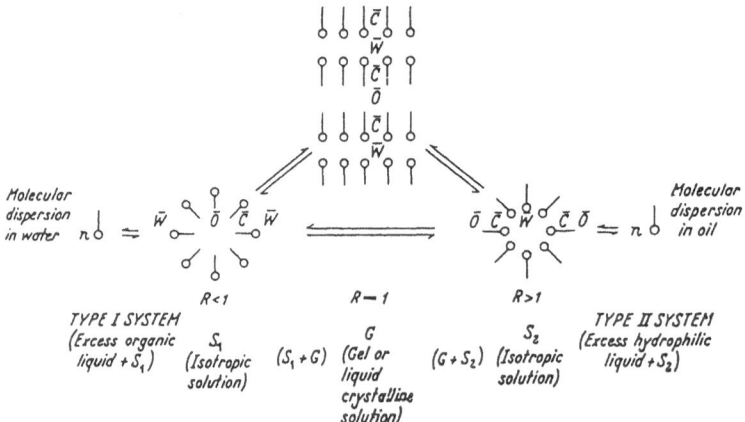

Figure 9. *Intermicellar equilibrium and associated phase changes
shown by amphiphilic systems according to Winsor*[2]. *This picture was
later extended to include reversed and normal hexagonal phases*[37].

either an ionic surfactant in an amphiphilic solvent or of aggregates of the surfactant and the amphiphile in non-polar solvents.

Microemulsions

It is tempting to suggest that as more water is added, the reversed micelles will always swell to accommodate it; the result
would be an aqueous droplet surrounded by a surfactant membrane.
This model was suggested in 1942 by Hoar and Schulman[100]. In the
system water/potassium oleate/pentanol/benzene they observed the
formation of clear isotropic solutions with large amounts of benzene and water and relatively small amounts of oleate and pentanol.
The condition was that the weight ratio oleate/pentanol was kept
within certain rather narrow limits. To explain this phenomenon,
Schulman proposed that an "oleopathic hydromicelle" was formed with
a size that would be limited by the requirement that all of the surfactant mixture should be adsorbed at the oil/water interface. The
1942 model (Figure 10) suggests a very small emulsion droplet but
it was not until considerably later[101] that Schulman et al actually
used the term "microemulsions". In the original paper[100] Schulman
refers to this aggregate as a "reversed micelle, the analog to the
hydrophilic swollen soap ionic micelle containing enclosed oil",
but at the same time he speaks of a stable oil-in-water dispersion.
Later Friberg et al[102] showed that the compositions described by
Schulman in Reference 100 do fall within a thermodynamically stable
solution region in the four-component phase diagram.

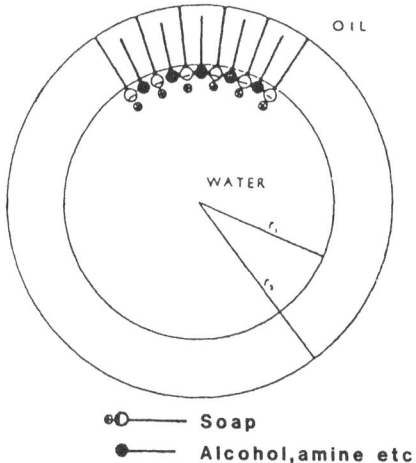

Figure 10.   "The oleapathic hydro-micelle" according to Schulman[100].

An additional source of confusion certainly is the fact that
already in the early thirties optically transparent solutions of
oil, water and surfactants were used in technical products[103]. In
these systems, of course, the distinction between thermodynamic
stability and stability for practical purposes was not a matter of
great concern. Thus, many practical systems that have been called
microemulsions are kinetically stabilized emulsions with small
droplets.

A very important contribution of Schulman[104] was the realization that the requirement for stabilization of microemulsions was
a low solubility of the surfactant (or surfactant mixture) in the
oil as well as in the water phase. It can be shown by very simple
arguments that only in such cases can one expect that adsorption
will lead to zero interfacial energy. If the surfactant forms micelles in aqueous solution or reversed micelles in oil, the state
of zero interfacial energy will represent a state of higher total
free energy. It is indeed borne out by many experiments that typical solutions that contain large amounts of water and oil form
three-phase equilibria with almost pure water or almost pure oil.
Some typical situations are illustrated in Figure 11-13.

In the last two decades studies of many phase equilibria[12,76-78]
have shown that in a three-phase equilibrium with oil and water, one
may have either a liquid crystalline phase (which in the majority of
cases is lamellar) (Figure 13) or a solution containing large amounts

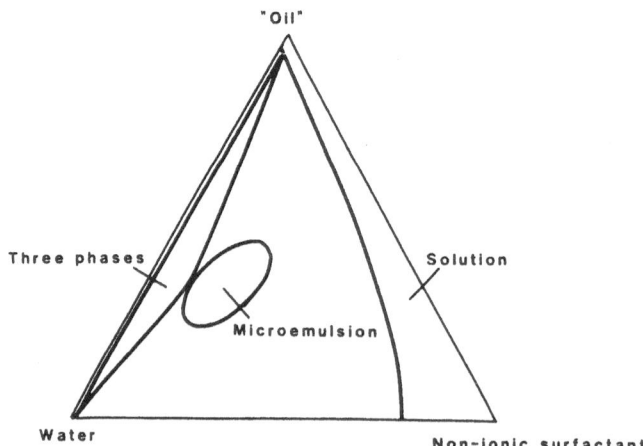

Figure 11. *(Schematic) Equilibrium between aqueous solution, oil
phase and microemulsion in typical three-component system of water,
oil and surfactant at the phase inversion temperature. Liquid
crystalline areas not shown.*

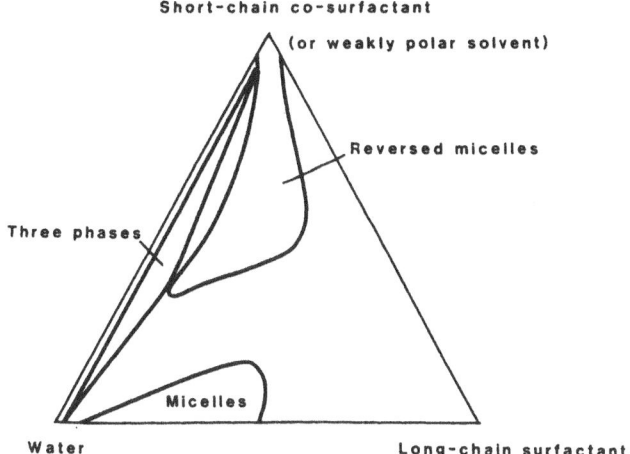

Figure 12. *(Schematic) Equilibrium between aqueous solution, dilute organic solution and concentrated organic solution in system consisting of water, long-chain surfactant and short-chain co-surfactant. Considerable amounts of completely non-polar solvent may usually be dissolved into the organic solution; the distribution of the co-surfactant between oil and surfactant membrane then becomes very important*[106]

of both oil and water (Figures 11,12). A phenomenological rule that clearly emerges from these studies is that if the chains of the surfactant or co-surfactant are able to form an "ordered layer", the third phase is a liquid crystal. The tendency to form such "ordered" layers can be decreased by choosing surfactants and co-surfactants with widely different sizes of the hydrocarbon moiety, by choosing the temperature so that there is a balance between hydrophilic and hydrophobic properties of the surfactant[105], or by choosing the oil in such a way that the co-surfactant is distributed towards the interface[106]. The result of such "disturbances" is a tendency to form oil/water solutions. Thus, the qualitative requirements for the formation of highly concentrated, stable oil/water mixtures are

(i) low equilibrium concentrations of the surfactant or surfactant/ co-surfactant mixture in oil and water; this reduces interfacial tensions to very low values, and

(ii) high fluidity and flexibility of the surfactant membrane.

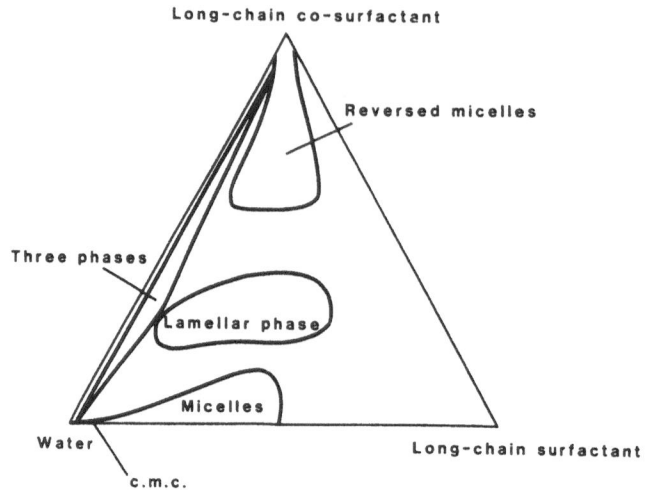

Figure 13. *(Schematic) Equilibrium between aqueous solution, organic solution on lamellar phase in a system consisting of water, long-chain surfactant and long-chain co-surfactant.*

These qualitative conclusions from phase equilibria have been confirmed by the nmr and esr determinations of order parameters and correlation times[76-107]. Recent diffusion measurements by Lindman, Stilbs and co-workers[108,109] have shown that on the timescale probed by nmr in many cases well-defined aggregates are not formed in typical microemulsions. It is obvious that even in aqueous solutions, transitions between this situation and typical micelles may occur[110,111].

The problem is that these principles are very difficult to quantify. First of all, we really need more experimental information on solution structure. Promising calculations of phase equilibria, based on electrostatic interactions between the surfactant bilayers in surfactant systems have recently been published[112,113]. But we also require a much better theoretical understanding of the conditions leading to increased flexibility of the surfactant membrane. Several such attempts have been published[114,118]. a stimulating review has been given by de Gennes and Taupin[119].

The "biological and technological relevance of reversed micelles and other amphiphilic structures in apolar media" is very obvious and is certainly the driving force that has produced the massive literature on amphiphilic systems. Fundamental research on

reversed systems certainly concerns the really most important unresolved problems in our knowledge of surfactant systems.

REFERENCES

1. G.S. Hartley, "Aqueous Solution of Paraffin – Chain Salts",
   Hermann et Çie, Paris (1936).
2. P.A. Winsor, "Solvent Properties of Amphiphilic Compounds",
   Butterworths, London (1954).
3. M.E. Laing, M.E.L. McBain and E. Hutchinson, "Solubilization
   and Related Phenomena", Academic Press, New York (1955).
4. K. Shinoda, T. Nakagawa, B. Tamamushi and T. Isemura,
   "Colloidal Surfactants", Academic Press, New York (1963).
5. M.J. Schick, ed., "Nonionic Surfactants", Marcel Dekker,
   New York (1966).
6. K. Shinoda, ed., "Solvent Properties of Surfactant Solutions",
   Marcel Dekker, New York (1967).
7. E. Jungermann, ed., "Cationic Surfactants", Marcel Dekker,
   New York (1975).
8. B. Lindman and H. Wennerström, Topics in Current Chemistry,
   87 : 1 (1980).
9. H. Wennerström and B. Lindman, Phys. Rep., 52 : 1 (1979).
10. P. Ekwall, I. Danielsson and P. Stenius in "Surface Chemistry
    and Colloids", MTP Int. Rev. Sci. Phys. Chem., Ser. I, M. Kerker,
    ed., Vol. 7, p. 97, Butterworths, London (1972).
11. P. Ekwall and P. Stenius in "Surface Chemistry and Colloids"
    MTP Int. Reve. Sci. Phys. Chem., Ser. II, M. Kerker, ed.,
    Vol. 7, p. 215, Butterworths, London (1975).
12. P. Ekwall, Adv. Liq. Cryst., 1 : 1 (1975).
13. G. Tiddy, Phys. Rep., 57 : 1 (1980).
14. H.-F. Eicke, Topics in Current Chemistry, 75 : 85 (1980).
15. K.L. Mittal, ed., "Micellization, Solubilization and Microemul-
    sions", Plenum Press, New York (1977).
16. G.C. Kresheck in "Water, a Comprehensive Treatise", F. Franks,
    ed., Vol. 4, p. 95, Plenum Press, New York (1975).
17. J.W. McBain and A. Taylor, Ber., 43 : 321 (1910).
18. J.W. McBain, Trans. Farad. Soc., 9 : 99 (1913).
19. J.W. McBain and A. Taylor, Kolloid-Z., 12 : 256 (1913).
20. J.W. McBain and C.S. Salmon, J. Amer. Chem. Soc., 42 : 426
    (1920).
21. A. Reychler, Kolloid-Z., 13 : 252 (1913).
22. W.D. Harkins and H.H. Kind, J. Amer. Chem. Soc., 41 : 970
    (1919).
23. J. Grindley and C.R. Burg, J. Chem. Soc., 679 (1929).
24. J. Grindley and C.R. Burg, Phil. Mag., 4 : 841 (1927).
25. P. Ekwall, Acta Acad. Aboensis Math. Phys., IV : 6 (1927).

26. G.S. Hartley, Kolloid-Z., 88 : 22 (1939).
27. G.S. Hartley, Rep. Progr. Chem., 45 : 33 (1948).
28. A. Lottermoser and F. Püschel, Kollioid-Z.,63 : 175 (1933).
29. C. Tanford, "The Hydrophobic Effect", 2nd ed., Wiley, New York (1978).
30. P. Mukerjee and K.J. Mysels, "Critical Micelle Concentrations of Aqueous Surfactant Systems", NSRDS-NBS 36, US Govt. Printing Office, Washington, D.C. (1971).
31. G.S. Hartley, Nature, 163 : 787 (1949).
32. P. Mukerjee, J. Phys. Chem., 76 : 565 (1972).
33. M. Corti and V. Degiorgio, J. Phys. Chem., 85 : 711 (1981).
34. P. Mukerjee, Adv. Colloid Interface Sci., 1 : 241 (1967).
35. I. Danielsson, S. Backlund, J.B. Rosenholm and P. Stenius, Progr. Colloid Polymer Sci., 61 : 1 (1976).
36. J.B. Rosenholm, P. Stenius and I. Danielsson, J. Colloid Interface Sci., 57 : 551 (1976).
37. P.A. Winsor Chem. Rev., 68 : 1 (1968).
38. F. Reiss-Husson and V. Luzzati, J. Phys. Chem., 68 : 3504 (1964).
39. K.G. Götz and K. Heckmann, J. Colloid Sci., 13 : 206 (1958).
40. H.A. Sheraga and J.K. Backus, J. Amer. Chem. Soc., 73 : 5108 (1951).
41. P. Ekwall, L. Mandell and P. Solyom, J. Colloid Interface Sci., 35 : 519 (1971).
42. J. Ulmius and H. Wennerström, J. Magn. Resonance, 28 : 309 (1977).
43. U. Henriksson, J.C. Eriksson and L. Ödberg, J. Phys. Chem., 81 : 76 (1977).
44. G. Lindblom, B. Lindman and L. Mandell, J. Colloid Interface Sci., 42 : 400 (1973).
45. L. Johnsson, G. Lindblom and B. Nordén, Chem. Phys. Lett., 39 : 128 (1976).
46. P. Mukerjee, J. Phys. Chem., 76 : 565 (1972).
47. D.C. Poland and H.A. Sheraga, J. Colloid Interface Sci., 21 : 273 (1966).
48. C. Tanford in "Micellization, Solubilization and Microemulsions", K.L. Mittal, ed., Vol. 1, p. 119, Plenum Press, New York (1977).
49. C. Tanford, J. Phys. Chem., 78 : 2469 (1974).
50. R. Nagarajan and E. Ruckenstein, J. Colloid Interface Sci., 60 : 221 (1979).
51. J.N. Israelachvili, J.J. Mitchell and B.N. Ninham, J. Chem. Soc., Faraday II, 72 : 1525 (1976).
52. N.A. Mazer, M.C. Carey and G.B. Benedek in "Micellization, Solubilization and Microemulsions, K.L. Mittal, ed., Vol. 1, p. 359, Plenum Press, New York (1977).

53. F. Eriksson, J.C. Eriksson and P. Stenius, Colloids Surfaces, 3 : 339 (1981).
54. G. Gunnarsson, B. Jönsson and H. Wennerström, J. Phys. Chem., 13 : 3114 (1980).
55. E.A.G. Aniansson, Ber. Bunsenges, 82 : 981 (1978).
56. E.A.G. Aniansson and S.N. Wall, J. Phys. Chem., 79 : 857 (1975).
57. E.A.G. Aniansson, S.N. Wall, H. Hoffman, J. Kielmann, W. Ulbricht, R. Zana, J. Land and C. Tondre, J. Phys. Chem., 80 : 905 (1976).
58. J.W. McBain and W.W. Lee, Oil and Soap, 17 (Febr. 1943).
59. J.W. McBain, C.C. Brock, R.D. Vold and M.J. Vold, J. Amer. Chem. Soc., 60 : 1870 (1938).
60. V. Luzzati and F. Husson, J. Coll. Biol., 12 : 207 (1962).
61. V. Luzzati, H. Mustacchi and A.E. Skoulios, Nature, 180 : 600 (1957).
62. V. Luzzati, H. Mustacchi, A.E. Skoulios and F. Husson, Acta Cryst., 13 : 660 (1960).
63. V. Luzzati and F. Reiss-Husson, Nature, 210 : 1351 (1966).
64. V. Luzzati, A. Tardieu and R. Gulik-Krzywicki, Nature, 217 : 1028 (1968).
65. J.W. McBain and M.C. Field, J. Phys. Chem., 30 : 1454 (1926).
66. J.W. McBain and M.E.L. McBain, J. Amer. Chem. Soc., 58 : 2610 (1936).
67. S.S. Marsden and J.W. McBain, J. Phys. Chem., 52 : 110 (1948).
68. E.C. Lumb, Trans. Farad. Soc., 47 : 1049 (1951); see also Reference 2.
69. R. Friman, I. Danielsson and P. Stenius, J. Colloid Interface Sci., 86 : 501 (1982).
70. K. Fontell, J. Colloid Interface Sci., 43 : 156 (1973).
71. J.C. Eriksson and G. Gillberg, Acta Chem. Scand., 20 : 2019 (1966).
72. J. Ulmius, J. Colloid Interface Sci., 65 : 88 (1978).
73. P. Mukerjee and J.R. Cardinal, J. Phys. Chem., 82 : 1620 (1978).
74. G. Lindblom, B. Lindman and L. Mandell, J. Colloid Interface Sci., 42 : 400 (1973).
75. P. Stilbs, J. Colloid Interface Sci., 87 : 385 (1982).
76. A.M. Bellocq, J. Biais, B. Clin, P. Lalanne and B. Lemanceau, J. Colloid Interface Sci., 70 : 524 (1979).
77. J. Biais, B. Clin, P. Lalanne and B. Lemanceau, J. Chim. Phys., 74 : 1197 (1977).
78. P. Lalanne, J. Biais, B. Clin, A.M. Bellocq and B. Lemanceau, J. Colloid Interface Sci., 74 : 311 (1980).
79. T.M. Kathopoulis, Thesis, USTL, Montpellier (1979).
80. I. Danielsson, M.R. Hakala and M. Jorpes-Friman, in "Solution Chemistry of Surfactants", K.L. Mittal, ed., Vol. 2, p. 659, Plenum Press, New York (1979).

81. S. Friberg and I. Buraczewska, Progr. Colloid Polymer Sci., 63 : 1 (1978).
82. E. Sjöblom and S. Friberg, J. Colloid Interface Sci., 67 : 16 (1978).
83. K. Shinoda and S. Friberg, Adv. Colloid Interface Sci., 4 : 281 (1975).
84. H. Saito and K. Shinoda, J. Colloid Interface Sci., 32 : 647 (1970).
85. K. Shinoda, H. Kunieda and N. Obi, J. Colloid Interface Sci., 80 : 305 (1981).
86. J.C. Lang and R.D. Morgan, J. Chem. Phys., 73 : 5849 (1980).
87. P. Ekwall, L. Mandell and K. Fontell, J. Colloid Interface Sci., 33 : 215 (1970).
88. H.F. Eicke, Pure Appl. Chem., 52 : 1349 (1980).
89. M. Zulauf and H.F. Eicke, J. Phys. Chem., 83 : 480 (1979).
90. R.C. Little and C.R. Singleterry, J. Phys. Chem., 68 : 3453 (1964).
91. F.M. Batson and C.A. Kraus, J. Amer. Chem. Soc., 65 : 2017 (1934).
92. C.W. Hoerr and A.W. Ralston, J. Amer. Chem. Soc., 64 : 2824 (1942).
93. C.R. Singleterry, J. Amer. Oil. Chem. Soc., 32 : 446 (1955).
94. H.F. Eicke, Topics in Current Chemistry, 87 : 86 (1981).
95. K. Kon-No and A. Kitahara, J. Colloid Interface Sci., 35 : 636 (1971).
96. A. Kitahara and K. Kon-No, ACS Symp. Ser., 9 : 225 (1975).
97. F.M. Fowkes, in "Solvent Properties of Surfactant Solutions, K. Shinoda, ed., p. 65, Marcel Dekker, New York (1967).
98. P. Ekwall, I. Danielsson and P. Stenius, In MTP Intern. Rev. Sci. Phys. Chem. Ser. 1, M. Kerker, ed., Vol. 7, p. 97, Butterworths, London (1972).
99. P. Ekwall and P. Stenius, in MTP Intern. Rev. Sci. Phys. Chem. Series 2, M. Kerker, ed. Vol. 7, p. 215, Butterworths, London (1975).
100. T.P. Hoar and J.H. Schulman, Nature, 152 : 102 (1943).
101. W. Stockenius, J.H. Schulman and L.M. Prince, J. Phys. Chem., 63 : 1677 (1959).
102. G. Gillberg, H. Lehtinen and S. Friberg, J. Colloid Interface Sci., 33 : 40 (1970).
103. L.M. Prince in "Microemulsions", L.M. Prince, ed., p. 1, Academic Press, New York (1977).
104. J.H. Schulman and J.B. Montagne, Ann. N.Y. Acad. Sci., 92 : 366 (1961).
105. K. Shinoda and H. Kuneida, in "Microemulsions", L.M. Prince, ed., p. 57, Academic Press, New York (1977).
106. E. Sjöblom and U. Henriksson, in "Surfactants in Solution", K.L. Mittal, ed., Plenum Press, New York (to be published).

107. M. Dvoilaitzky, R. Ober and C. Taupin, C. R. Acad. Sci. Paris, 296 : II : 27 (1979).

108. B. Lindman, P. Stilbs and M.E. Moseley, J. Colloid Interface Sci., 83 : 569 (1981).

109. P.-G. Nilsson and B. Lindman, J. Phys. Chem., 86 : 271 (1982).

110. P. Stilbs, J. Colloid Interface Sci., 87 : 385 (1982).

111. J.B. Rosenholm and P. Stenius, in "Surfactants in Solution", K.L. Mittal, ed., Plenum Press, New York (to be published).

112. G. Gunnarson, B. Jönsson and H. Wennerström, J. Phys. Chem., 84 : 3114 (1980).

113. B. Jönsson, H. Wennerström and B. Halle, J. Phys. Chem., 84 : 2179 (1980).

114. D. Mitchell and B.W. Ninham, J. Chem. Soc. Farad. Trans. II, 77 : 601 (1981).

115. E. Ruckenstein and J.C. Chi, J. Chem. Soc. Farad. Trans. II, 71 : 1690 (1975).

116. Y. Talmon and J. Prager, J. Chem. Phys., 69 : 2984 (1978).

117. J. Overbeek, Farad. Disc., 65 : 7 (1978).

118. W. Helfrich, Phys. Lett. A., 43 : 409 (1973).

119. P.G. De Gennes and C. Taupin, J. Phys. Chem., 86 : 2294 (1982).

REVERSE MICELLES AND AQUEOUS MICROPHASES

H.-F. Eicke and P. Kvita

Institute of Physical Chemistry
University of Basel
4058 Basel, Switzerland

The structure and physical properties of reverse
micelles formed by alkali (and ammonium) bis-(2-
ethylhexyl)sulfosuccinates are reviewed. The
transition to aqueous microphases and phenomena
related to the possible critical state of water-
in-oil microemulsions are also discussed.

Unlike the study of micelles in water, the study of reverse
micelle formation has suffered from the lack of a guiding principle,
such as the 'hydrophobic effect' which is considered to be responsible for the formation of micellar aggregates in aqueous solutions
of surfactants. This fact, together with the rather limited number
of well-defined and easily obtainable surfactants[1], frequently led
people to concentrate their efforts on one particular surfactant
system.

The study presented here is accordingly devoted to the well-
known, and in many respects particularly well-defined, surfactant
Aerosol OT (AOT), i.e. the sodium salt of bis-(2-ethylhexyl)sulfo-
succinate. This surfactant has a counterion that can easily be replaced by other alkali and alkaline earth ions and in this way,
details about the structure and properties of this surfactant can
be inferred. Although this amphiphilic compound has been known for
decades with respect to its special efficiency in reducing the
interfacial free energy and to its solubilizing capacity, more
systematic investigations are rare.

To start with, Figure 1 shows a typical conformation of AOT
from which it might be concluded, according to the considerations

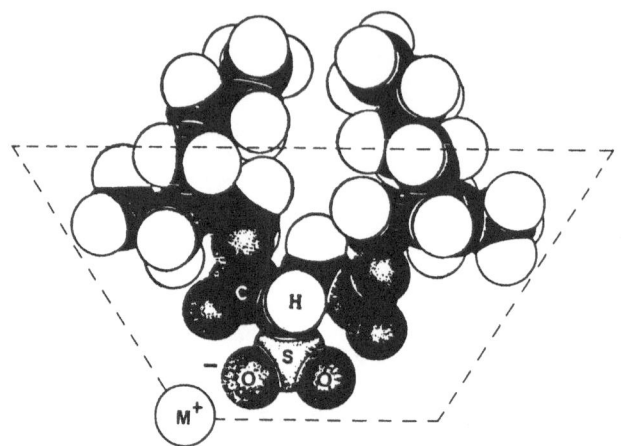

Figure 1. Structure of alkali bis-(2-ethylhexyl)sulfosuccinates.

of Mitchell and Ninham[2] and Israelachvili et al.[3], that reversed
structures, i.e. micelles and aqueous microphases, are favored in
binary surfactant/oil or ternary water/surfactant/oil systems. It is
expected that hydrophobic tails would exert steric constraints if
spherical structures are assumed, so that the size of the spherical
shaped micelles would be independent of the particular counterion.
Zundel[4] reports that IR investigations of polystyrene sulfonates
reveal that there is asymmetric attachment of the counterions with
respect to the sulfonate head group. The polarization of the S-O
bond in the direction of the counterion bond minimizes the energy
of the system. In the case of sulfosuccinates, chelate structures
may form, and IR measurements strongly support this assumption[5].
Such structures could certainly contribute to the pronounced stability of some members of the alkali and alkaline earth salts of
bis-(2-ethylhexyl)sulfosuccinates, and could also explain the poor
solubilizing power of Li-bis-(2-ethylhexyl)sulfosuccinate. Chelate
structures for mono and divalent counterions are shown in Figure
2, where the coordination numbers of the ions are four and six.

Our investigations show that the hydrodynamic radius of the
micelles (as determined from quasi-elastic light scattering experiments) plotted against the counterion charge/radius-ratio q/r_i
(q = ionic charge, r_i = radius of the i-th ion) displays two distinct regions (Figure 3) : above about $q/r_i \simeq 1$, the mean hydrodynamic radii ($\bar{r}_h$) of the micellar aggregates being independent
of the counterions (in agreement with the above mentioned suppo-
sition), while below this particular value a pronounced dependence
of $\bar{r}_h$ on the charge/radius-ratio of the counterions is observed.
We believe this indicates that the micellar aggregates have a non-
spherical shape. Small angle light scattering and vapor pressure

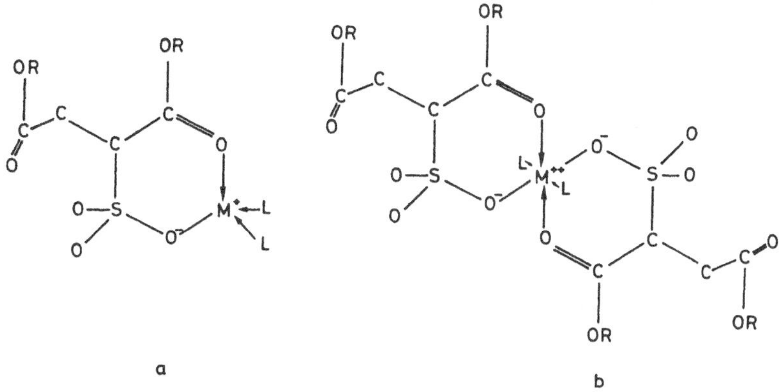

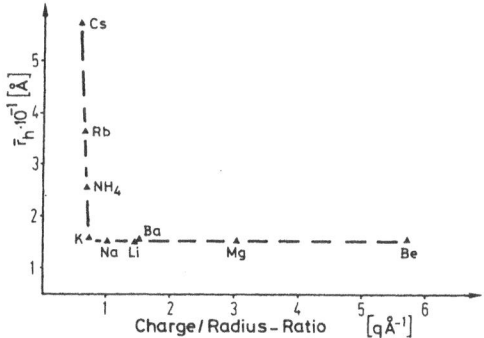

*Figure 2.   Suggested chelate structures of mono- (a) and divalent
(b) (alkali and alkaline earth respectively)ions with bis-(2-ethyl-
hexyl)sulfosuccinate.*

osmometry data obtained as a function of the weighed-in concentration
of surfactant (Figure 4) support the results shown in Figure 3, i.e.
a concentration-dependent aggregation number for q/r_i values below
approximately 1, while larger q/r_i values of the counterions correspond to concentration-independent micellar aggregation numbers.

On the one hand these effects are attributed to the competition
between hydration (i.e. hydrogen-bonding interactions) and chelate
formation, due to the solubilizing capacities of the different bis-
(2-ethylhexyl)sulfosuccinates, and on the other, to the simultaneous
decrease of both phenomena with decreasing q/r_i ratios. The latter
observation is of course not surprising, since both phenomena depend
strongly on the charge/radius ratio of the counterions. Hence, the

*Figure 3. Mean hydrodynamic radii of micellar aggregates against
charge/radius ratio of alkali and alkaline earth counterions(298 K).*

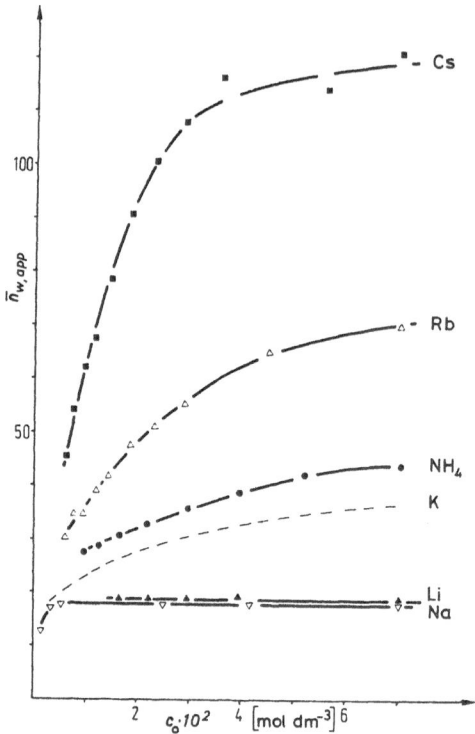

Figure 4. Weight-average aggregation numbers of alkali (and ammonium) bis-(2-ethylhexyl)sulfosuccinates in iso-octane (298 K) plotted against their weighed-in concentrations.

poor solubilization of Lithium-bis-(2-ethylhexyl)sulfosuccinate is due to the strong polarization of the S-O bond by Li^+ and the concomitant formation of a chelate structure in which -C=O and -S-O bonds of the sulfosuccinate participate.

    The chelate formation is believed to prevent the building up of even the first hydration layer in Lithium-bis-(2-ethylhexyl)sulfo-succinate. The tendency of the sodium ion to form a chelate structure is less pronounced, which gives rise to the well-known large solubi-lization capacity for water or aqueous solutions. The potassium ion seems to mark some kind of transition state between the two types of monovalent counterions mentioned above, having neither particular interactions nor pronounced chelate formation. This is supported by the concentration-dependent aggregation numbers (see Figure 4).The experimental data were evaluated according to a procedure of Reering[6] who assumed cylindrically or prolate ellipsoidally shaped aggregates in order to derive weight-average molecular weights and second virial coefficients from apparent molecular weights of the micellar

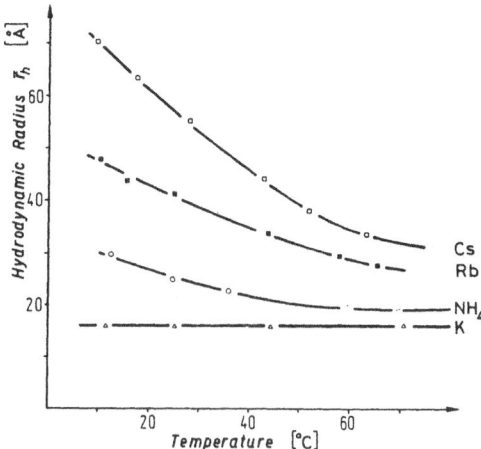

Figure 5. Temperature-dependence of micellar hydrodynamic radii of alkali (and ammonium) bis-(2-ethylhexyl)sulfosuccinates in iso-octane; concentration: 6.7·10⁻² mole dm⁻³.

aggregates. The dotted curve for the potassium salt indicates that the evaluation is questionable in this case.

The weight-average aggregation numbers agree reasonably well with the number – average aggregation numbers obtained from vapor pressure osmometric measurements. The plot (Figure 4) predicts a limiting aggregate size with increasing concentration for counter-ions with a charge/radius ratio smaller than one. If this pronounced concentration dependence is forming monomers, the aggregate sizes (expressed by $\bar{r}_h$) should be temperature dependent. The corresponding experimental results are displayed in Figure 5. The plots indicate that the aggregate sizes approach limiting values with increasing

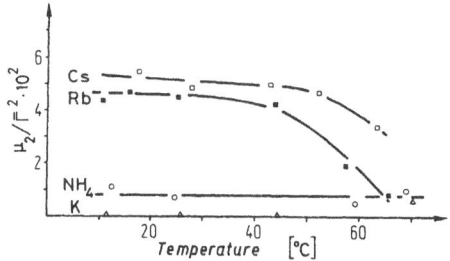

Figure 6. Temperature-dependent polydispersity of alkali (and ammo-nium)bis-(2-ethylhexyl)sulfosuccinates:concentration: 6.7·10⁻²mole dm⁻³.

temperature. The temperature dependence of the polydispersity of these aggregates (Figure 6) parallels these findings.

We were therefore prompted to apply a procedure of Mazer and Benedek[7] which allows the aggregate shapes to be approximately evaluated from their temperature-dependent, apparent micellar masses. A direct determination of aggregate shapes from rotational diffusion coefficients via quasi-elastic light-scattering experiments was not possible, since the particle sizes ranged from $\lambda/20$ to $\lambda/10$. These authors start from the idea that it is in fact possible to relate the intensity ratio $I(T)/I_0$ (where I_0 is the intensity scattered by a reference spherical micelle) to the temperature-dependent micellar mass ratio times a form factor which is also a function of temperature. A connection with the hydrodynamic radii of the particles is obtained by converting the micellar masses into volumes with the aid of partial specific volumes and Perrin's relations for ellipsoidally shaped particles. The results are shown in Figure 7, where the theoretical curves refer to the left-hand ordinate, the experimental plots to the right-hand ordinate. Considering the approximations involved in this procedure, the experimental data seem to fit satisfactorily with the theoretical scattering curve for cylindrically shaped particles.

AQUEOUS MICROPHASES

The extension of the system to three components, i.e. by adding water to these non-polar micellar solutions, produces a new set of phenomena which justifies the introduction of new concepts. Figure 8 shows weight-average apparent molecular weights of aggregates (determined by light-scattering and ultracentrifuge measurements) plotted against the added aliquots of water at constant temperature. Two regions of the plot can clearly be distinguished. For small amounts of water a straight line indicates a "swelling" of the micellar aggregates by the uptake of water. For larger aliquots of water a non-linear increase describes a coalescence of droplets. This is due to the fact that with a constant amount of surfactant, increasing quantities of water are accommodated by the system. The transition between the linear and non-linear parts of this plot is dependent on the counterion, as is seen in Figure 9. This has already been discussed in connection with the chelate formation and was seen to depend on the competition between chelation and hydration for monovalent alkali ions with $q/r_i \simeq 1$ (see Figure 3). There seems to exist a certain regularity regarding decreasing hydration interaction in proceeding from K^+ to Cs^+.

The shift of the transition region between the micellar state (where the "swelling" is observed) and the microphase domain parallels this order. This transition can also be quite suitably followed by other means, e.g. ^{1}H-NMR spectroscopy (Figure 10),

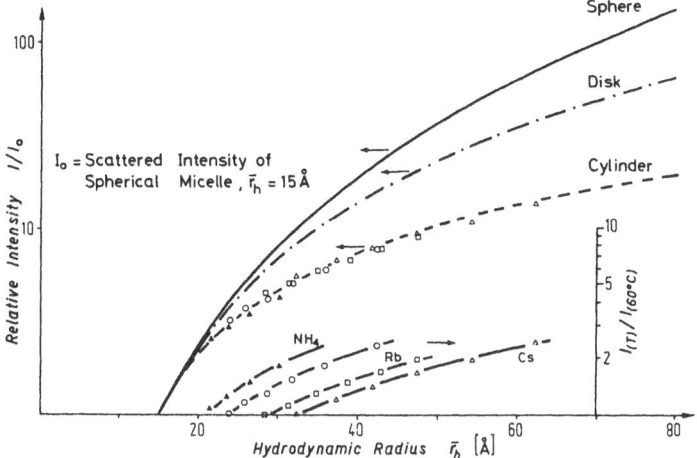

Figure 7. Estimation of micellar shapes of aggregated alkali (and ammonium) bis-(2-ethylhexyl)sulfosuccinates in iso-octane[7].

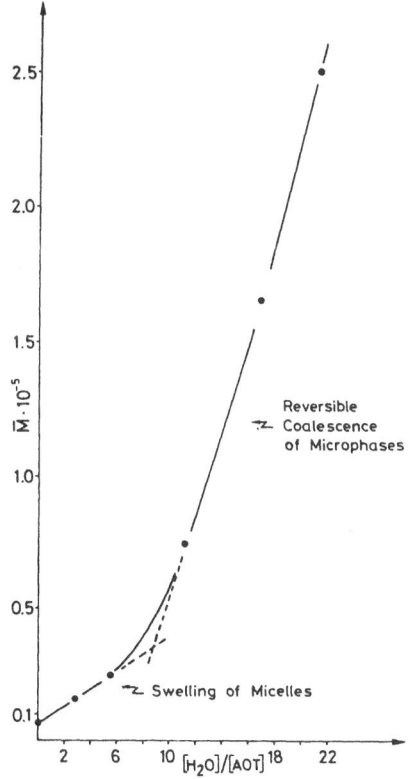

Figure 8. Mean apparent molecular weight of sodium bis-(2-ethyl-hexyl)sulfosuccinate (AOT) aggregates as function of aliquots of water in iso-octane (298 K).

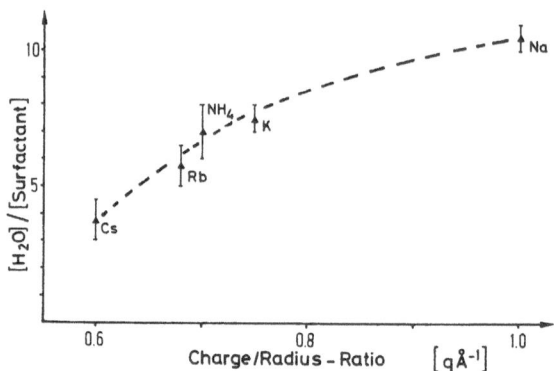

Figure 9. Aliquots of water ((H₂O/(surfactant)) which induce the transition between micellar and microphase states in iso-octane (298 K) against charge/radius ratio. Surfactants: alkali (and ammonium) bis-(2-ethylhexyl)sulfosuccinates.

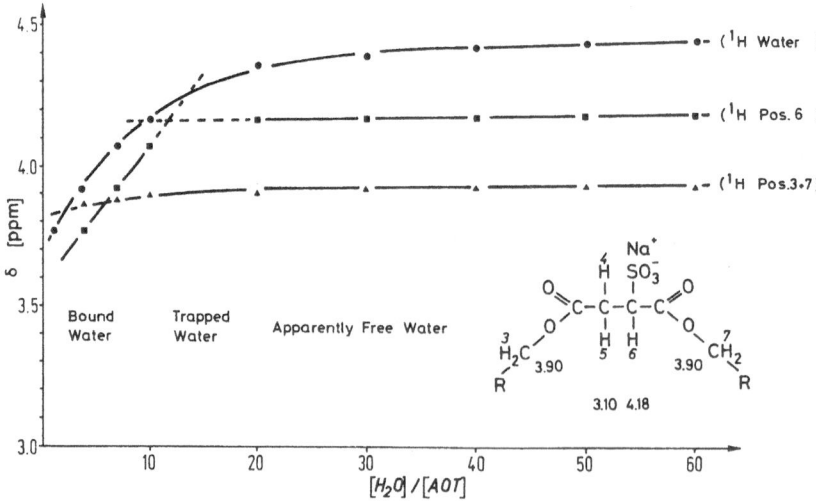

Figure 10. Dependence of proton chemical ·shifts in water (AOT)/ iso-octane (W/O) microemulsions on the added aliquots of water (298 K)[22].

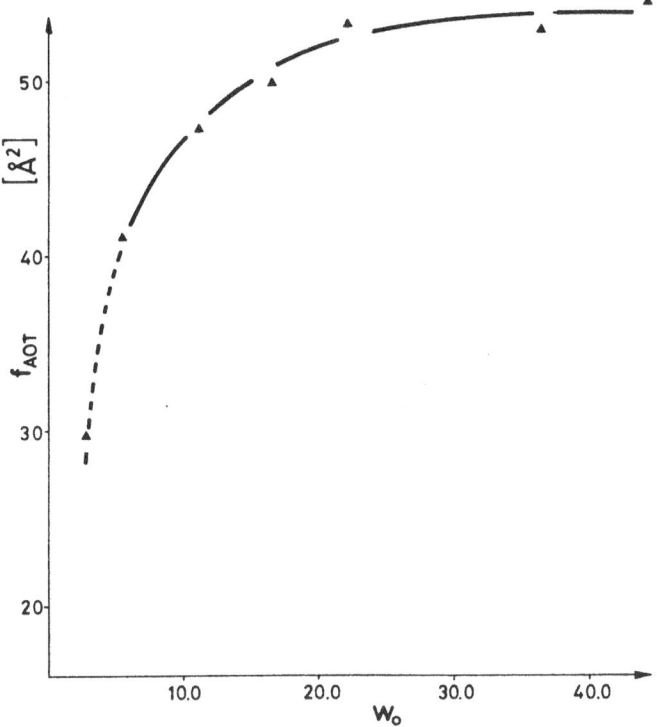

Figure 11. *Average area of the hydrocarbon water interface covered by one AOT molecule ($\overline{f}_{AOT}$) plotted against aliquots of added water (298 K).*

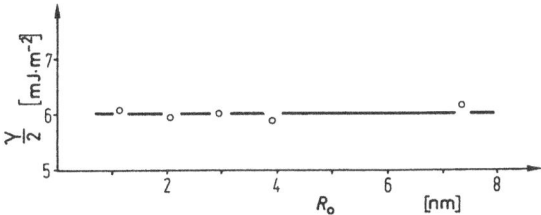

Figure 12. *Electrostatic contribution to the interfacial free energy of an aqueous microphase plotted against its radius R_0.*

where the shifts of the water proton and those of the two different
protons of the AOT molecule, one in the geminal position to the sulfonate group and the other two (not resolvable) within the first
$-CH_2$ groups of the 2-(ethylhexyl)esters are traced. Such a plot of
the chemical shift of the water proton against w_o (H_2O/AOT) indicates
that the environment of the water proton in the microphase appears
to be almost the same as that in the bulk water. The proton geminal
to the sulfonate group in AOT displays behavior analogous to the
water proton, while the two ester protons are apparently unaffected
by this transition.

Another strong indication of this transition is obtained from
the determination of the variation of the average area of the water/
hydrocarbon interface covered by one AOT molecule (= f_{AOT}),(Figure
11). The apparent transition is again located at an (H_2O)/(AOT) ratio
of about 15. It is obvious from this plot that f_{AOT} approaches a
constant value (this apparently being true for all surfactants and
combinations of surfactants and cosurfactants)[8]. It can actually be
shown that a maximum value of $f_{surface\ active\ material}$ (abbreviated f
surf.act.mat.) is to be expected from thermodynamic considerations
which are compatible with the conditions of zero surface tension[9].
In order to demonstrate this one can start from a single (plane,
i.e. no curvature energy considerations) interface which is composed
of a bulk contribution, an oil/water interfacial contribution and a
free energy of the surfactant, including the mutual repulsion between
surfactant molecules in the interface. Minimization of the total free
energy with respect to $f_{surf.act.mat.}$ at constant number of surfactant
molecules yields $f_{surf.act.mat.}^{max}$. It should however be emphasized
that the condition of zero surface tension is not at all restricted
to this particular f-value (as could be inferred from[9]. On the
contrary, it was shown[13] that this condition prevails for all $f_{surf.act.mat.}$
values ecept for the micellar state where the definition
of a macroscopic interface appears questionable.

It can reasonably be assumed in the present case,where ionic
surfactants have been applied, that there are essentially two contributions to the interfacial free energy that determine the state
of zero surface tension, i.e.$\gamma_{uncharged}$ and $\gamma_{electrostat}$,[11,12].
Solving the non-linear Poisson-Boltzman equation approximately with
appropriate boundary conditions for aqueous microphases[10], it became possible to evaluate $\gamma_{electrostat}$, and in particular its
dependence on the microphase radius. As can be seen in Figure 12,
$\gamma_{electrostat}$ is independent of the curvature of the spherical microphase[13].This could have already been inferred from the fact that
$\gamma_{uncharged}$ and $\gamma_{electrostat}$ almost compensate each other (the contribution of entropy of mixing of the microphases with oil has been
neglected). However, since $\gamma_{uncharged}$ is determined by nearest
neighbor interactions it is expected to be independent of curvature
to a first approximation. Hence, also $\gamma_{electrostat}$ has to be independent of curvature.

Zero surface tension and spherically shaped aqueous microphases seem at first sight to be incompatible properties of these systems. It can however be argued[11] that even small deformations of the minimum interfacial area will increase the interfacial tension, which is immediately counteracted by the Gibbs-Maragoni effect driving the shape of the microphases back to spherical. Hence, it is not surprising that the Percus-Yervick[14] and Carbahan-Starling[15] hard sphere approximations were successfully applied to these systems[16]. Accordingly, the microphases might be called "isodispersive" particles[11].

Also, within the scope of the above calculations, the degree of polydispersity of the microphase ensemble (= the so-called microemulsion) might be estimated. The surprisingly small polydispersity of these systems is shown in Figure 13.

Although the three-component water-Aerosol OT-oil system has been selected as one of the simplest examples that exhibits all features typical of an aqueous microphase ensemble, temperature variations show a variety of phenomena which still present difficulties for unambiguous interpretations.

Quasi-elastic light scattering experiments with this system display an almost constant mean hydrodynamic particle radius for the investigated temperature region, while above a certain amount of added water, these 'radii' start to increase non-linearly. The slopes become steeper the greater the amount of water added. The experiments were evaluated under the assumption that the viscosity of the dispersion medium essentially determines the underlying hydrodynamic processes. The same is true with electro-optical (pulsed) Kerr effect studies which permitted one to follow the temperature-dependent processes over a larger temperature interval without being disturbed by increasing turbidity. The purely induced dipole moments were attributed to the polarization of the electric double layer of the aqueous microphase which is covered by a monolayer of ionic surfactants in the present case. Two relaxation times were derived from the experimentally observed decay of the double refraction by a computerized evaluation procedure[18]. Recent experimental information led us to conclude that one probably has to attribute the two relaxation times observed within a particular concentration region to two rotational diffusion structures: i.e. the spherical microphases*) (derived from light- and small angle neutron scattering[23]) and rod-(pearl chain) like particles. The latter structure would support observations of percolation-like phenomena and field-free birefringence of one of the oil-continuous phases which separates from an almost pure oil phase after prolonged equilibration of the turbid mixture. The rod-like structures are formed by a continuous

---

Footnote: *) They are probably slightly deformable in an electric field.

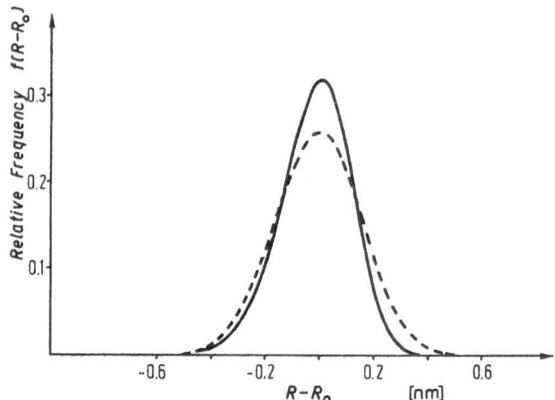

Figure 13. *Polydispersity of aqueous microphases at 298 K calculated according to the considerations in Reference 13.*

phase transition (second order process?) from originally spherical
to more extended (reverse hexagonal(?) or lamellar[17]) structures.
The particular type of mesophase which is built up during this
process depends also on the initial concentrations of the aqueous
microphases. The two τ-values can be correlated remarkably well.
A typical plot of both relaxation times is shown in Figure 14.

An interesting result is obtained (Figure 15) by studying the
temperature-dependent birefringence detected in different solvents
with constant aliquots of water. The diagrams seem to indicate that
the transition from aqueous spherical microphases to "lamellar"
structures is sensitively influenced by different organic solvents
corresponding to their polarizabilities and/or space requirements
within the surfactant monolayer.

Although it could be shown by small angle neutron scattering[23]
that there exists a temperature-dependent growth of the aqueous
microphases, the latter can nevertheless not interpret the strong
temperature dependence of the relaxation times. The assumption that
critical phenomena are occurring[19,20] appears more appropriate. In
this way the relation of the observed maxima in the relaxation
time - temperature diagrams to critical points and their shift with
the applied electric field is straightforwardly explainable. Also
the above mentioned percolation-like increase of the specific conductivity (Figure 16) and the corresponding variation of the self-
diffusion coefficient of water with a temperature increase[24] are in
line with this view.

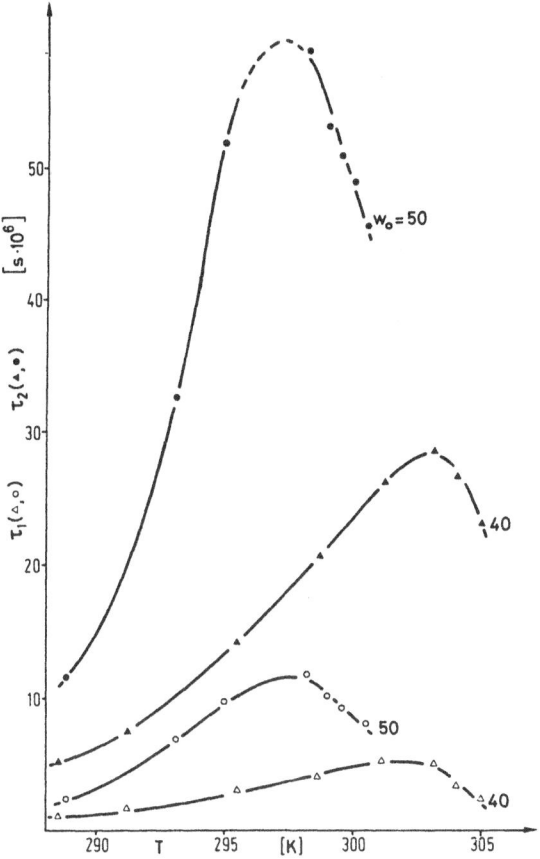

Figure 14. *Temperature-dependent relaxation times of electric bi-refringence in water/0.18 mol.dm^3 AOT/iso-octane. Parameter: $(H_2O)/(AOT)$.*

The different positions of the turbid-end-points of the curves (see Figure 15) with respect to the temperature abscissa would then indicate the distance to the critical (demixing) point. This may then be considered to be a measure of the solubilities of the micro-phases in different dispersion models.

The work reported herein is part of a project (No. 2.025-0.81) of the Swiss National Science Foundation.

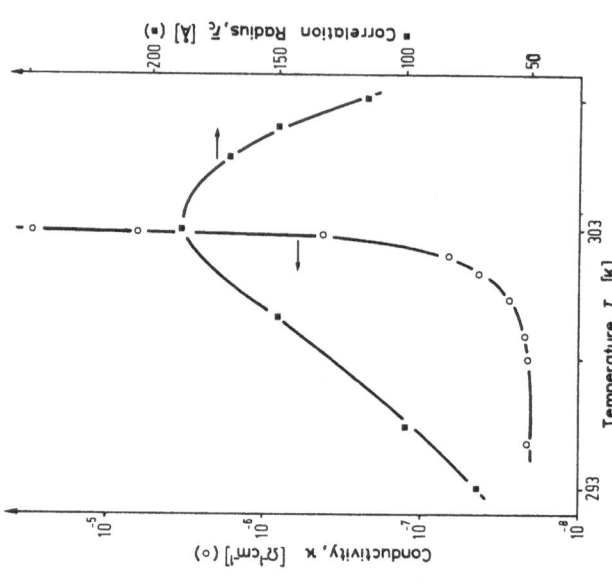

Figure 16. Temperature-dependent mean correlation radii and specific conductivity of water/0.18 mol. dm^{-3} AOT/iso-octane.

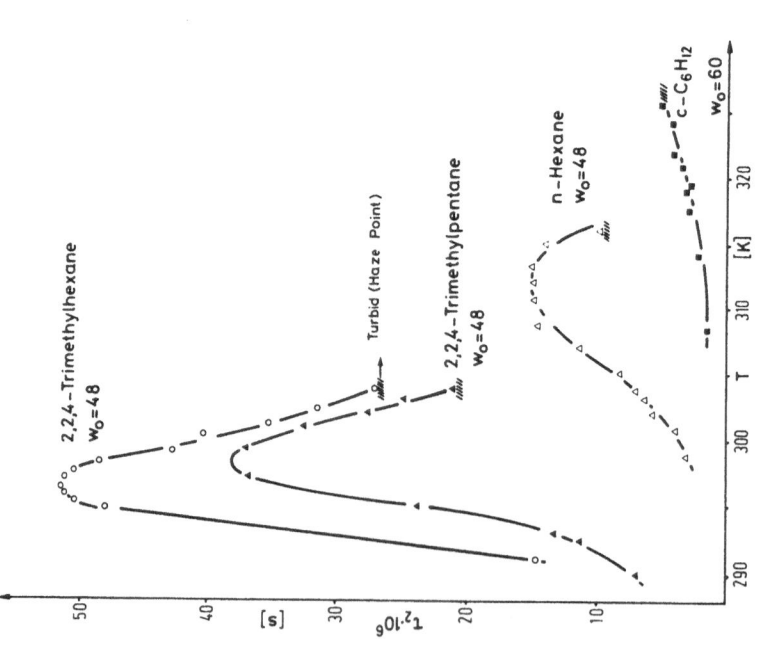

Figure 15. Temperature-dependent relaxation times of electric birefringence in water/0.18 mol. dm^{-3} AOT/apolar solvents. Parameter: organic solvents[21].

REFERENCES

1. H.F. Eicke, "Surfactants in Nonpolar Solvents", Topic in Current Chem. 87 : 86-145 Springer, Berlin (1980).
2. D.J. Mitchell and B.W. Ninham, J. Chem. Soc. Faraday Trans. II, 77 : 601 (1981).
3. J. Israelachvili, S. Marcelja and R.G. Horn, Rev. Biophys., 13 : 121 (1980).
4. G. Zundel, "Hydration and Intermolecular Interaction" Academic Press, New York (1969).
5. P. Kvita, Ph.D. Thesis, University of Basel, 1982.
6. H. Reering, J. Coll. Int. Sci., 20 : 217 (1965). ·
7. N.A. Mazer, G.B. Benedek and M.C. Carey, J. Phys. Chem., 80 : 1075 (1976).
8. H.F. Eicke and J. Rehak, Helv. Chim. Acta, 59 : 2883 (1976).
9. P.G. de Gennes and C. Taupin, J. Phys. Chem., 86 : 2294 (1982).
10. B. Jönsson and B. Wennerström, J. Coll. Int. Sci., 80 : 482 (1981); B. Jönsson, Ph.D. Thesis, University of Lund, 1981.
11. J.T.G. Overbeek, Faraday Disc. Chem. Soc., 65 : 7 (1978).
12. J. Frenkel, "Kinetic Theory of Liquids" Dover Publ. New York p. 362 f. (1955).
13. R. Kubik, H.F. Eicke and B. Jönsson, Helv. Chim. Acta, 65 : 170 (1982).
14. J.K. Percus and G.J. Yecick, Phys. Rev., 110 : 1 (1958).
15. N.F. Carnahan and K.E. Starling, J. Chem. Phys., 51 : 635 (1969).
16. W.G.M. Agterof, J.A.J. van Zomeran and A. Vrij, Chem. Phys. Letters, 43 : 363 (1976).
17. B. Tamamushi and N. Watanabe, Colloid & Polymer Sci., 258 : 174 (1980).
18. H.F. Eicke and Z. Markovic, J. Coll. Int. Sci., 85 : 198 (1982).
19. A.M. Cazabat, D. Langevin, J. Meunier and A.J. Pouchelon, Phys. Letters, 43 : 89 (1982).
20. A.M. Cazabat, D. Chatenay, P. Guering, D. Langevin, J. Meunier, O. Sorba, J. Lang, R. Zana and M. Paillette, in "Surfactant Chemistry in Solution" Lund (1982).
21. D. Dünnenberger, unpublished results, Inst. for Phys. Chem., Univ. Basel (1982).
22. H.F. Eicke, Chimia, 36 : 241 (1982).
23. M. Kotlarchyk, S.H. Chen and J.S. Huang, J. Phys. Chem. Lett., 86 : 3273 (1982).
24. Private communication from Dr. M. Holz, University of Karlsruhe.

NATURALLY OCCURRING AMPHIPHILES: ASPECTS OF THEIR PHASE BEHAVIOR

H. Hauser

Laboratorium für Biochemie
Eidgenössische Technische Hochschule
CH-8092 Zürich (Switzerland)

INTRODUCTION

My discussion of naturally occurring amphiphiles will be confined to the class of phospholipids. Phospholipids are major constituents of biological membranes and serum lipoproteins. They usually make up 25-50% of the dry weight, the exception being myelin, where the phospholipid content amounts to ∿80%. Phospholipids are also commercially important, their technological applications ranging from emulsifiers to ingredients of foodstuffs, cosmetics, lubricants and medicinal products. The chemical structures of the most common phospholipids are presented in Table 1.

Many reviews and monographs have dealt adequately with the nomenclature, chemistry and physics of phospholipids[1-6]. This survey is addressed to some aspects of phospholipid phase behavior which are of utmost importance in the industrial application of phospholipids.

POLYMORPHISM OF PHOSPHOLIPIDS

As amphipathic molecules, phospholipids have a great tendency to aggregate and, like fatty acids, to form more than one structure in the crystalline (solid) state as well as in aqueous dispersions. This property, referred to in the literature as polymorphism, is illustrated in Figure 1. For a detailed discussion of the various structures, the reader is referred to reviews by Williams and Chapman[5] and Luzzati[6]. Since this volume is devoted to the discussion of reversed micelles, it should be pointed out that phospho-

Table I.

Phospholipids   X Equals

CH₂——CH——CH₂—O—P—X   Phosphatidic acid, −OH
| | ‖   Phosphatidylcholine, −O−CH₂−CH₂−N⁺Me₃
O   O   O   Phosphatidylethanolamine, −O−CH₂−CH₂−NH₂
| | ⁻   Phosphatidylserine, −O−CH₂−CH−NH₂
CO   CO |
| | COOH
(CH₂)ₙ (CH₂)ₙ (n = 12-24) Phosphatidylthreonine, −O−CH−CH−NH₂
| | | |
CH₃   CH₃   CH₃ COOH

                                                      OH
                                                      |
                           Phosphatidylglycerol, −O−CH₂−CH−CH₂OH

                                                      OH
                           Phosphatidylinositol, HO⟨ring⟩O−
                                                  with OH, OH OH

Plasmalogens   X Equals

CH₂−CH−CH₂−O−P−X   Choline plasmalogen, −O−CH₂−CH₂−N⁺Me₃
| | ‖   Ethanolamine plasmalogen, −O−CH₂−CH₂−NH₂
O   O   O
| |
CH   CO
‖ |
CH   (CH₂)ₙ
| |
R   CH₃

Cardiolipin

            O CH₂−CH−CH₂−O
            ‖ |
      O−P−O   OH   O=P−O⁻
      | |
CH₂——CH−CH₂−O   O−CH₂−CH——CH₂
| | | |
O   O   O   O
| | | |
CO   CO   CO   CO
| | | |
(CH₂)ₙ (CH₂)ₙ   (CH₂)ₙ (CH₂)ₙ
| | | |
CH₃ CH₃   Diphosphatidylglycerol CH₃   CH₃

Sphingolipids   X Equals

HO−CH———CH−CH₂−OX   Ceramide, H
   | | Sphingomyelin, −PO₂−O−CH₂−CH₂−N⁺Me₃
   CH   NH   Cerebroside, β-D-galactosyl-(1→4)-β-D-glucosyl-
   | | lactosyl-
   CH   CO   Galactosylceramide, β-D-galactosyl-
   | |
   (CH₂)₁₂ (CH₂)₁₄
   | |
   CH₃   CH₃

Lipids : Phase Behaviour

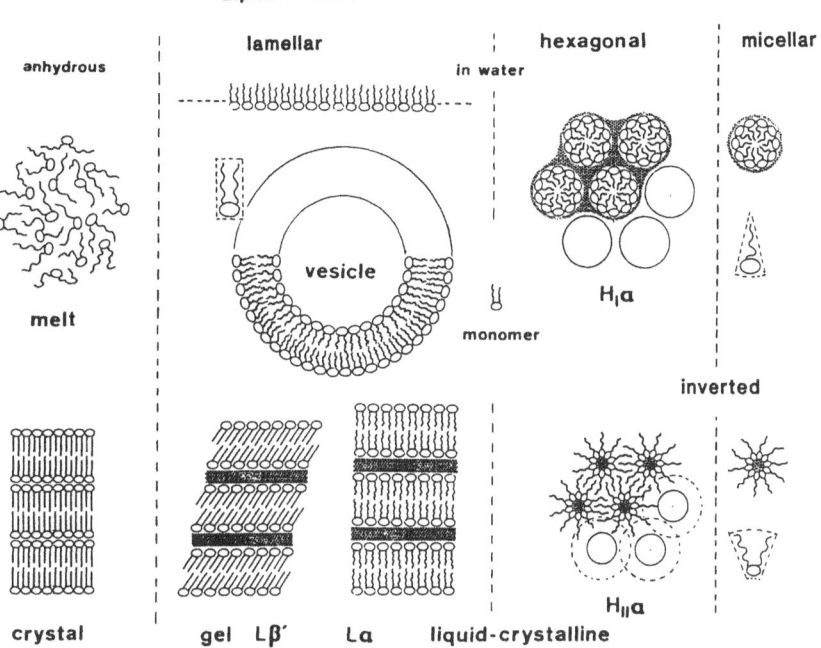

Figure 1. *Schematic diagram showing various lipid phases in the solid state and in aqueous dispersion.*

On heating a lipid crystal the lipid molecules do not pass directly from the crystal to the isotropic melt, but usually there are a number of liquid crystalline states intermediate between the crystal and the isotropic liquid (not shown). In the presence of water phospholipids are known to form various hydrated, usually liquid crystalline phases (center and right-hand panel), which are in equilibrium with a certain concentration of monomers. However, this critical micellar concentration is extremely small ($\lesssim 10^{-10}M$) for diacyl phospholipids with long hydrocarbon chains ($\gtrsim 14$ C atoms) indicating an overwhelming preference of the lipid molecule for the aggregated state. For a detailed discussion of the various lipid structures the reader is referred to Ref. 5 and 6. (The diagram was kindly provided by Dr. I. Pascher, University of Göteborg, Sweden).

lipids can form inverted micelles, at least under certain experimental conditions (bottom right hand panel, Figure 1). Phospholipids have also been reported to form inverted hexagonal phases ($H_{II}\alpha$) which are structurally related to inverted micelles (bottom right, Figure 1). The hexagonal phase $H_{II}\alpha$ consists of infinite cylinders of water which are lined by phospholipid polar groups.

The axes of the water cylinders are packed in a hexagonal array and
the space between the cylinders is taken up by liquid crystalline
hydrocarbon chains. This phase was first described for a human brain
lipid water system and has since been reported for several other
phospholipids[5]. To the best of my knowledge there is no systematic
study of phospholipd reversed micelles or of reversed hexagonal
phases.

CRYSTAL STRUCTURES OF PHOSPHOLIPIDS

In the crystal and in aqueous dispersions, phospholipids preferentially aggregate to form bilayers. Several single-crystal structures have been solved up to now, a summary of which is presented
in References 7-9. As a representative example, the single-crystal
structure of dilauroyl phosphatidylethanolamine[10] is depicted in
Figure 2. Some general points emerging from this and other crystal
structures of phospholipids can be summarized as follows:

1) The phospholipid molecules are packed as bilayers; in one excep-
tional case they have been reported to be arranged as a monolayer.

2) The parallel stacking of the hydrocarbon chains is a prominent
feature of phospholipid crystals. It leads to the well-known tuning
fork conformation of diacyl phospholipids. This arrangement of the
hydrocarbon chains is accomplished by a specific conformation about
the C1-C2-C3-O glycerol group (torsion angles θ_3 and θ_4; for the
atom numbering and notation of torsion angles see Figure 2). As a
result of this conformation the glycerol group (C1-C2-C3) is oriented approximately perpendicular to the plane of the bilayer. As
shown in Figure 2, the three glycerol C-atoms form a continuous
zig-zag with the fatty acyl chain esterfied to glycerol C-3.

3) The orientation of the polar head group is determined by the
conformation about the phosphodiester group. This conformation is
synclinal/synclinal (torsion angles α_2/α_3), producing a 90° bend at
the phosphodiester group (Figure 3). As a result of this kink about
the phosphodiester group the polar head group attains an almost
parallel orientation with respect to the bilayer plane.

LYOTROPIC PHASE BEHAVIOR OF NEUTRAL AND ISOELECTRIC PHOSPHOLIPIDS

When phospholipid crystals are heated to a certain temperature,
the Krafft point or T_c temperature, the hydrocarbon chains melt. The
phospholipids undergo a phase transition from the crystal to the
liquid crystalline state at this temperature. A number of intermediate states are found to exist between the crystal and the isotropic liquid (not included in Figure 1) and this phenomenon is re-

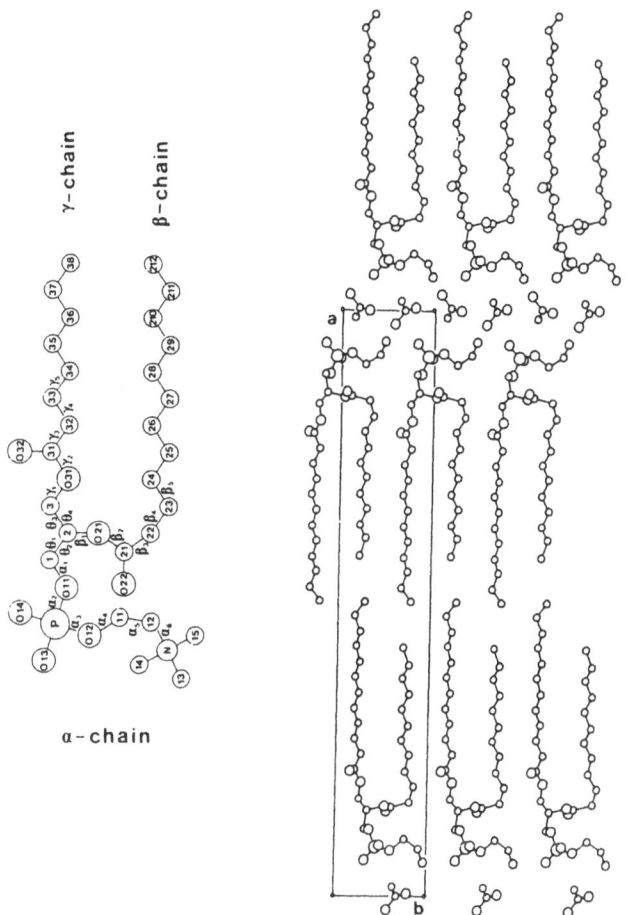

Figure 2. *Single-crystal structure of dilauroyl phosphatidyletha-*
nolamine acetic acid. The rectangle represents the unit cell con-
taining four lipid molecules. The atom numbering and notation for
torsion angles is included[10,7].

ferred to as thermotropic mesomorphism[5]. Here we are more interested in phospholipid-water systems and the phenomenon of lyotropic
mesomorphism. As with thermotropic mesomorphism, where phospholipids do not pass directly from the crystal to the isotropic liquid, phospholipids do not usually pass directly from the crystalline state to a solution when water is present. Various hydrated
phases are encountered before solution of the phospholipid in water
occurs. This behavior is called lyotropic mesomorphism. Figure 4
shows that for pure phospholipids, the transition temperature T_c
decreases when water is added. For the isoelectric phospholipid

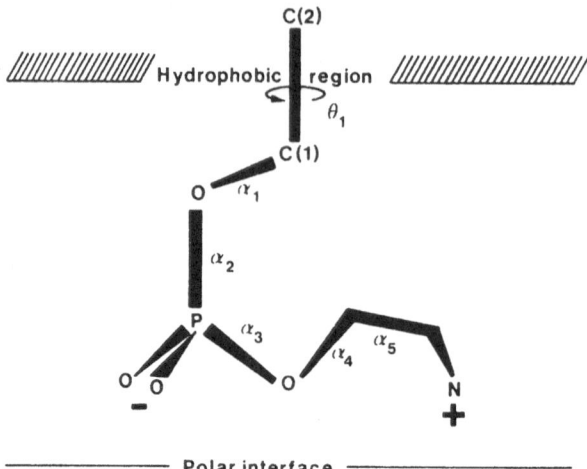

Figure 3. *Schematic diagram showing the orientation of the phos-*
phatidylcholine or phosphatidylethanolamine polar head group with
respect to the bilayer plane. There is an approximate 90° bend at
the phosphodiester group[30].

dipalmitoyl phosphatidylcholine, the transition temperature decreases
from 93°C to a limiting value of 41°C which is reached at maximum
hydration, i.e. at about 30% H_2O (Figure 4A). This temperature of
41°C is the minimum temperature at which water will readily penetrate between the crystalline lipid bilayers. Above the T_c line
(Figure 4A) dipalmitoyl phosphatidylcholine forms a mesomorphic lamellar phase with liquid crystalline hydrocarbon chains. On cooling
the phospholipid/water system below the T_c line, a lamellar gel forms
with all the water being retained between the bilayers. In the gel
state the hydrocarbon chains pack in a hexagonal subcell[5]. Above the
T_c line addition of water in excess of 40% gives rise to a two-phase
system consisting of multilamellar liposomes dispersed in excess
water (Figure 4A). Below the T_c line, the two-phase system consisting of gel and excess water forms at water contents lower than 30%.
The phase behavior shown for dipalmitoyl phosphatidylcholine (Fig. 4A)
can be generalized as follows: above the limiting T_c temperature
(plateau of the T_c line in Figure 4A) phosphatidylcholines in water
swell by accommodating water between their bilayers. The swelling
continues until the maximum hydration is reached, at which point a
two-phase system forms. The swelling of phospholipids is readily
determined by X-ray diffraction. Smectic (lamellar) phases give an
X-ray diffraction pattern at low angles consisting of a series of
equally spaced lines in the ratio $1:(\sqrt{2}):(\sqrt{3}):(\sqrt{4})...(1/n)$, from

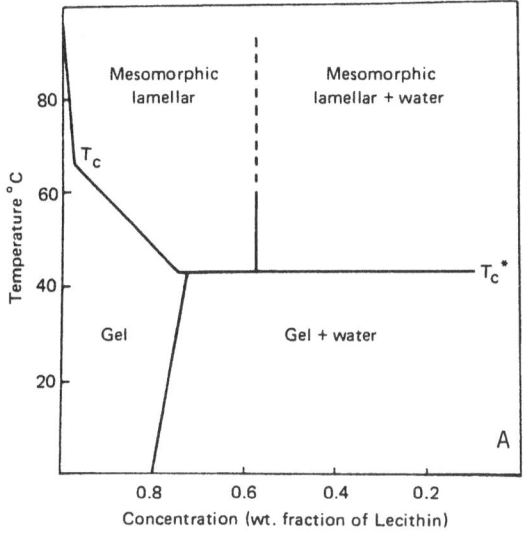

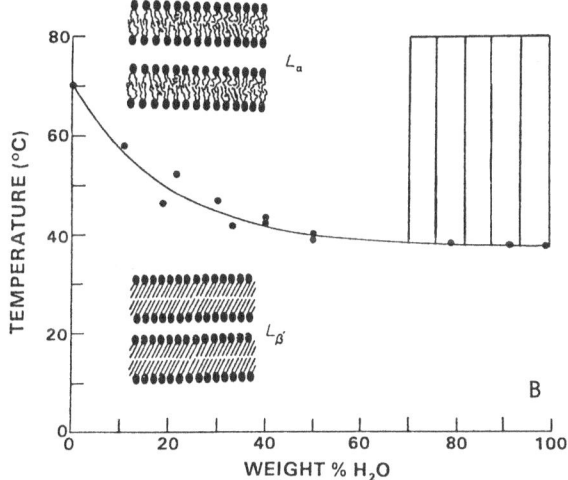

Figure 4. *Phase diagram of (A) the system 1,2-dipalmitoyl-sn-phosphatidylcholine in water[12]; (B) NH_4^+ salt of 1,2-dimyristoyl-sn-phosphatidylserine. The solid line represents the T_C line separating the mesomorphic lamellar phase L_α with liquid crystalline hydrocarbon chains from the gel $L_{\beta'}$. The schematic drawing indicates that the hydrocarbon chains have a small molecular tilt in the gel. The hatched area (right) indicates that at water contents exceeding about 70% there is a transition from the liquid crystalline, fully swollen lamellar phase to a two-phase system[13].*

which the lamellar repeat distance d (= sum of the thickness of the lipid and the water layer) is obtained.

LYOTROPIC PHASE BEHAVIOR OF CHARGED PHOSPHOLIPIDS

There is a marked difference between the phase behavior of the isoelectric and neutral lipids discussed above and that of charged lipids. As an example, the phase diagram of a negatively charged phospholipid, dimyristoyl phosphatidylserine as the NH_4^+ salt, is shown in Figure 4B. The vertical line separating the one-phase from the two-phase region is shifted to the right. This implies that a single lamellar swelling phase exists up to much higher water contents, at least up to 70% H_2O. The precise position of the border between the one-phase and the two-phase system is difficult to determine and is unknown. At high hydration levels the low-angle reflections become diffuse, indicating that the stacking order in the lamellar phase decreases with increasing water content. There may be a gradual rather than a sharp transition from the one-phase to the two-phase region somewhere in the hatched area of the phase diagram (Figure 4B).

The difference in the swelling in water between isoelectric phosphatidylcholines and charged lipids is illustrated in Figure 5. Figure 5A shows the limited swelling of egg phosphatidylcholine and dipalmitoyl phosphatidylcholine. For both lipids the lamellar repeat distance d increases with water content, reaching a limiting value at a water content of 45% and 27%, respectively. Up to these limits a single, swelling lamellar phase exists and above these limits two-phase systems are observed. For comaprison, the swelling of a commercial dipalmitoyl phosphatidylcholine is included. In contrast to the two pure phosphatidylcholines, d increases continuously with increasing water content, indicating a single lamellar swelling phase, at least to a water content of about 60%. The swelling characteristics of charged lipids is shown in Figures 5B-D. The NH_4^+ salt of dimyristoyl phosphatidylserine and the Na^+ salt of ox brain phosphatidylserine swell in water, both below and above the T_c up to water contents of about 50% and 70%, respectively. At higher hydration the low-angle reflections become too diffuse to allow the determination of d. The continuous swelling of the salts of phosphatidylserine is contrasted by that of isoelectric phosphatidylserine at pH $\sim$2. At this pH the carboxyl group of phosphatidylserine is fully protonated and hence the lipid has no net charge. At pH = 2 and 20°C (below T_c), d = 42.0 Å and d = 45.7 Å for dimyristoyl and dipalmitoyl phosphatidylserine, respectively. These d-values are invariant over the total hydration range investigated (horizontal solid lines, Figure 5B). Furthermore, Figure 5B shows that the swollen gel of NH_4^+ dimyristoyl phosphatidylserine at 50% water

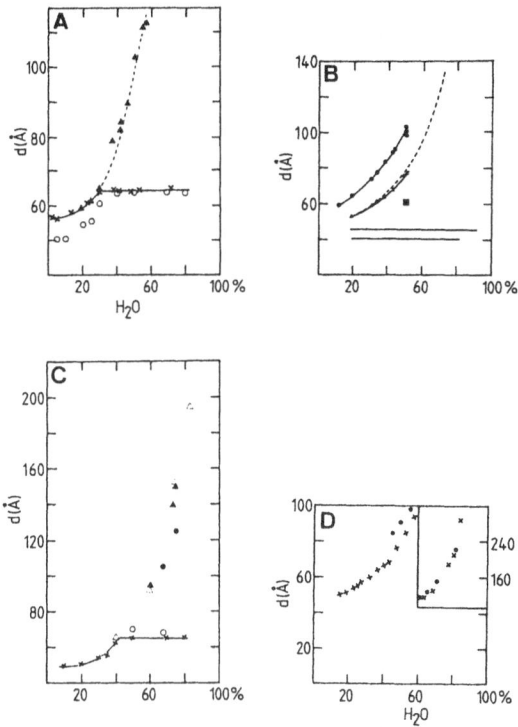

Figure 5. *Swelling of diacyl phospholipids in water. (A) 1,2 di-palmitoyl-sn-phosphatidylcholine at 25°C in the gel state (x)[12], egg phosphatidylcholine at 24°C in the liquid crystalline state (o)[14]. For comparison the swelling of a commercial 1,2-dipal-mitoyl-rac-phosphatidylcholine at 25°C is included (▲).*
(B) NH_4^+ 1,2-dimyristoyl-sn-phosphatidylserine at 25°C (below T_c = 39°C) (●) and at 50°C (above T_c) (▲). For comparison the swelling of 1,2-dimyristoyl and 1,2-dipalmitoyl-sn-phosphatidyl-serine, both in the protonated form, is shown (solid lines at d = 42 Å and 45.7 Å, respectively); Na^+ salt of ox brain phos-phatidylserine (dotted line); at 50% H_2O, 20°C and [NaCl] ⩾ 0.5M, d for NH_4^+ 1,2-dimyristoyl-sn-phosphatidylserine is given (■)[13];
(C) egg phosphatidylcholine (x) and mixtures of egg phosphati-dylcholine and charged lipids: egg phosphatidylcholine and cetyl-trimethylammonium bromide at 0.6% (o); 0.9% (●) and 1.5% (Δ), egg phosphatidylcholine and 3.5% sodium oleate (▲)[15]. (D) lipid extracts of beef heart mitochondria at 25°C (x)[31] and of human erythrocytes at 0°C (●)[32].

content shrinks when NaCl is added. NaCl at 0.5 M apparently ex-
trudes most of the water between the bilayers. That the almost infinite swelling of charged lipids is a general phenomenon is shown
in Figure 5C. By adding small amounts of a positively charged detergent or Na$^+$ oleate to egg phosphatidylcholine the swelling behavior of this phospholipid is converted to that typical for charged lipids. In the light of this experiment the continuous swelling
of a commercial sample of dipalmitoyl phosphatidylcholine is readily
explained. The presence of a charged contaminant at a level of a few
percent could account for the continuous swelling. "Infinite" swelling as shown in Figures 5B and C has been reported for other negatively charged phospholipids, e.g. for the sodium salt of phosphatidic acid, phosphatidylglycerol and phosphatidylinositol[11]. On the
basis of their swelling, lipids can therefore be divided into two
classes: 1) neutral and isoelectric lipids showing no swelling in
the crystal, and limited swelling in the liquid crystalline state
(above T_c);   2) charged lipids showing almost infinite swelling
with increasing water content. The thickness of the water layer may
be a multiple of the bilayer.

EXCESS WATER REGION OF THE PHASE DIAGRAM OF CHARGED LIPIDS:
FORMATION OF LARGE UNILAMELLAR VESICLES

A question of some practical interest is what happens when
dispersions of charged lipids are diluted to the extent that the
low-angle reflections become diffuse or are lost altogether (see
Figure 4B, hatched region). X-ray experiments with samples which
fall in this region of the phase diagram show that the sharp low-
angle reflections, typical for multilamellar structures, are replaced by a broad scattering peak, illustrated in Figure 6A-D. The
scattering curve of a 2.5% unsonicated dispersion of Na$^+$ ox brain
phosphatidylserine in H$_2$O (Figure 6A) consists of a broad diffuse
peak in the range $0.05 < h < 0.15$ (Å)$^{-1}$. This scattering curve resembles that obtained from sonicated phospholipid dispersions
(Figures 6B and D), known to consist of small vesicles (diameter
~ 250 Å) surrounded by a single bilayer. For comparison, the X-ray
diffraction pattern of a dilute unsonicated egg yolk phosphatidylcholine dispersion is shown in Figure 6C. The sharp diffraction
lines at 66 Å and 33 Å are the first and second order diffractions
corresponding to the lamellar repeat distance d of the maximally
hydrated egg phosphatidylcholine dispersion. The evidence presented
in Figures 5 and 6 suggests that the fundamental structure of charged lipids in excess water is the unilamellar closed vesicle. The
following rule can be derived from these two figures: charged lipids
forming smectic phases in water swell continuously up to at least 70%
to 80% water (Figure 5). At a certain dilution at the boundary bet-

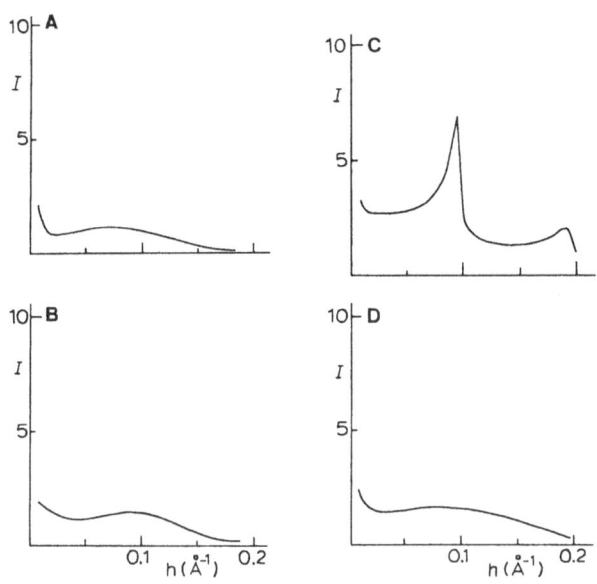

Figure 6. *X-ray scattering curves for aqueous dispersions of phos-*
phatidylserine; (A) 2.5 weight % unsonicated dispersion of the Na$^+$
salt of ox brain phosphatidylserine in H$_2$O; (B) 5 weight % sonicated
dispersion of the same lipid as in (A); (C) for comparison the X-ray
diffraction pattern of a 5 weight % unsonicated egg phosphatidyl-
choline dispersion; (D) 5 weight % sonicated dispersion of egg phos-
phatidylcholine, fractionated by gel filtration on Sepharose 4B
prior to the X-ray experiment. I is the experimental scattering in-
tensity, h = (4π sin θ/λ), where 2θ is the scattering angle and λ
the wavelength[29].

ween the one-phase and two-phase system, the multilamellar structures
break up and the bilayer sheets seal to form large closed unilamellar
vesicles. It is important to note that continuous swelling is not
only a property of pure, charged lipids but also of mixed lipid bilayers of relatively low surface charge density (Figure 5C). The nature of the charge is unimportant indicating that the swelling is a
purely electrostatic phenomenon. The experiment (■) shown in
Figure 5B supports this notion. The addition of sufficient counterions, in this case 0.5 M Na$^+$, makes a fully swollen dimyristoyl phosphatidylserine dispersion shrink. Equilibrium is attained between
two opposing forces: the repulsive forces of the double-layer potential and the attractive van der Waals forces between bilayers. The
equilibrium lamellar repeat distance is d = 62 Å (Figure 5B). Under
these conditions charged lipid bilayers form multilamellar struc-

tures as uncharged or isoelectric lipids do. As with phosphatidylcholine, the swelling of charged lipids in water is limited in the
presence of Na^+ and the two-phase system in excess water consists
of multilamellar lipid structures and water. To summarize, the
phase behavior of charged lipids in water differs significantly in
two points: 1) charged lipids swell continuously up to high water
contents (>70%) forming a single phase; 2) at even higher water
contents a two-phase system forms which consists of large unilamellar vesicles dispersed in water. The property of forming large unilamellar vesicles in excess water appears to be a general feature
of the phase behavior of charged lipids. The vesiculation occurs in
the hatched region on the right-hand side of the phase diagram
(Figure 4B). It should however be made clear that this phenomenon
is not a special property of naturally occurring charged lipids, but
one which is also observed with the large group of synthetic, charged
detergents. The area of the phase diagram of charged lipids discussed above is important in terms of liposome technology. Although the
"dispersing" effect of charged lipids has been known for a long time
and widely used in the past, the principle behind it may not have
been fully appreciated.

EXCESS WATER REGION OF THE PHASE DIAGRAM: CHARGED LIPIDS WITH
CHARACTERISTIC STRUCTURAL FEATURES UNDERGO SPONTANEOUS VESICULATION.
FORMATION OF SMALL UNILAMELLAR VESICLES OF DIAMETER 200 - 600 Å

In terms of liposome technology, three aspects of the phase behavior of charged lipids seem to be important:

1) In excess water, in the two-phase domain of the phase diagram,
charged lipids forming large unilamellar vesicles can revert back
to a multilamellar ordered packing according to the following
scheme:

$$\text{(large unilamellar vesicles)} \quad \underset{\substack{\text{removal of ions, e.g. } Ca^{2+} \\ \text{by adding EDTA}}}{\overset{\text{addition of ions, e.g. } Ca^{2+}}{\rightleftarrows}} \quad \text{(multilamellar structures)}$$

From this scheme is is obvious that the structural changes are reversible. Depending on the experimental conditions, on the nature of
the ions present and on their amount (ionic strength) dispersions of
charged lipids in excess water consist of multilamellar liposomes or
unilamellar vesicles (see discussion under 3).

2) The phase behavior of neutral and isoelectric lipids can easily
be converted to that of charged lipids by doping the neutral lipid
with a small quantity (one to a few %) of a charged lipid or amphiphile.

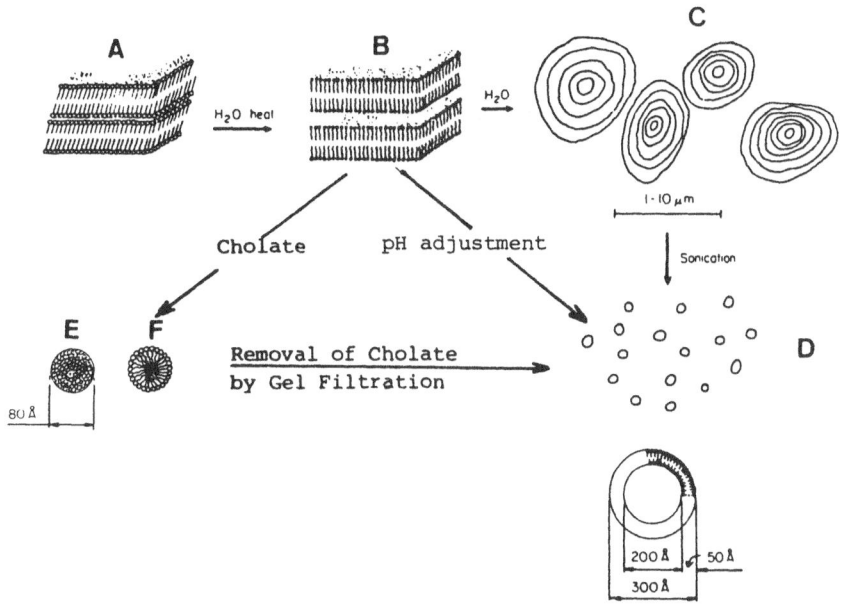

Single Bilayer Vesicle

Figure 7. (A) Section of a phosphatidylcholine bilayer in the solid state (crystal) or in the gel phase with crystalline hydrocarbon chains, except that in this case the polar groups are hydrated and their degree of motional freedom is greater than in the crystal, (B) represents the liquid crystalline phase with the water being incorporated between the hydrocarbon chain layers. When the water content exceeds 23 mol of H_2O/mol of phosphatidylcholine, a two phase system is observed consisting of multilamellar particles and water. (C) Each line in (C) represents a bilayer, arranged like the layers of an onion. When a dispersion of liposomes (C) is subjected to ultrasonication, the bilayers disintegrate and the small, unstable fragments reseal to vesicles surrounded by a single bilayer (D). In (E) and (F) (cross section) small spherical, mixed micelles consisting of phospholipid and detergent are shown. Upon detergent removal small unilamellar vesicles are formed (D). Spontaneous vesiculation of smectic lamellar phases of phosphatidic acid or mixtures of this with phosphatidylcholine occurs when the pH of the lipid dispersion is raised transiently to 10-12.

3) Another important aspect of the phase behavior of charged lipids in excess water is depicted in Figure 7. Most of what is shown in this figure is applicable to neutral and isoelectric lipids. Water readily penetrates the polar group lattice, if the crystal (A) is heated above the T_c temperature. The limited swelling of the liquid

crystalline bilayers in the one-phase system is indicated in B. In
excess water (two-phase region) the bilayers form onion-like multilamellar liposomes (C). Subjecting these multilamellar structures
to ultrasonic irradiation produces small unilamellar vesicles of
200-600 Å outer diameter (D). These single-bilayer vesicles may also
be produced by another milder treatment. Detergents such as cholate,
deoxycholate, Triton X-100 and others added to the fully hydrated,
liquid crystalline phases B or C (Figure 7) solubilize the bilayers
to small mixed micelles (E and F). Upon detergent removal by dialysis or by gel filtration small unilamellar vesicles form. In contrast to those vesicles produced by ultrasonication the vesicle
population formed by detergent removal is homogeneous with respect
to particle size[16]. We have recently shown[17] that lamellar phases
of phosphatidic acid form small unilamellar vesicles of 200-600 Å
diameter almost spontaneously when the pH of the lipid dispersion
is raised transiently to 10-12. The method is also applicable to
mixed phospholipid dispersions containing phosphatidic acid, provided the pH of the lipid dispersion is transiently raised so that
the phosphatidic acid is fully ionized[17].

The method of spontaneous vesiculation by pH adjustment is included in Figure 7. The term "spontaneous vesiculation" is meant to
indicate that small unilamellar vesicles form readily, whereby te-
dious standard procedures such as sonication or detergent removal
widely used to produce these vesicles are avoided. In the case of
spontaneous vesiculation, the smectic lamellar phases depicted in
B and C (Figure 7) have to be envisaged to consist of phosphatidic
acid or mixtures of this with other charged or neutral lipids. That
this method is of wider applicability can be readily demonstrated.
In smectic lamellar phases of almost any neutral or isoelectric
lipid and lipid mixtures, charged or uncharged, containing phosphatidic acid, a transient increase in pH rendering the phosphate group
of phosphatidic acid fully ionized will induce spontaneous vesiculation. The resulting phospholipid dispersion is usually a mixture
of small unilamellar vesicles of a narrow size distribution (average
diameter 200-300 Å) and large unilamellar vesicles of a wide size
distribution ranging between about 0.1μ and several microns. The
relative proportion of the two populations depends on experimental
conditions, such as the rate of the pH increase, the pH to which
the dispersion is transiently exposed, and in mixed lipid dispersions, also on the proportion of phosphatidic acid. The active reagent inducing the spontaneous vesiculation (i.e. the formation of
small unilamellar vesicles)is phosphatidic acid, a molecule having
two negative charges at pH > 10 and two hydrocarbon chains with
more than 12 C-atoms. It can be shown that lysophosphatidic acids
such as lauroyl, myristoyl and other long-chain lysophosphatidic
acids are as effective in inducing vesiculation as phosphatidic

acid itself under otherwise identical experimental conditions. In the fully ionized state these compounds have two negative charges that are counterbalanced by only one long hydrocarbon chain. Testing a variety of structurally different amphiphiles as to their ability to induce spontaneous vesiculation has led us to postulate the following rule (H. Hauser and N. Gains, unpublished results): The minimum requirement for a compound to be active and induce spontaneous vesiculation in smectic lamellar phases is one charge, positive or negative, per hydrocarbon chain of 12 or more C-atoms. There are many amphiphiles, e.g. detergents, matching this description which, when dissolved in water, do not form smectic lamellar phases but micellar solutions. However, mixed with appropriate lipids, e.g. diacyl phospholipids, they may form smectic lamellar phases. In this case, one would predict that in the phase diagram of such three-component systems, diacyl phospholipid, detrgent and water, there will be a region which consists of small unilamellar vesicles. Furthermore, according to the rule proposed above, diacyl lipids are expected to vesiculate or to induce vesiculation, provided they have two equal charges in their polar group when present in lipid mixtures. This is true for phosphatidic acid at pH > 10. A number of structurally different compounds meeting this criterion have been tested as to their ability to induce vesiculation; all have been found to be active supporting the above rule (H. Hauser and N. Gains, unpublished).

CONCLUDING REMARKS

The method of spontaneous vesiculation will be an important supplement to the existing liposome technology. It has a number of significant advantages over the methods presently used for the production of small unilamellar vesicles. It is easy and quick, it does not require elaborate laboratory equipment and avoids time-consuming procedures such as ultrasonication and detergent removal by dialysis or gel filtration. Another important advantage is the versatility of the method. As mentioned, the method produces a mixture of small (< 600 Å) and large (diameter > 0.1μ) unilamellar vesicles. The ratio of the two populations depends on a number of factors, foremost on the pH and the rate of pH increase and furthermore, on the ionic strength and nature of the ions present. By varying these parameters it is possible to control not only this ratio but also to vary the surface charge density of the bilayer and hence the surface potential of the resulting vesicles within wide limits. Considering these advantages, the method is potentially useful for the encapsulation of drugs. The unilamellar vesicles produced by this method have characteristics which should make them particularly suitable as drug delivery systems.

The discussion of phospholipids presented here is centered around certain aspects of lipid phase behavior that are important from the viewpoint of industrial applications. Consequently other properties, though equally important for the understanding of phospholipid behavior, had to be neglected. There is however a justification for presenting what, at first sight, might appear to be an unbalanced survey of phospholipids. Some of the basic principles of the phase behavior of charged lipids discussed here and their implications in lipid or liposome technology have only recently been recognized. As far as I am aware, they have not been the subject of any review article. However, those phospholipid properties which could not be discussed here have all been adequately reviewed in the past, for instance, the molecular packing and motion in various lipid aggregates, particularly bilayers below and above T_c[4,18-20], the conformation and segmental motion of phospholipid hydrocarbon chains[4,18-20], the hydration[22], and the conformation and segmental motion of the phospholipid polar group in gels and liquid crystalline phases which have been the subject of several recent reviews [7,23-26]. This also applies to phospholipid-lipid (cholesterol)[4,18-27], phospholipid-ion[28] and phospholipid-protein[21,28] interactions. A good understanding of the phase behavior, as well as those properties not discussed here is required if the behavior of phospholipids in bilayers and biological membranes is to be appreciated.

REFERENCES

1. D. Chapman, "The Structure of Lipids", Methuen, London (1965).
2. A.D. Bangham, Prog. Biophys. Mol. Biol. 18 : 29 (1968).
3. G.G. Shipley, in: "Biological Membranes", D. Chapman and D.F.H. Wallach, eds., Academic Press, London and New York (1973).
4. M.K. Jain and R.C. Wagner, "Introduction to Biological Membranes", John Wiley, New York (1980).
5. R.M. Williams and D. Chapman, in: "Progress in the Chemistry of Fats and other Lipids", R.T. Holman, ed., Pergamon Press, Oxford (1970).
6. V. Luzzati, in: "Biological Membranes", D. Chapman, ed., Academic Press, London and New York (1968).
7. H. Hauser, I. Pascher, R.H. Pearson and S. Sundell, Biochim. Biophys. Acts 650 : 21 (1981).
8. I. Pascher, S. Sundell and H. Hauser, J. Mol. Biol. 153 : 791 (1981).
9. I. Pascher, S. Sundell and H. Hauser, J. Mol. Biol. 153 : 807 (1981).
10. P.B. Hitchcock, R. Mason, K.M. Thomas and G.G. Shipley, Proc. Natl. Acad. Sci. USA 71 : 3036 (1974).

11. H. Hauser, in: "Liposome Letters", A.D. Bangham, ed., Academic Press, London (1984).
12. D. Chapman, R.M. Williams and B.D. Ladbrooke, Chem. Phys. Lipids 1 : 445 (1967).
13. H. Hauser, F.Paltauf and G.G. Shipley, Biochemistry 21 : 1061 (1982).
14. D.M. Small, J. Lipid Res. 8 : 551 (1967).
15. T. Gulik-Krzywicki, A. Tardieu and V. Luzzati, Mol. Cryst. Liquid Cryst. 8 : 285 (1969).
16. J. Brunner, P. Skrabal and H. Hauser, Biochim. Biophys. Acta 455 : 322 (1976).
17. H. Hauser and N. Gains, Proc. Natl. Acad. Sci. USA 79 : 1683 (1982).
18. D. Marsh and A. Watts, in: "Liposomes: from Physical Structure to Therapeutic Applications", C.G. Knight, ed., Elsevier, Amsterdam (1981).
19. D.A. Wilkinson and J.F. Nagle, in: "Liposomes: from Physical Structure to Therapeutic Applications", C.G. Knight, ed., Elsevier, Amsterdam (1981).
20. J.N. Israelachvili, S. Marčelja and R.G. Horn, Q. Rev. Biophys. 13 : 121 (1980).
21. J. Seelig and A. Seelig, Q. Rev. Biophys. 13 : 19 (1980).
22. H. Hauser, in: "Water, a Comprehensive Treatise", F. Franks, ed., Plenum Press, New York and London (1975).
23. J. Seelig, Q. Rev. Biophys. 10 : 353 (1977).
24. J. Seelig, Biochim. Biophys. Acta 515 : 105 (1978).
25. J.L. Browning, in: "Liposomes: from Physical Structure to Therapeutic Applications", C.G. Knight, ed., Elsevier, Amsterdam (1981).
26. G. Büldt and R. Wohlgemuth, J. Membr. Biol. 58 : 81 (1981).
27. M.C. Phillips, Progr. Surface Membr. Sci. 5 : 139 (1972).
28. H. Hauser and M.C. Phillips, Progr. Surface Membr.Sci. 13 : 297 (1979).
29. D. Atkinson, H. Hauser, G.G. Shipley and J.M. Stubbs, Biochim. Biophys. Acta 339 : 10 (1974).
30. S. Sundell, PhD Thesis, Dept. of Structural Chemistry, Univ. Göteborg (1980).
31. T. Gulik-Krzywicki, E. Rivas and V. Luzzati, J. Mol. Biol. 27 : 303 (1967).
32. R.P. Rand and V. Luzzati, Biophys. J. 8 : 125 (1968).

THE ASSEMBLY OF AMPHIPHILES: MOLECULAR AND PHENOMENOLOGICAL MODELS

Peter Fromherz

Department of Biophysics
University of Ulm
D-7900 Ulm-Eselsberg, FRG

Non-bilayer assemblies are necessarily some form or other of transformations of lipid bilayers. A direct experimental analysis of their structure is not possible, due to their transient nature. A theoretical treatment without invoking arbitrary constraints is not feasible at the present. On the other hand the mere assignment of symbols is not sufficient for an understanding of the biological processes in membranes and for guiding the adaptation of these processes to technological and medical purposes.

The present paper defines some simple chemical and physical concepts in regard to the molecular structure and to the thermodynamics of non-bilayer assemblies in membranes. Narrow convex and concave curvatures in assemblies of amphiphiles are built-up according to a block architecture. Prototypes are the biconvex micelle, the straight-convex bilayer-edge, the concave-convex semi-inverted-micelle within bilayers and the biconcave cylindrical inverted micelle within bilayer junctions. The energetics of the simplest case – the edge of bilayers – is treated by a thermodynamical approach. Several implications of the models are discussed: Synthesis of the droplet and of the bilayer model of micelles, the magnitude of the edge-tension of membranes, the hysteresis of the closure and opening of vesicles, the mechanism of intra-bilayer lipid exchange and the nature of lipid particles within bilayer attachment sites.

55

INTRODUCTION

The planar lipid bilayer is usually symbolized by parallel aligned knobbed sticks representing the hydrocarbon chains and the polar headgroups of lipids (Figure 1). There is general agreement on the basis of a wide range of experimental data as to how this symbol can be interpreted in terms of a true molecular structure with respect to the conformations within chains and headgroups and with respect to rotational and translational dynamics of the molecules[1].

The lipid bilayer, however, is subjected to processes, such as intra-bilayer lipid exchange, bilayer lysis, bilayer attachment and fusion, which are beyond the range of possibilities expressed by the usual symbol. Since the non-bilayer structures involved in these processes are transient, they are not amenable to the same experimental techniques as the bilayer itself. The only approach which remains at the present moment is to study stable non-bilayer assemblies of amphiphiles and to derive from their structures possible non-bilayer defects in membranes. In this regard, until now, the usual starlike symbols for micelles and inverted micelles (Figure 1) have been assigned to presumed non-bilayer events[2]. As a further step, concepts are required that refer to the true physical nature of non-bilayer defects in order to promote experimental and theoretical investigations on model systems and as guidelines for the biological research. Such heuristic concepts may refer to the geometry of packing on a molecular level and to the energetics of the presumed defect assemblies on a phenomenological level.

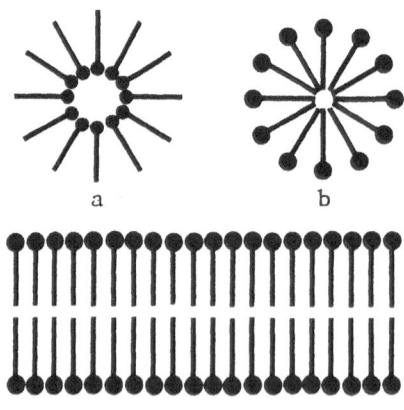

Figure 1. *The conventional symbols for (a) an inverted micelle, (b) for an isometric surfactant micelle and (c) for a planar lipid bilayer.*

It is the purpose of the present paper to discuss and implement an architectural principle for molecular packing and a thermodynamic framework for the energetics of non-bilayer assemblies that have recently been proposed[3-7]. The starting point is the structure of surfactant micelles[3,6]. Their block-architecture is assigned to the edge of bilayers as well[4,5]. The structure of edges is crucial in the processes of closure and lysis of lipid vesicles which is treated phenomenologically: The control of vesiculation is described by a Gibbs-isotherm with respect to edge-active solutes binding to the bilayer-edge[7]. Then block-architecture is applied to describe elementary edge-defects within a closed bilayer. The implications of these defects are discussed with respect to pertinent non-bilayer processes[4,5]. Finally a compact inverted block-micelle is proposed which may exist within bilayer junctions as an initial stage of membrane fusion[4,5].

SURFACTANT MICELLES

Surfactants like sodium dodecylsulfate aggregate in aqueous solution to form rather well defined assemblies micelles[8,9]. In classical micelle research, two concepts were invoked to describe their nature: Mc Bain considered them to be solid fragments of a bilayer composed of extended amphiphiles[10-12] and Hartley considered them to be disordered clusters[13,14] (Figure 2). The energetics of the system was treated by Stigter and Tanford using the concept of a compact sphere-like droplet, taking into account only the hydrocarbon/water interface and the headgroup repulsion[15,16], disregarding the structure of the liquid[17].

Recent experimental data referring to the nature of the hydrocarbon conformation[9] and of the hydrocarbon/water contact[18] indicate that the concept of the bilayer model holds for describing the structural features of a micelle, if the cylindrical bilayer fragment is considered to be punched out from an infinite bilayer in its liquid-crystalline state[3,4,6]. It appears that a bilayer and a droplet are antithetic models of micelle structure. The first does not account for the energetics due to its extreme headgroup interaction and its arbitrary size, the second does not account for structure, as it refers only to thermodynamic, i.e. average features. On the other hand, however, the two models are complementary: Together they describe all the structural and energetic features of the micelle.

To resolve the inherent contradictions between the two classical concepts, the surfactant-block model was proposed (Figure 2). The amphiphiles aggregate to isometric assemblies which resemble the bilayer in that there is a parallel organization of the chains in blocks whose width corresponds to the length of their chains, the

blocks being perpendicular to each other. The surfactant-block model
does not contradict the droplet model, but interprets it in terms of
a particular structure in cases where a structural assignment is required. Considering classical sketches of micelles (Figure 2), namely
of the solid bilayer[12], the disordered cluster[13], and the strict
droplet[15], it is apparent that the new model is a synthesis. It
implies local disorder, allowing for segmental motion, molecular rotation and lateral motion as we'l as total rearrangement, yet the
disorder is restricted by the partial parallel alignment of the
chains. On an average it resembles the compact droplet. It may be

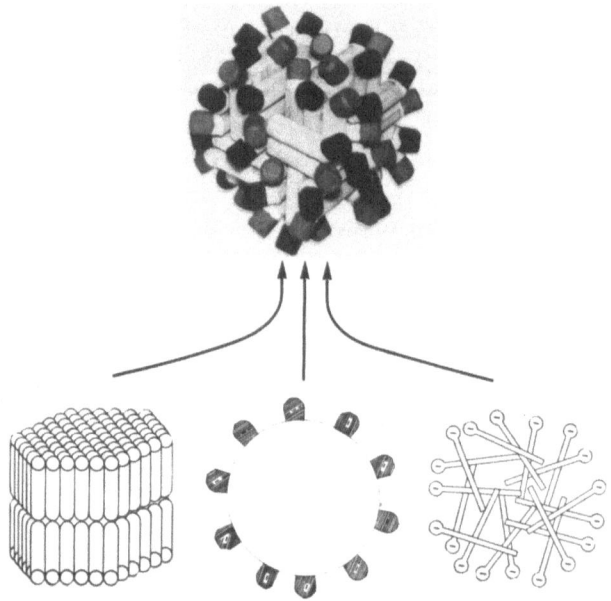

Figure 2. *Surfactant-Block model of a micelle of sodium-dodecyl-
sulfate. The term block refers to the regions of parallel alignment of the molecules. The range of parallel correlation equals
the length of a hydrocarbon chain. The wood models chosen represent
the volume occupied by a hydrocarbon chain in its liquid crystalline state. They account for kinks and voids in the packing. The
dynamics of the system is implicit in this design. The kinks near
the heads indicate how bulky heads may be accommodated in the
packing. Half of the headgroups are marked with a dark color to
indicate charge neutralization in ionic micelles by bound counterions. At the bottom of the figure three classical representations
of a micelle are drawn which reflect partial aspects of block-
packing: The solid bilayer of McBain/Philippoff, the structure-
free droplet of Stigter/Tanford and the disordered cluster of
Hartley.*

necessary to repeat here that the "wood-paint-glue" models used to
depict the surfactant-block model[3,4,6] (Figure 2) consist of sticks
which represent the space as filled on an average by a hydrocarbon
chain in a liquid-crystalline state. Thus these stiff-looking figures
of the block-model account for the kink formation and voids that are
present in a liquid crystal. The dynamics of the assembly is described implicitly by those wood models. Dynamics is not equivalent
to disorder.

The most peculiar features of the surfactant-block model are:
a) A spheroid compact core of a constant density and with no extensive coiling. b) Stoichiometry that is determined by molecular
dimensions, as by the ratio of chain length to chain width. c) The
average radius of the compact hydrocarbon core may exceed the chain
length. d) Distinct sites have differing physical environments
(they are averaged for time constant above 1 ns). e) There is a
fraction of molecules located tangentially along the surface.
f) There is segregation of the surface into fatty patches and zones
rich in headgroups.

Experimental data which would confute the model have not been
reported. Considering the successful rationalization with experimental data, it is the clarity of the model and the simplicity of its
design as derived from the well-known lipid bilayer which make it
a useful model for structural studies, even at the present semiquan-
titative level[6].

MEMBRANE EDGE

Considering that the micelle structure derives from the bilayer
configuration, it is most natural to use the block-architecture to
describe an open edge of a bilayer (Figure 3). Since the width of a
liquid-crystalline paraffin chain is about equal to the length of
four extended methylene groups, a block consists of four chains of
the usual membrane lipids, i.e. of one molecule cardiolipin, two
molecules lecithin or four molecules lysolecithin.

The hydrocarbon/water area of the edge is essentially unchanged for any reorganization of the edge. (This holds true even for a
structure-free semicyclindrical liquid-rim model of the edge, if
the hydrocarbon maintains constant density and is water-free). Thus
the area is determined by the thickness of the bilayer, $d = 4$ nm[19].
Considering the hydrophobic energy $\epsilon_{HP} = 1.75 \times 10^{-20}$ J/nm[2][16], the
work required to form a unit length of an edge, i.e. the edge ten-
sion due to the hydrophobic effect, is $\gamma_{HP} = 7 \times 10^{-20}$ J/nm[7].

Figure 3. *Block architecture of a bilayer edge. The sticks indicate dipalmitoyl-chains in their liquid crystalline state, the black bricks indicate phosphatidylcholine groups. Note the hydrocarbon/ water contact which is distributed between the headgroups on edge sites.*

The total edge energy is obtained by adding the change in head-group interaction. This is due to the transformation from an intact bilayer to an open edge. Block assembly distributes the open area of the edge in such a way that the space available for the headgroups of molecules located on edge sites is doubled. In this way, lipids with repulsive headgroups (bulky or electrically charged) are accom-modated. A compact structural organization of the open edge is at-tained with amphiphiles having straight stems and bulky heads. Wedge-shaped molecules are not required to match the edge structure.

VESICLE CLOSURE

Stable membranes are characterized by a high edge energy, which is dominated by the hydrophobic component. Open planar fragments may exist, however, if resealing of the edge by vesicle formation implies too large an increase in the elastic energy caused by the curvature. A solution of these planar fragments may lower its energy by lateral fusion, which reduces the total edge length. Above a certain size of the fragments, vesiculation occurs as the decrease in edge energy overcomes the elastic stress.

The total energy in a dispersion of 5.4×10^4 nm^2 of egg-lecithin bilayer arising from curvature and from open edges is shown in Figure 4 as a function of the fragment shape for various degress of fragmentation. For high fragmentation, the shape that would have a global energy minimum is a disc. Growth leads to a decrease in energy of the system. For 100 fragments (diameter 26.5 nm), the local minima for the closed vesicular state have the same energy as the disc state. The minimum corresponding to the disc disappears after fusion down to 25 fragments: The disc diameter of 53 nm is the limit for vesiculation. So, a dispersion of egg-lecithin bilayer equili-

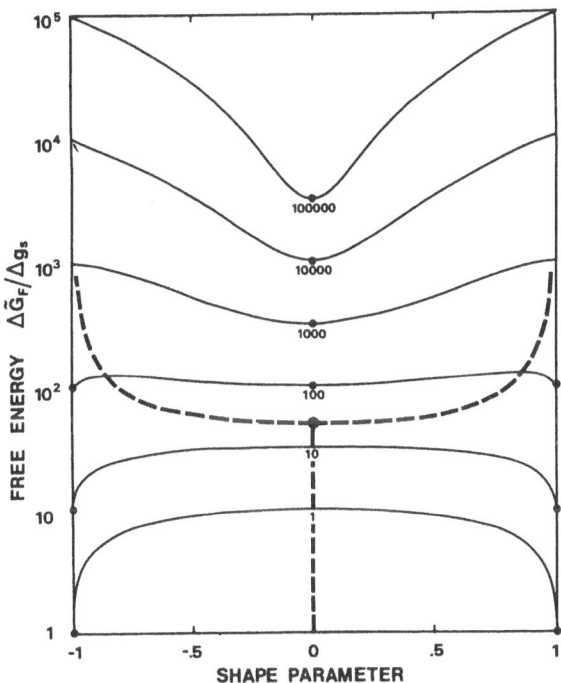

Figure 4.   Free energy, $\tilde{\Delta G}_F$, of a solution of monodisperse bilayer fragments versus the shape parameter Ω. The energy is calculated relative to the elastic energy of a single closed vesicle Δg_s. The shape parameter indicates the relative height of the curved shell formed by a fragment. It is $\Omega = 0$ for planar discs, and $\Omega = \pm 1$ for closed vesicles. The potential profiles are drawn for various degrees of fragmentation as indicated. The total bilayer area is 5.4×10^4 nm^2, the elastic modulus is 2.3×10^{-19} J, the edge tension 7×10^{-20} J/nm (cf ref. (7)). The bifurcated broken line connects the energy maxima of the profiles.

brates first along the energy minimum of the disc state, then jumps
into one of the local energy minima of the closed vesicles. Further
decreases in energy may be attained by vesicle fusion until a single
lamella is attained. As soon as the vesicular state is attained,
however, the dispersion is usually metastable, because of the high-
energy intermediates of vesicle fusion.

Considering the energy profiles in Figure 4 it is apparent that
just after vesiculation the energy of the dispersion is in a deep
minimum with respect to a burst of the vesicles reverting to the
disc state. To reverse the energy profile so that the disc state
again becomes more stable for a constant size of the fragments, the
edge energy has to be drastically lowered. Closure and opening of
vesicles does not occur under identical physical conditions: This
hysteresis of the closure/burst transition is the fundamental principle of the existence of closed membrane structures.

EDGE - ACTANTS

The critical size of membrane fragments necessary for spontaneous closure to vesicles is determined by the magnitude of the edge
tension. Partial compensation of the hydrophobic edge energy neces-
sitates larger fragments[7]. Thus control of edge tension implies control of vesicle diameter.

The edge tension is lowered by solutes in the bulk water or in
the bulk membrane which accumulate at the edge of the bilayer. The
term "edge-actant" is introduced for these solutes[7]. Their action
is described by an equation analogous to Gibbs' adsorption isotherm
for surfactants and surface tension. Edge-active agents are detergents which form the typical micelle, such as lysolecithin and
doubly charged phosphatidic acid; and also solvents which accumulate at a hydrocarbon/water interface, such as ethanol and ether.

Addition of an edge-actant to a lipid/water suspension lowers
the edge tension of the system, until, at a certain critical concentration, the edge-tension vanishes. Further addition of edge-
actant leads to the spontaneous fragmentation of stable closed membranes into stable planar open discs, i.e. into mixed micelles[20].
The equilibrium states of the infinite planar bilayer and of finite
mixed micelles as well as the threshold of the disc/vesicle transition are indicated in Figure 5.

Vesiculation may be described as a cyclic process: 1) Edge-
actant is added to a lipid/water dispersion. 2) The system relaxes, forming micellar fragments. 3) The concentration of the

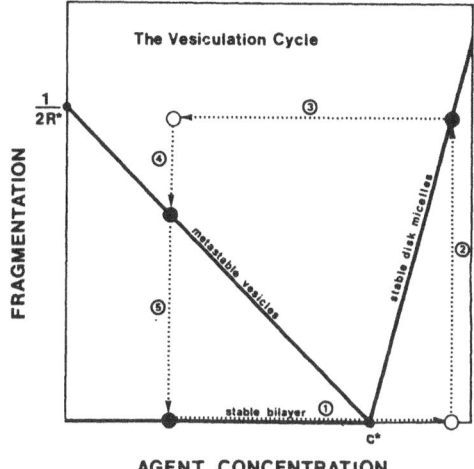

Figure 5. *Fragmentation of a bilayer dispersion (reciprocal diameter 2 R_S of vesicles or reciprocal radius of planar discs) versus the concentration of an edge-actant (schematic drawing). Below
a critical concentration c* the planar edge-free bilayer is thermodynamically stable. Above c*, where the edge tension vanishes,
emulsification of the bilayer by the agent leads to stable mixed
micelles. The range of unstable finite bilayer fragments is divided by the limit for vesiculation of the disc fragments. The vesicles formed are metastable with respect to the planar edge-free
state. Vesicle formation in a lipid dispersion is drawn as a cyclic
process as discussed in the text.*

edge-actant is decreased below a critical concentration c*. 4) The
fragments grow until the size for vesiculation is attained. 5) Further growth by vesicle fusion up to the stable planar edge-free bilayer is kinetically hindered. The cycle is realized by the addition of detergents and their removal by dialysis or dilution[21-23],
by the addition of organic solvents and their removal by evaporation
or dilution[24,25] or by titration and back-titration with an agent
that transforms lipid components into edge-actants, as in the case
of phosphatidic acid treated with acid and base[26].

NON-BILAYER DEFECTS

At a finite edge-tension, tiny nuclei of edges may occur as
fluctuations in an intact bilayer. Due to their transient nature a
direct experimental analysis of their structure is not possible.
Thus suitable models of those defects are necessary.

By bending the block-model of an edge (Figure 3) in the bilayer plane to a concave pore and by reducing the radius, one attains a compact radial aggregate of rotated blocks (Figure 6). Compactness is attained by solvating the radial blocks with bilayer material. The assembly is matched to the geometry of a bilayer. The (convex) micellar character of this assembly is reflected in the exposed hydrocarbon. The (concave) inverted micellar character is reflected in the headgroup contacts. The defect may be called a "semi-inverted micelle" or SIM-defect.

The frequency of SIM-defects is determined by their energy, according to Boltzmann's law. The hydrophobic surface of the rotated blocks leads to a high energy, suppressing SIM-formation. A stabilization may involve (A) the micellar character due to (A1) agents that lower the hydrophobic energy (alcohol, ether, polyethyleneglycol) or to (A2) agents that expose hydrophobic surfaces (such proteins as phospholipases[27] and cholates which are bound parallel to the membrane plane[28]) or to (A3) edge-actants which have repulsive headgroups which accumulate on bilayer sites adjacent to the rotated blocks (lysolecithin, detergents). On the other hand stabilization may involve (B) the inverted micellar character by (B1), agents that have attractive headgroups, which are not necessarily

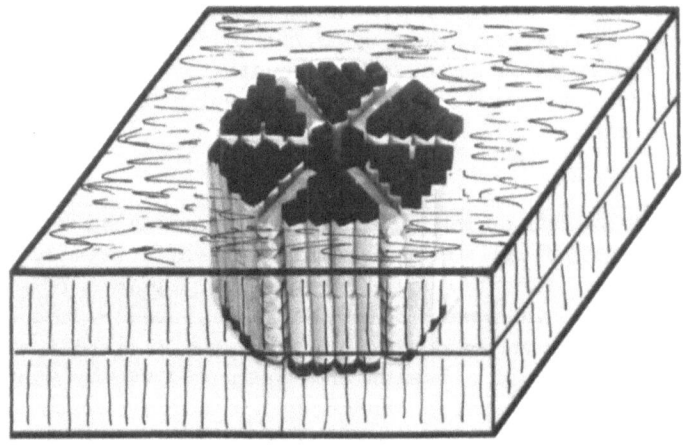

Figure 6. *Semi-inverted micelle (SIM) in block architecture within a lipid membrane. The compact packing is attained by solvating the radial rotated blocks with bilayer material. Note the exposed hydrocarbon chains reflecting the micellar character of the assembly and note the headgroup contacts in the center reflecting the inversed micellar character. This SIM-defect is matched to the bilayer geometry. Its existence is due to certain agents and leads to certain observable effects, as discussed in the text.*

cone shaped, (carboxylic acids with hydrogen bonds, phosphatidylethanolamine with ionic bonds, cardiolipin/calcium with salt bridges) or by (B2), an electric field across the bilayer which is parallel to the headgroup dipoles of the rotated blocks[29].

The SIM-structure implies a series of observable effects:
(a) The rotation of the blocks perpendicular to the membrane and
the tumbling of the rotated blocks in the membrane lead to a topology of headgroup motion which is identical with the topology of
motion in hexagonal phases or in isotropic structures, although a
hexagonal phase or a spherical structure is not actually present.
(b) The periphery of the SIM-defect interrupts the bilayer packing.
Mechanical stress leads to a preferential rupture across the monolayers along the cylindrical periphery of the SIM-defect, producing
monodisperse lumps of about 6 nm diameter. (c) Lipids are brought
from one side of the bilayer to the other side: Rotation of blocks
orients molecules parallel to the membrane. Alignment of rotated
blocks creates a pathway for lateral hopping of the molecules across
the bilayer. Resealing of the defect completes the process (ROTation-
HOPping Mechanism). (d) Hydration of the central core of the SIM-
defect creates a hydrophilic zone across the membrane. Ionic conductivity across this channel is blocked when the radial defects within
the two monolayers are displaced by lateral motion.

Considering the data of P-31 NMR spectroscopy[2], the observations of intramembrane lipid exchange[30], of freeze-fracture particles[31] and of ion permeation[32] it is noteworthy that so many features of "non-bilayer" events can be interpreted in terms of the
SIM-defects. This coincidence indicates that the SIM-defect serves
at least as a heuristic basis for the discussion of non-bilayer
structures in membranes.

MEMBRANE JUNCTIONS

Lipid vesicles may close and open not only in their free state
but also when they are in contact with a second bilayer[33]. In such
a process of fusion the change in elastic energy may be seen as
similar to the case of free burst. However, the change of edge
energy is different, since both the initial and final states are
edge-free and since the edge is composed of two bilayers. An estimation of the magnitude of edge-tension, which dominates the initial
stages of fusion, requires a molecular model, so that predictions
about the fusion process can be made.

The prerequisite for bilayer fusion is, of course, a close contact of the two bilayers and this in turn requires attractive forces

between the headgroups. Block architecture provides for the next
step which is the axial rotation of blocks in the two outer monolayers (Figure 7). The formation of such a cylindrical inverted
micelle - a CIM-defect - is an immediate consequence of bilayer
contact: The conditions of headgroup attraction are fulfilled and
the lateral contact of the rotated blocks across the junction (contact region) prevents the exposure of hydrocarbons to water. The
CIM-defect is a compact inverted assembly. On the other hand it is
clearly different from the radial symbols used for inverted micelles within bilayer junctions[34],[35] and on the other it is the most
direct molecular representation of those mere symbols. Compactness
of the reversed state is attained by the parallel and perpendicular
alignment of the molecules, which is analogous to the block-model
for micelles.

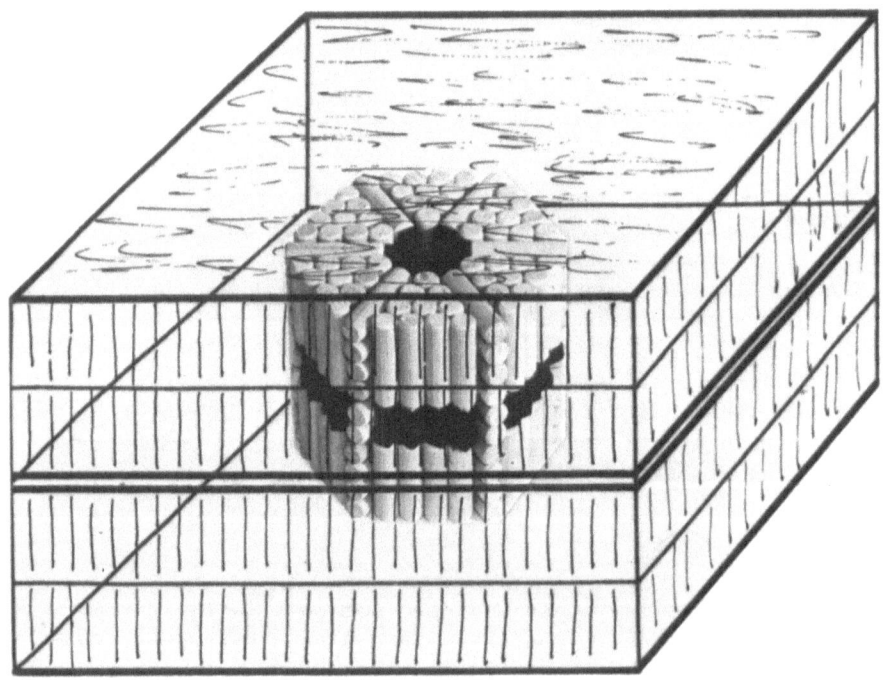

Figure 7. *Cylindrical-inverted micelle (CIM) in block architecture
within the two contacting monolayers of a bilayer attachment site.
Note the compact packing attained by parallel and perpendicular
alignment of the chains and the mutually saturated headgroup contacts. No hydrocarbon is exposed to water. The CIM-defect derives
from a close contact of the bilayers and is the first stage towards
membrane fusion.*

The observable consequences of transient or stable CIM-defects are isotropic motion of headgroups, inter-membrane lipid exchange and monodisperse particles under freeze-fracture[34,35]. In asymmetric bilayers the cylindrical defect may induce conical deformations leading to mutual attraction of the defects in analogy to the case of intrinsic membrane proteins[36].

Within block architecture the third step towards fusion is a propagation of the defect across the whole junction, i.e. the formation of two stacked SIM-defects (cf Figure 6), where the central part is the CIM (Figure 7). In this stage, additional agents are required to stabilize the exposed hydrocarbon moities. Permeation of hydrophilic materials across the junction is an additional observable effect[37]. Further steps in the pathway of fusion are then 4) enlargement of the hydrophilic core, 5) insertion of a cylindrical bilayer between the membranes, 6) smoothing of all edges until complete fusion is achieved. A thermodynamic elaboration of the fusion process along these structural intermediates, analogous to the closure/burst transition, can be attempted.

The pathway 1 to 6 appears to be consistent with the scarce data available about fusion. This will be discussed elsewhere. Here it is sufficient to note that the mechanism based on block-architecture may provide at least a model to direct and test experimental work.

REFERENCES

1. J. Seelig and A. Seelig, Quart. Rev. Biophys.13 : 19 (1980).
2. B. De Kruijff, P.R. Cullis and A.J. Verkleij, Trends Biochem. Sci. 5 : 79 (1980).
3. P. Fromherz, Chem. Phys. Lett. 77 : 460 (1981).
4. P. Fromherz, Ber. Bunsenges. Phys. Chem. 85 : 891 (1981).
5. P. Fromherz, Biophys. Struct. Mech. 7 : 297 (1981).
6. P. Fromherz, in: Proceedings of the International Symposium on Surfactants in Solution, Lund 1982, eds. K.L. Mittal, B. Lindman, Plenum Press New York, in press.
7. P. Fromherz, Chem. Phys. Lett. 94 : 259 (1983).
8. A. Reychler, Kolloid Z. 12 : 277 (1913).
9. H. Wennerström and B. Lindman, Phys. Rept. 52 : 1 (1979).
10. J.W. McBain, J. Chem. Educ. 6 : 2115 (1929), p. 2121.
11. J.W. McBain, Colloid Science, Boston 1950, p. 255.
12. W. Philippoff, Kolloid Z. 96 : 255 (1941).
13. G.S. Hartley, Aqueous Solutions of Paraffinic Chain Salts, Paris (1936).
14. G.S. Hartley, Quart. Rev. Chem. Soc. 2 : 152 (1948).

15. D. Stigter, J. Phys. Chem. 78 : 2480 (1974).
16. C. Tanford, J. Phys. Chem. 78 : 2469 (1974).
17. J.P. Hansen and I.R. McDonald, Theory of Simple Liquids, London (1976).
18. F.M. Menger, Acc. Chem. Res. 12 : 111 (1979).
19. M.C. Phillips, E.G. Finer and H. Hauser, Biochim. Biophys. Acta 290 : 397 (1972).
20. N.A. Mazer, G.B. Benedek and M.C. Carey, Biochemistry 19 : 601 (1980).
21. J. Brunner, J. Skrabal and H. Hauser, Biochim. Biophys. Acta 455 : 322 (1976).
22. O. Zumbühl and H.G. Weder, Biochim. Biophys. Acta 640 : 252 (1981).
23. P. Schurtenberger, N.A. Mazer, W. Känzig and R. Preisig, in: Proceedings of the International Symposium on Surfactants in Solution, Lund 1982, eds. K.L. Mittal, B. Lindman, Plenum Press New York, in press.
24. S. Batzri and E.D. Korn, Biochim. Biophys. Acta 298 : 1015 (1973).
25. D. Deamer and A.D. Bangham, Biochim. Biophys. Acta 443 : 629 (1976).
26. H. Hauser and N. Gais, Proc. Natl. Acad. Sci. USA 79 : 1693 (1982).
27. G.H. deHaas, Presentation at Rigi-Kaltbad 1982.
28. J. Ulmius, G. Lindblom, H. Wennerström, L.B.A. Johansson, K. Fontell, O. Södermann and G. Arvidson, Biochemistry 21 : 1553 (1982).
29. R. Benz and U. Zimmermann, Biochim. Biophys. Acta 640 : 169 (1981).
30. R.P. Kornberg and H.M. McConnell, Biochemistry 10 : 1111 (1971).
31. A.J. Verkleij and P.J.Z. Ververgaert, Biochim. Biophys. Acta 515 : 303 (1978).
32. V.F. Antonov, V.V. Petrov, A.A. Molnar, D.A. Prevoditelev and A.S. Ivanov, Nature 283 : 585 (1980).
33. D. Papahadjopoulos, G. Poste and W.J. Vail, in: Methods of Membrane Biology vol. 10 ed. E.D. Korn, Plenum Press, New York 1979 p. 1.
34. S.W. Hui and T.P. Stewart, Nature 290 : 427 (1981).
35. A.J. Verkleij and B. de Kruijff, Nature 290 : 427 (1981).
36. H. Gruler, Z. f. Naturforsch. 30c : 608 (1975).
37. J. Wilschut, N. Düzgünes and D. Papahadjopoulos, Biochemistry 20 : 3126 (1981).

PURITY OF AEROSOL-OT(AOT):   EFFECT ON PROCESSES IN REVERSED

MICELLES AND WATER-IN-OIL MICROEMULSIONS

Paul D.I. Fletcher, Neil M. Perrins, Brian H. Robinson
and Christopher Toprakcioglu

Chemistry Laboratory
University of Kent at Canterbury, U.K.

The presence of impurities in AOT can have very dramatic effects on the kinetics of reactions in microemulsions and to a lesser extent on other properties, e.g. solubilization capacity. If the AOT contains a water-soluble impurity at the 0.1% level (mole per cent), a water-in-oil microemulsion prepared using 10^{-1} mol dm^{-3} AOT and containing 1% by volume of water will have an impurity concentration in the water phase of 10^{-2} mol dm^{-3} of water. In a kinetic experiment, this is likely to be in excess of the reactant concentrations. On the other hand, impurities which partition into the oil phase are normally of little consequence.

Impurities in the AOT at the 0.1% level are not easy to identify and their subsequent removal can pose considerable problems.

There are two classes of impurity in AOT: those present as a result of the manufacturing process and those produced after preparation of reversed micellar/microemulsion systems. AOT is often prepared by di-esterification of maleic or fumaric acids with 2-ethyl hexanol. The di-ester is then sulfonated with sodium bisulfite. Consequently, possible impurities are sodium bisulfite, the parent dicarbolic acids, various acidic monoesters present as a result of incomplete esterification and 2-ethylhexanol. The alcohol has a characteristic odor and is readily dedected in some commercial samples of AOT. Water is also present, but generally at a H$_2$O:AOT mole ratio of <1. The existence of an aromatic species and a reducing impurity is also indicated. In a water-in-oil microemulsion, the main impurities are carboxylic acids and 2-ethylhexanol formed as a result of self-hydrolysis of AOT in the dispersion. Acid impurities are by far

69

the most significant in most applications since they act in the
water phase and can exert a very significant buffering effect on
the 'pH' of the system.

The hydroxide-promoted hydrolysis of AOT results in the production of a carboxylate ion (the conjugate-base anion of either an α
or β substituted sulfonate monoester carboxylic acid) and 2-ethyl
hexanol.

Therefore we have:

$$R'OCR'' + OH^{\ominus} \longrightarrow R'OH + R''CO_2^{\ominus} \qquad \qquad Equation\ 1$$

There are two possible products of the reaction, depending on
which of the two carboxyl groups on AOT is attacked. In water the
rate of hydrolysis has been measured at concentrations of AOT<cmc[1]
and in the presence of excess OH⁻. The reaction is first order in
hydroxide and AOT is found to have a half-life of ∿3 days at pH∿12
and a temperature of 25°C. The decomposition of AOT in water is also
acid-catalyzed. When the AOT is located at the oil-water interface
of a microemulsion droplet, it can still be decomposed by hydroxide
according to Equation 1. The result is that the 'pH' of the pool
system, which is initially at high pH, decreases. The kinetics have
been studied utilizing the color change of various water-soluble
pH indicators, e.g. sulfo-orange, alizarin yellow, to monitor the
progress of the reaction[2]. The rate of reaction is quite similar to
that in bulk water, with $-d[AOT]/dt \sim 5 \times 10^{-8}$ mol $dm^{-3}s^{-1}$ at
$[H_2O]/[AOT] = 20$, $[AOT] = 0.1$ mol dm^{-3} and a fixed $[OH^-] = 10^{-2}$ mol
dm^{-3} (of water). For a constant concentration of OH⁻ of ∿10^{-2} mol dm^{-3}
the half-life of AOT would be ∿1 week.

In practice, AOT is in great excess over hydroxide so that in
a microemulsion at a hydroxide concentration of 10^{-2} mol dm^{-3} (of
water), the 'effective' AOT concentration (at 0.1 mol dm^{-3} overall
in a microemulsion containing 1% water) is 10 mol dm^{-3} of water.
Hence only 0.1% of the AOT would be decomposed and the reaction rate
will decrease rapidly as OH⁻ is consumed, since $-d[AOT]/dt$ is proportional to $[OH^-]$. The time for 90% of the hydroxide to be consumed
is typically of the order of one hour, a more rapid rate being obtained as the water content is decreased. Hence stock solutions containing base must be used immediately after preparation to avoid a
change in 'pH' unless a large excess of a buffering agent is present.
It is also clear that if a pH-dependent reaction is involved in basic
microemulsions, it is essential to study the reaction in the presence
of a high concentration of buffer.

Microemulsions containing water at neutral pH are stable to decomposition (at the 0.01% level) over a period of several days and stock solutions of AOT in a hydrocarbon such as heptane (no added water) are probably stable over a much longer period. In general, care must be taken in solvents other than alkanes (e.g. halogen-containing solvents) where water-soluble impurities (e.g. HCl) might be present at significant concentrations in the solvent.

The pKa's of the carboxylic acids present are unknown, but might be expected to be in the region 3-6. They will effectively buffer the aqueous phase of the microemulsion in this pH region.

Hence, the implication is that acid- and base-containing microemulsions are inherently unstable and this may possibly be a factor in explaining the instability of AOT microemulsions over long time periods[3]. Also an attempt to prepare acid-containing droplets at e.g. pH = 2 may be thwarted by the buffering effect of any acid impurity present. To prepare OH^--containing microemulsions, it would be first necessary to titrate the acid impurity. If this is not done the 'pH' of the microemulsion may well be lower than that predicted from the weighed-in amount of base.

The 'pH' of the microemulsion system based on AOT has been measured using the indicator 4-nitrophenol-2-sulfonate(sodium salt). The pKa of this indicator in water is $6.80(\pm0.05)$ at $25^{\circ}C$ (I → 0) reducing, as the ionic strength is increased, to ∿6.0. The pKa is not thought to be much perturbed on solubilization in the microemulsion (pKa ∿6.2 (±0.2))[4]. Both forms of the indicator (charges -1 and -2) are thought to be located in the water phase since they are totally insoluble in heptane and very soluble ($>10^{-2}$mol dm^{-3}) in water. Other advantages are the small size of the probe and the relatively large extinction coefficient of the di-anion, which means that low concentrations can be used. Another method we have used to probe the 'pH' of the acid pools is to add the water-soluble indicator anion murexide (ammonium purpurate). From the rate of fading of the indicator (λ_{max} = 520nm), which is acid-base catalyzed, the 'pH' can be obtained. In droplets at neutral pH, the indicator is stable.

The presence of 'acid' impurities will very significantly affect reactions in microemulsions where pH is important. Enzyme reactions will be particularly sensitive to the presence of such impurities since the pH of the aqueous pool will be uncertain. It will then be necessary to introduce a titration procedure or a large additional buffering capacity if misleading results are not to be obtained.

The main effect of the presence of alcohols (2-ethylhexanol) is to influence the rate of exchange of reactants between droplets[5] through perturbation of the surfactant layer. This effect would only be of significance in the case of reactions for which the overall rate is controlled by communication. The size of the droplets, as determined by small angle neutron scattering, is not much affected by the named impurities (at $[imp]/[surf]<0.05$) provided that the composition of the system is not near to phase boundaries (i.e. R/R_{max} <0.5 (where $R = [H_2O]/[AOT]$) and $T_c-T >10^oC$ (where T_c is the temperature corresponding to the upper phase transition)).

It would seem that it is difficult to purify AOT using procedures described in the literature. Clearly the initial purity of the supplied material is very important. Previously, the quality from different suppliers has been very variable, and even different batches from the same supplier have contained widely varying amounts of the acid impurity. However, recently the quality of our purchased AOT (from Sigma) has been much improved. We would strongly recommend the purchase of a large amount of a single batch for a whole series of experiments.

REFERENCES

1. E.F. Williams, N.T. Woodberry and J.K. Dixon, Journal of Colloid Science, 12 : 452 (1957).
2. P.D.I. Fletcher, A.M. Howe, N.M. Perrins, B.H.Robinson, C. Toprakcioglu and J.C. Dore, in "Proceedings of the Third International Symposium on Surfactants in Solution", ed. K.L. Mittal, Plenum, 1983.
3. A. Assih, P. de Lord and F.C. Larché, this publication.
4. D.C. Steytler, 'Ph.D. Thesis', University of Kent, 1981.
5. S.S. Atik and J.K. Thomas, J. Amer. Chem. Soc., 103 : 3543 (1981).

THE PROBLEM OF CONCENTRATION AND REACTIVITY IN REVERSED MICELLES

AND WATER-IN-OIL MICROEMULSIONS

Paul D.I. Fletcher, Andrew M. Howe, Brian H.Robinson
and David C. Steytler

Chemical Laboratory
University of Kent at Canterbury, U.K.

Microemulsions have aroused much interest as a novel medium for chemical reactions. It is usual to attempt to compare reactivity in the dispersed system with homogeneous solvent media. For water-soluble reactants, insoluble in the dispersion medium, the aim would be to compare reactivity in the aqueous droplets of the microemulsion with bulk aqueous solution. In the case of first-order reactions the measured rate constants (units $-s^{-1}$) may, in most cases, be compared directly in the two types of system, (e.g. k_{cat} for an enzyme reaction). However, in the case of a bi-molecular process, the calculation of the second-order rate constant (units $-dm^3 mol^{-1}s^{-1}$) involves consideration of the concentration of species in the microemulsion, and the question arises as to whether the overall concentration (moles dm^{-3} of total solution) or the concentration in the water should be employed. In the latter case there are two possibilities: a) the reactant concentration in a discrete droplet and b) the concentration in moles dm^{-3} of the dispersed aqueous phase. In principle it is possible to define and use any concentration scale. The problem is to decide on the most applicable units in order to best interpret the data in terms of a comparison of reactivity in the microemulsion and bulk solvent media.

Some kinetic parameters are more easily obtained than others. For example, reaction enthalpies ($\Delta H^\ominus$) and activation enthalpies ($\Delta H^\ddagger$) are independent of the concentration units employed so they may be directly measured and compared. Reaction entropies ($\Delta S^\ominus$) and activation entropies ($\Delta S^\ddagger$) involve consideration of standard state definitions in the microemulsion. $\Delta V^\ominus$ and $\Delta V^\ddagger$ may also be measured directly and these parameters may be useful in the assignment of mechanism. A problem with these parameters is that they may be influenced by a change in state of the microemulsions, and

73

so changes in e.g. droplet concentration with temperature or
pressure may need to be considered.

The discussion in this paper will be restricted to reacting
species which are totally confined within the aqueous domains of the
water-in-oil microemulsion system. Considerations of the most
appropriate concentration units to be employed are then determined
predominantly by the time domain of the reaction under study compared
with the time required for redistribution of the reactant species
between the aqueous droplets of the microemulsion dispersion. Three
cases can be envisaged: (a) - (c).

(a) Distribution is slow relative to the reaction rate :

For very fast reactions, the distribution of reactants among
the water pools is frozen on the time scale of reaction, i.e. no
exchange of reactants between droplets takes place. In this case,
the concentration of reacting species which is meaningful in the
interpretation of kinetic data is obtained by dividing the number
of moles (molecules) of that species in a droplet by the volume of
water in the discrete droplet. This situation is commonly observed
for fluorescence-quenching reactions in microemulsion systems of the
type water/Aerosol-OT/heptane. The time scale of the reaction is
typically 10^{-8} - 10^{-6}s and exchange of ionic quenchers occurs in
the 10msec - 100μsec time range[1,2]. For the situation where there
is more than one quenching ion per pool, a complex convolution of
the rates arising from pools containing one fluorescer and different
low, integral numbers of quencher molecules is observed. As the
quencher concentration is reduced a situation is reached when the
fluorescence decay is made up of two processes corresponding to the
natural fluoresence lifetime and the shorter lifetime when one
quencher is present in the pool with the fluorescer. The rate of
reaction corresponding to pools containing a single quencher, and
the statistical distribution of quencher ions between pools can be
readily extracted from the data. The distribution is usually found
to be of the 'Poisson' type which implies a random, non-cooperative
distribution. The second-order rate constant for quenching by a
single quencher in a pool can be calculated (in units of dm^3 (of
water) $mol^{-1}s^{-1}$) by expressing the quencher concentration, equal
to one molecule per pool volume, as N_{AV}^{-1}/(volume of water per
pool in dm^3). The rate constant obtained is then directly comparable
with the second-order rate constant measured in bulk water. Quenching is often a process controlled by diffusion so that ion motion
within a droplet and local viscosity effects can be investigated.

(b) Distribution is the rate-determining step of the chemical reaction :

If the aqueous reactants are initially contained in separate
pools and reaction is induced by mixing the pool systems, then it

is necessary for the reactants (or one reactant) to exchange between pools prior to reaction. For intrinsically fast chemical reactions e.g. proton transfer, electron transfer, labile metal-ion/ligand substitution, the inter-droplet exchange process may be the rate-limiting step in the overall reaction. The rate constant for exchange then sets a limit for the fastest possible reaction in the system, and is analogous to the diffusion-controlled rate constant in a homogeneous medium. The observed chemical reaction then serves simply as an indicator for the exchange process. Exchange rates have been measured in this way for a variety of fast chemical reactions mainly in water/AOT/alkane systems[3]. For the process:

$$\text{(A)} \; + \; \text{( )} \quad \xrightarrow{k_{ex}} \quad \text{( )} \; + \; \text{(A)} \qquad\qquad \textit{Equation 1}$$

exchange can readily occur on collisions between droplets. The rate constant is rather insensitive to the actual species A being transferred (at least when A is small) and second-order rate constants are typically in the range 10^6–10^8 dm^{3}mol^{-1}s^{-1}. The volume dm^3 refers to the total volume of solution and mol to one mole of A or one mole of droplets since they are equivalent. Exchange occurs following energetic collisions between the droplets. If every collision were effective, k_{ex} could be calculated from the Brownian motion of the droplets and application of the Smolochowski Equation; a value of about 10^{10}dm^{3}mol^{-1}s^{-1} would be obtained for a low viscosity solvent. Hence exchange takes place in about 1 in 10^3 of the collisions. Since droplet concentrations are typically in the range 0.01–1mmol dm^{-3} of total solution, the exchange process, as already indicated, takes place on the 10ms to 100μs time scale. This would seem to be the case for a variety of surfactant systems.

However the situation is more complex when we consider the case of multiple occupancy of the pools. Representing a droplet containing n molecules of species A as $\text{(A}_n\text{)}$, the stoichiometric equation for the reaction may be represented as:

$$\sum \text{(A}_n\text{)} \; + \; \sum \text{(B}_m\text{)} \; \rightleftharpoons \; \sum\sum\sum \text{(}^{A_x}_{\substack{B_y\\C_z}}\text{)} \qquad\qquad \textit{Equation 2}$$

$$
\begin{array}{lll}
 & & x = 0, \; \infty \\
n = 0, \; \infty & m = 0, \; \infty & y = 0, \; \infty \\
 & & z = 0, \; \infty
\end{array}
$$

where the chemical process occurring within the droplets is A + B $\rightleftharpoons$ C.

Equation 2 represents the sum of an 'infinite' number of exchange processes, in some of which reaction occurs, in others A, B and C are simply redistributed. To calculate the rate constant for exchange, it is reasonable to make the assumption (for small species and low occupancies) that the exchange rate is independent of the pool contents, which is equivalent to the statement that the species A, B and C are distributed randomly throughout the pools (i.e. according to the Poisson distribution equation). The same rate constant is applied to all species being transferred; this can be shown to be the case by direct experiment. Even so, Equation 2 is thought to be insoluble analytically. However, by restricting the reactant concentrations it is possible to produce a situation corresponding typically to Equation 3:

$$\sum \left(A_n\right) + \sum \left(B_m\right) \rightleftharpoons \sum \sum \sum \left(\begin{matrix} A_x \\ B_y \\ C_z \end{matrix}\right) \qquad Equation\ 3$$

$$n = 0,4 \qquad m = 0,1 \qquad \begin{matrix} x = 0,4 \\ y = 0,1 \\ z = 0,1 \end{matrix}$$

Writing down all the separate rate processes results in 74 rate equations with identical rate constants after introduction of a step-dependent statistical factor which is readily calculated. A computational integration procedure may then be used to simulate the exchange process. The simulations successfully model the random distribution of species in the absence of reaction and also the observed experimental kinetic data, lending support to the general concept and the Poisson distribution. In the presence of a large excess of 'empty' pools, Equation 3 reduces simply to Equation 4 as this is the only process occurring of significance in the kinetic analysis.

$$\left(A_1\right) \quad + \quad \left(B_1\right) \longrightarrow \left(C_1\right) \quad + \quad \bigcirc \qquad Equation\ 4$$

The concentrations of $\left(A_1\right)$, $\left(B_1\right)$, are then simply given by the overall concentrations (moles per dm^3 of microemulsion). The observed second-order rate constant for reaction may then be directly equated with the exchange rate constant.

To summarize, under conditions of rate-determining pool exchange, the physically-significant concentrations to extract the exchange rate constant are $[\bigcirc]$, $[\left(A_1\right)]$, $[\left(A_2\right)]$ etc; that is the discrete occupancy of the droplets must be considered.

(c) Rate-limiting chemical reaction :

Most slow reactions studied in microemulsions fall into this category. The main questions which arise are whether overall or aqueous concentrations should be used, and which derived rate constant is most readily compared with the rate constant measured in bulk water.

Consider the reaction:

$$A \ + \ B \longrightarrow C$$

In a dispersion, the reaction scheme can be written as:

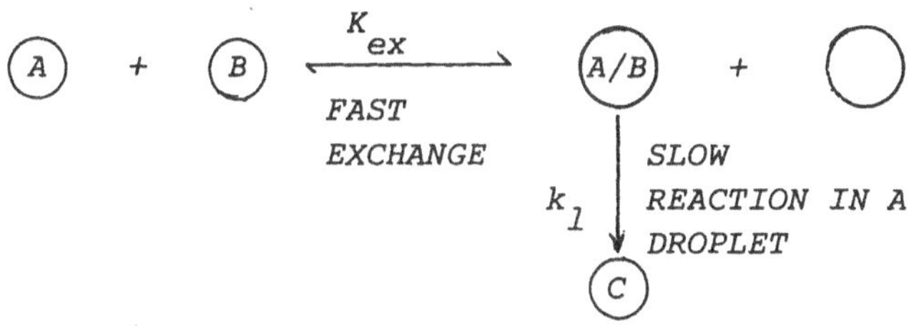

Equation 5

Consider a system such that $[\bigcirc] \gg [\text{\textcircled{A}}]$, $[\text{\textcircled{B}}]$. Then no pools will contain more than one molecule of A, B or C.

The fast exchange process can be expressed as a fast pre-equilibrium with equilibrium constant K_{ex}:

i.e. $\quad K_{ex} = \dfrac{[\text{\textcircled{A/B}}][\bigcirc]}{[\text{\textcircled{A}}][\text{\textcircled{B}}]}$

The experimental second-order rate constant may then be expressed as:

$$\frac{d[\text{\textcircled{C}}]}{dt} = k_2{}^{ov}[\text{\textcircled{A}}][\text{\textcircled{B}}]$$

where $k_2{}^{ov}$ has units dm^3 (of total solution)$mol^{-1}s^{-1}$, and $[\text{\textcircled{C}}]$ represents the number of moles of C/dm^3 of total solution.

Also $\quad \dfrac{d[\text{\textcircled{C}}]}{dt} = k_1[\text{\textcircled{A/B}}]$

$$= \frac{K_{ex}k_1[\text{\textcircled{A}}][\text{\textcircled{B}}]}{[\bigcirc]} \qquad\qquad \text{Equation 9}$$

Hence $\qquad k_2{}^{ov} = \dfrac{K_{ex} k_1}{[\bigcirc]}$ $\qquad\qquad$ *Equation 6*

For a random distribution of reactants in the pools in the fast exchange process, $K_{ex} = 1$. If the concentration of pools is known then $k_1(s^{-1})$ can be calculated[4].

From another viewpoint, because of rapid exchange, the dispersed aqueous phase can be envisaged as a 'pseudo-continuous' phase on the (slow) time scale of the reaction. Then we have:

$$\frac{d[C]_{aq}}{dt} = k_2{}^{aq.pool} [A]_{aq}[B]_{aq}$$

where $[C]_{aq}$ represents the number of moles of C/dm^3 of dispersed water, and

$$[\bigcirc] = [C]_{aq} f_w \ etc.$$

where f_w is the volume fraction of water in the system.

Then it is clear that:

$$k_2{}^{ov} = \frac{k_2{}^{aq.pool}}{f_w} \qquad\qquad \textit{Equation 7}$$

and so either concentration scale can be used.

$k_2{}^{aq.pool}$ may differ from the value of k_2 measured in bulk water for three reasons. Firstly the reactants may prefer to stay in different water pools, which is equivalent to the statement that K_{ex} is < 1. Secondly, the reactants may partition within the same pool between the water core and the surfactant interface. Thirdly, the rate of the reaction may be altered by a specific effect due to the nature of the dispersed water. The following discussion is concerned with the first effect. The treatment leading to Equation 7 ignores the discrete nature of the pools, and it is of interest to demonstrate, for $K_{ex} = 1$, that $k_2{}^{aq.pool}$ then only differs from k_2 measured in bulk water by the second and third effects mentioned (i.e. $k_2{}^{aq.pool}$ reflects reactivity changes caused by intra-pool effects rather than inter-pool effects).

Considering the actual reaction step within the individual pools and defining a second order rate constant k_{INT} within a pool as follows:

$$A + B \xrightarrow{k_{INT}} C$$

Then

$$\frac{d[C]_{INT}}{dt} = k_{INT}[A]_{INT}[B]_{INT}$$

The subscript INT (for intrinsic) now refers to the concentration of species in the water pools actually occupied by the species. Hence this volume does not include the volume of water in 'empty' pools. k_{INT} can therefore measure intra-pool effects on the chemical reaction. How is k_{INT} related to the directly measurable quantities $k_2{}^{ov}$ or $k_2{}^{aq.pool}$?

In one liter of a microemulsion, the number of moles of C formed in unit time is given by:

$$\frac{dC}{dt} = k_{INT}[A]_{INT}[B]_{INT} \ f_w \ \frac{[\ (A/B)\ ]}{[pools]} \qquad \text{Equation 8}$$

$f_w [\ (A/B)\ ]/[pools]$ is the volume of the water droplets containing both A and B/dm^3 of microemulsion.

$$[A]_{INT} = [B]_{INT} = 1 \ molecule \ per \ water \ pool \ volume = [pools]/f_w$$

Equating Equation 8 and Equation 9, substituting for $[A]_{INT}$ etc. and remembering that $[pools] = [\ (\bigcirc)\ ]$, we have:

$$k_1 = \frac{k_{INT}[pools]}{f_w}$$

From Equation 6:

$$k_1 = \frac{k_2{}^{ov}[\ \bigcirc\ ]}{K_{ex}}$$

Therefore

$$k_{INT} = (k_2{}^{ov} \ f_w)/K_{ex} = k_2{}^{aq.pool}/K_{ex}$$

)

In the case of random distribution, $K_{ex} = 1$ and $k_{INT} = k_2^{aq.pool}$. Therefore, for random distribution, the measured value of $k_2^{aq.pool}$ is not influenced by the discrete nature of the droplets. The dispersed phase behaves as a separate continuous phase within the total volume of the system.

The available experimental evidence supports the view that small solutes (and ions) are indeed randomly distributed. However, this may not be the case for large solubilized species such as enzymes. The reactivity parameters for enzyme reactions[5] may involve a non-unity value for K_{ex}.

In general, intra-pool effects are determined by the nature of the water pools and not the number of pools. Many reactions show an invariance of $k_2^{aq.pool}$ under experimental conditions where the concentration of pools is varied without changing their nature[6]. Therefore, the use of aqueous concentrations would seem to be preferred as any alteration in the reactivity of the chemical system in the solubilized water as compared to bulk water is directly revealed.

The treatment shown in this paper is developed for special conditions of low pool occupancy, but we expect the general conclusions to remain valid when multiple occupancy occurs.

When reactants partition significantly into the oil phase, the situation is further complicated. Provided that partitioning processes are fast relative to chemical reaction, the situation may be accommodated using equations similar in form to those developed by Berezin[7] to describe reactivity in aqueous micellar systems.

REFERENCES

1. S.S. Atik and J.K. Thomas, J. Amer. Chem. Soc., 103 : 3543 (1981).
2. N.J. Bridge and P.D.I. Fletcher, J. Chem. Soc. Farad. Trans. 1, to be published.
3. P.D.I. Fletcher and B.H. Robinson, Ber. Bunsenges. Phys. Chem., 85 : 863 (1981).
4. B.H. Robinson, D.C. Steytler and R.D. Tack, J. Chem. Soc. Farad. Trans. 1, 75 : 489 (1979).
5. S. Barbaric and P.L. Luisi, J. Amer. Chem. Soc., 103 : 4329 (1981).
6. P.D.I. Fletcher, unpublished work.
7. I.V. Berezin, K. Martinek and A.K. Yatsimirski, Russ. Chem. Rev., 42 : 787 (1973).

ACIDITIES AND BASICITIES IN REVERSED MICELLAR SYSTEMS

Omar A. El Seoud

Instituto de Quimica, Univ. de São Paulo
C.P. 20.780
01000 - São Paulo, S.P., Brazil

INTRODUCTION

The determination of the "effective" acidities and basicities in reversed micelles (RM's) is fundamental for the rationalization of any pH-dependent interaction taking place in the micellar "water pool, WP". The subject also bears on the basic question of whether reversed micellar catalysis is due, at least partially, to a change in the physico-chemical properties of the reactants when they are solubilized in the WP. A nice example of an enhanced reactivity due to the substrate inclusion in the micellar pseudo-phase is provided by the case of the enzyme α-chymotrypsin. The reactivity of this enzyme in the WP of the anionic surfactant bis(2-ethylhexyl)sodium sulfosuccinate (AOT) is actually higher than it is in water, and the pK_a of the enzyme active site is shifted to a higher value[1,2].

The study of RM-mediated reactions is a relatively new subject[3], so that the question of the effective acidities and basicities in the WP's has only recently been examined[1,4-15]. Reaction rates have been determined in the presence of solubilized buffers, and the pH values used in the calculation of the kinetic data are those of the starting buffer solution (pH_{st}, i.e., that before its solubilization in the RM)[1,2,16]. Here the underlying assumptions are that the pK_a of the buffer in the WP is equal to its value in bulk water, and that the buffering capacity is maintained in the reversed micelle. However there is no guarantee, a priori, that these assumptions are valid for all buffer systems and all surfactants. Moreover, several techniques have shown that the properties of the RM-solubilized water are different from those of bulk water, even at higher R values (R = [water] / [surfactant] molar ratio)[3,17]. Since the solvent properties of the micellar water are different, it is plausible, at least in principle, that acid-base equilibria in the WP's may be displaced from their positions in bulk water. This realization has intensified the interest in a detailed examination of acid-base equilibria in RM's.

   The objective of the present review is to summarize the work that has been done in this area of research, and to show the relevance of the obtained results to the understanding of the RM-mediated interactions. Effective acidities in the micellar WP's can most easily be probed by studying the effect of the RM on the pK_a of suitable indicators. Since the system is complex (there are at least 4 components, the solvent, the surfactant, the solubilized water, buffer, and the probe) the experiments have to be meticulously planned and the results carefully examined. Therefore, some points which are important for a correct determination of the micellar pK_a values and for the interpretation of the obtained results will be emphasized.

EXPRESSIONS AND ABBREVIATIONS USED IN THIS WORK

RM : Reversed micelle or detergent aggregate in a non-polar solvent.
WP : The term micellar "water pool" describes the water droplet which
     is surrounded by a monolayer of the surfactant head-ions.
AOT: Is the anionic surfactant

$$\text{Na}^+ \ \bar{O}_3S \underset{\displaystyle \overset{|}{\text{CH-CO}_2\text{CH}_2\text{CH}(C_2H_5)C_4H_9}}{\overset{\displaystyle \text{CH}_2\text{CO}_2\text{CH}_2\text{CH}(C_2H_5)C_4H_9}{\rule{0pt}{0pt}}}$$

pH_{st} :Denotes the pH of an aqueous solution, as measured by a pH-
     meter, before its solubilization in the RM.
R   : [Water]/[surfactant] molar ratio.
pH_{wp} :Is the pH inside a WP. If an acid is added to a reversed mi-
     celle, the pH_{wp} is calculated by assuming that all the added
     acid concentrates in the already present WP. In this case the
     stoichiometric $[H_3O^+]$ should be multiplied by 55.55/molarity
     of the micelle-solubilized water. For eg., in the presence of
     0.55M water, $[H_3O^+]$ should be multiplied by 100; i.e. the
     stoichiometric pH should be decreased by a factor of two.
ΔpK_a :pK_a in bulk water - pK_a in the RM.
I   : The indicator ratio = [unprotonated form] / [protonated form].
Micellar periphery:
     Denotes the region of the RM that contains the head-ions plus
     the first hydration shell.

Micellar center:
    The rest of the micellar core.
pH_{eff}: Denotes the effective pH value at a certain region (periphery
    or center) of the RM.
pH_b :Is the pH value at the periphery of the WP, i.e. it is based on
    the protons in the first hydration shell.
pH_f :Is that at the center of the WP, i.e., it is a measure of the
    free protons in the center part of the pool.

The Use of Indicators

    Acid-base indicator equilibria for a series of colored indica-
tors (bromophenol blue, malachite green, methyl orange, and thymol
blue) were observed spectrophotometrically in the presence of the
RM's of the nonionic surfactant Igepal CO-530 (polyoxyethylene(6)
nonylphenol) in benzene[4]. No buffers were used and the pH's in the
WP were calculated by assuming that all the added acids (HCl or
$HClO_4$) are localized in the WP. All pK_a values were much lower than
those in bulk water, i.e., positive ΔpK_a's were obtained (ΔpK_a =
pK_a in water - pK_a in the RM). Thus in the presence of 0.5M surfactant, and 0.55M of solubilized water, the ΔpK_a values were 4.57,
5.46, 5.39 and 5.35 for bromophenol blue, malachite green, methyl
orange, and thymol blue, respectively. For malachite green, ΔpK_a
increases as a function of decreasing R, showing that the surfactant competes with the indicator for available protons[4].

    A study of the acid-base equilibria of several dinitrophenols,
bromophenol blue, bromophenol red, malachite green, and vitamin B_{12}
in the presence of the zwitterionic dodecylammonium propionate (DAP)
RM's in benzene has also been carried out[5]. No buffers were used.
However the case here is complex, because the added mineral acids
(HCl or $HClO_4$) react with the surfactant itself. By correlating the
extent of this reaction with the equilibrium of the indicators, it
was possible to determine the pK_a values. The ΔpK_a's were small
(< 0.4) except that for malachite green (1.3). Thus the water entrapped in the DAP micelles seems to have an acidity analogous to
that of bulk water[5].

Acid-Base Equilibria in AOT Reversed Micelles

    More attention has been focused on AOT, probably because of
its large solubilizing capacity, and since more is known about the
water solubilized in the RM's of this detergent[3,17]. The pK_a of
p-nitrophenol in the presence of AOT in heptane has been determined[6].
The pH values used in the Henderson-Hasselbach equation were pH_{st}.
Whereas the indicator ionized near pH=11 when NaOH was used, a

fraction of it ionized near its pH value in bulk water when imidazole was used as the base. The fraction of the ionized phenol increased as the concentration of the solubilized imidazole was increased, and the R value increased. These results were explained by
assuming that there are two distinct p-nitrophenol species within
the WP. The first is adsorbed at the AOT/heptane interface. This is
the "high pK_a" species ($\Delta pK_a \simeq -4.5$ units) whose ionization is im-
paired due to the electrostatic effect of the sulfonate groups of
the surfactant. Displacements of some of this indicator into the WP
by the solubilized imidazole produces the "low pK_a" species whose
ΔpK_a is between -0.4 and -0.8 units. Unlike imidazole, methanol,
n-butylamine, n-butylammonium chloride, and n-butanol seemed unable
to displace the indicator from the surfactant/solvent interface[6].

The pK_a of 2,4-dinitrophenol has also been determined in the
presence of AOT in octane, as a function of R, using $pH_{wp} = pH_{st}$[7].
The micellar pK_a value was higher than that in bulk water ($\Delta pK_a = -1.7$ units) and was found to be independent of R from 7 to 14. The
negative ΔpK_a was attributed to the fact that the polarity of the
AOT-solubilized water is lower than that of bulk water[7]. Indeed, it
has been shown that nitrophenols have higher pK_a values in solvents
of lower polarities, such as ethanol, and methanol[7,18].

The indicator ratio I (= [unprotonated form]/[protonated form])
for methyl red in the presence of AOT aggregates in isooctane has recently been studied[11]. It was found that (I) decreases as a function
of increasing the surfactant concentration, showing that the negative
RM stabilizes the conjugate acid of the indicator. The effect of imidazole, octylamine buffers, and of cetyltrimethylammonium bromide on
the (I) ratio was also examined. No change in (I) was observed as a
function of increasing the pH_{st} values of the latter additive. On the
other hand, (I) increased as a function of increasing the pH_{st}'s of
the imidazole and the octylamine buffers, showing that both can bind
competitively to the AOT sulfonate groups[11]. Although the results of
this work agree with those of Menger[6], we have found that methyl red
does not behave ideally in the AOT reversed micelle. Specifically,
both λ_{max} and the absorbance of the dye depend on R, the presence of
solubilized HCl and/or imidazole buffer[15]. Thus the results of this
study can only be considered if a rationale is first found for the
dependence of λ_{max} and of the absorbance on the experimental conditions.

The Use of ^{31}P-NMR

Another interesting technique to probe acid-base equilibria in
the WP is ^{31}P-NMR. It was found that the chemical shifts of the
phosphorous atom of certain species are pH-dependent and this pro-

perty was used to measure the local pH values in the interior and the exterior of cell membranes[19]. Thus a method is available to measure pH_{wp} by comapring the ^{31}P chemical shifts in the RM with those in aqueous solutions. The assumption which is made is that the pK_a of the probe (for eg., the phosphate buffer) will not change upon its solubilization in the RM. This is likely, since negatively charged ions like HPO_4^{2-} or PO_4^{3-} are very hydrophilic, and are not likely to interact with the anionic surfactants which are used[8,10,13].

In sodium octanoate RM's the ^{31}P chemical shifts of the phosphate buffer showed that $pH_{wp} > pH_{st}$, even after a correction (to account for the ionic effect of the surfactant) was applied. When an effective pH scale was established for the RM, it was used to calculate the pK_a of phenol red, and the ΔpK_a was found to be -0.5 units[8].

The same approach, i.e., to consider that the pK_a of the phosphate ion is the same in the RM and in water, was also extended to the AOT system[10]. Several organic-phosphate mixed buffers were used to establish a pH scale for the WP. The pH jumps (= $pH_{st} - pH_{wp}$) were found to depend on the composition of the mixed buffer, the total concentration of the phosphate ion in the buffer, but not on the value of R, nor on the presence of co-solubilized lysozyme in the WP. These pH_{wp} values were used to determine the pK_a's of two organic dyes as a function of the experimental variables. All micellar pK_a values were higher than the pK_a's in water. However the ΔpK_a's were found to be dependent on the buffer used. For example, in the case of phenol red, the ΔpK_a's were -1.1, and -3.2 for glycine-phosphate, and pyrophosphate buffers, respectively. The same dependence on the buffer was also observed for p-nitrophenol, which had a ΔpK_a that was $\lesssim$ -4.0 units, and which was dependent on concentration and R. These anomalies led the authors to criticize the use of dyes to probe the acidities of the micellar water pools[10].

More recently, the same Japanese group which studied the sodium octanoate RM's extended their ^{31}P-NMR studies to the AOT system[13]. The pH values in the RM's were calculated from the NMR chemical shift data of the phosphate ion after a correction (to account for the surfactant induced shift) was applied. In the pH_{st} range of 6.5 to 8.1 (R=50) the pH jump was found to be 0.4 units. Using these pH values the pK_a of phenol red was found to be 7.7, i.e., the ΔpK_a is only 0.1 units. The micellar dissociation constant for the colored indicator was then used to calculate pH_{wp} for a series of organic buffers (not containing phosphorous) solubilized in the WP. The determined values were found to be close to their pH_{st} counterparts. From the observed ΔpK_a for phenol red these authors calculated that the surface potential, ψ, of the AOT micelle is + 6 mV.

They concluded (by comparison with the sodium octanoate case, for which ψ = -30 mV) that the AOT aggregates ionize less than the sodium octanoate RM's. It is relevant, however, that the calculated value of ψ will depend on ΔpK_a, so that our results for bromocresol green (vide infra) indicate that the surface potential for AOT micelle is < -30 mV.

Effective pH Values in RM's

An important point is that in the preceding work the micellar interior is treated as one entity. As a consequence, and in spite of the awareness that the probes are localized at different sites in the RM[6,7,20], the pH values which were used in the Henderson-Hasselbach equation were either pH_{st} or pH_{wp}. The question which will now be addressed is: are the effective pH values where the probe is localized in the RM (pH_{eff}) always correctly represented by either one of these two pH's?

Effective pH values in the different regions of the RM have recently been calculated for AOT aggreages in heptane. The calculation was based on the theory of ion exchange between the ionic species in the WP and the surfactant counterion (Na$^+$ ion)[12]. The basis for this procedure is to divide the RM into a "periphery" and a "central" part, as shown in Figure 1. The former region includes

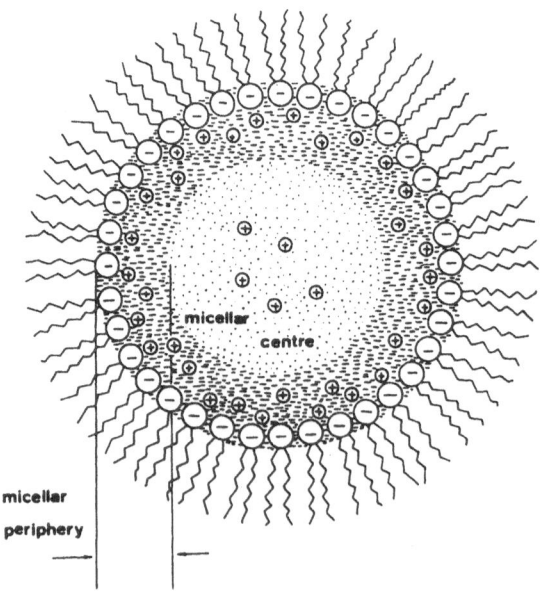

Figure 1. *A schematic representation of the RM of Aerosol-OT (AOT) showing the different regions of the micellar interior.*

the head ions plus the tightly associated first hydration shell
(ca. 6 water molecules/AOT)[13,16,21]. The rest of the aqueous solution is contained in the second region, or the center part of the
pool. This division is based on the many studies of water solubilization by AOT where a' sudden change in the magnitude of the physical
property that is being followed is usually observed at R ≈ 6[13,16,21].
The next steps are to localize the solubilization site of the probe
(i.e., at the center, or at the periphery of the WP), determine the
pH_{eff} values therein, and then use these values to calculate the
micellar pK_a's.

The simple case where no buffer was used in the pK_a determination will be examined first. The only ion exchange to be considered
is that between the hydronium ions in the pool (H_f) and the undis-
sociated Na^+ of the micelle (Na_b), as shown below

$$Na_b + H_f \rightleftharpoons Na_f + H_b$$

Here (H_b) refers to the concentration of the hydronium ions bound
in the first hydration shell and (Na_f) refers to the concentration
of the free sodium ions. The above equation permits the calculation
of H_f and H_b from which the corresponding pH values can be obtained[12]. Figure 2 shows the variation of pH_b and pH_f as a function
of pH_{wp} at two different R values. The most important point of
Figure 2 is that the variation of pH_b and pH_f parallels that of pH_{wp}
till pH ≈ 1. At higher acidities, i.e. when $[H_{wp}] > [AOT]$, pH_b decreases slowly as a function of decreasing pH_{wp} since there are only a
few remaining Na^+ counterions to be exchanged. Consequently, most of
the added protons concentrate in the center part of the pool, as
shown from the variation of pH_f. The differences between part A
(small pool) and B (large pool) of Figure 2 are due to the fact that
the degree of the dissociation of the RM, and the volume of the
center part of the pool, are larger for the latter case.

The significance of the preceding section for the subject of
the present review will now be considered. A good example is provided by malachite green. This hydrophobic indicator is adsorbed at
the AOT/heptane interface, i.e., pH_b's should be used in the pK_a determination, giving ΔpK_a of 0.4 ± 0.1 units. If pH_{wp} was used instead, one would have obtained a ΔpK_a of 1.4 ± 0.1 units[12]. Conver-
sely, for a hydrophobic probe whose $pK_a < 1$, a small ΔpK_a will be
obtained if pH_{wp} is used instead of pH_f, since the latter decreases
faster than the former. In summary, the use of the inappropriate pH
scale can lead to an apparent attenuation, or augmentation, of the
effect of the RM on acid-base equilibria.

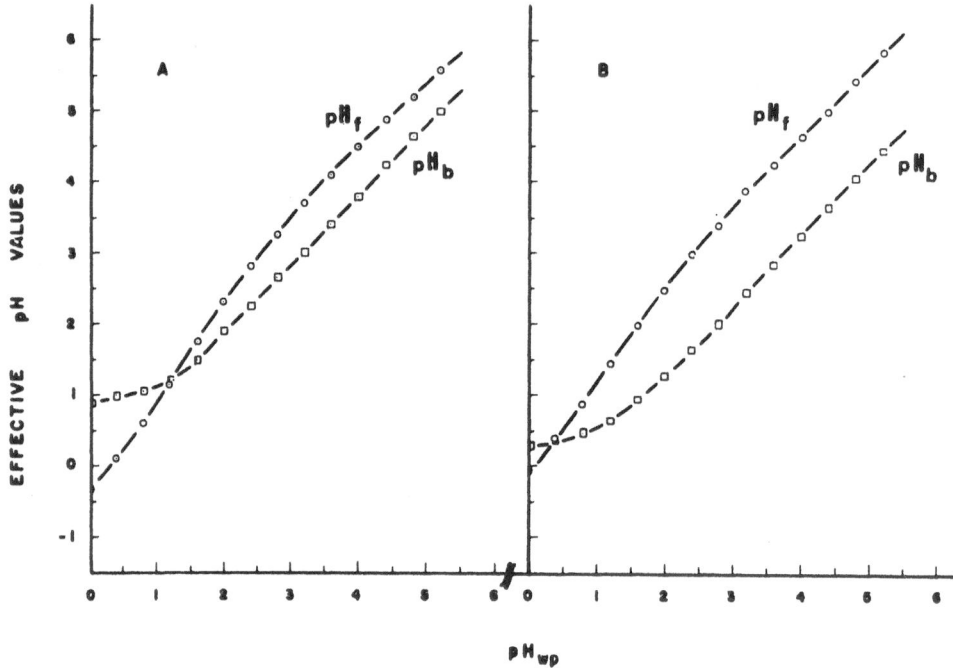

Figure 2. *Variations of the pH values at the micellar periphery*
(pH$_b$) and at the center part of the pool (pH$_f$) as a function of
pH$_{wp}$. Part A of the Figure is for a small pool (R=10), whereas
part B is for a large one (R=40). These effective pH values are
for AOT=0.1M, and were calculated based on the procedure outlined
in Reference 12.

The Use of Buffers in Reversed Micelles

We now consider pK_a determinations using aqueous buffers. Our
results[22], and those of others[6-10], show that the situation here is
more complex, and the results are sometimes anomalous. A worrisome
feature when buffers are used is that the ΔpK_a values depend on the
type and composition of the buffer, on its concentration (for a
constant pH_{st}) and on R[10,22]. The implication of these observations
for reaction kinetics in RM's is far-reaching, because it means
that the results of pH-dependent interactions can only be compared
under strictly the same conditions (i.e., using the same buffer
with the same concentration and R value). This is a consequence of
the observation that different buffers with the same pH_{st} values
produce, upon solubilization in the RM, different effective pH
values.

In the following discussion it will be shown how this apparently confusing situation can be handled if one considers the ion exchange processes between the buffer and the RM. Consider the simple
case where the pK_a of a hydrophilic probe is being studied. The buffer in the second hydration shell is relatively outside the domain
of the electrostatic field effect of the ionic groups of the surfactant. Moreover, in the case of AOT, negative buffers such as the
phosphate or borate buffers cannot exchange with the Na^+ counterion[22]. That is, these types of buffers are more or less intact in
the WP. Accordingly, the assumption that their pK_a values are not
changed upon solubilization in the RM[8,10,13] seems quite plausible.
We also recall here that the properties of the water in the center
part of the pool are closer to those of bulk water, in comparison
with the water molecules in the first hydration shell[3,16,21].
Menger did an interesting experiment by inserting a glass electrode
in an AOT solution in heptane containing solubilized phosphate or
borate buffers[1]. The pH readings that he obtained were fairly close
to the corresponding pH_{st} values. In the conference at Rigi-Kaltbad
I discovered that several people repeated Menger's experiment and
came up with similar results. Although the meaning of these readings
is open to debate, they can serve to support the idea that pH_{st} is
a good measure for the pH at the center of the WP. Another evidence
that a hydrophilic buffer remains largely unperturbed in the WP
comes from the determination of the first and second dissociation
constants of maleic acid in AOT aggregates in heptane. A small ΔpK_a
value was obtained for both ionizations (0.3 units) which was independent of R, although one expects a larger micellar effect on the
second pK_a (producing the dianion) than on the first pK_a[12,22].

A more complex case is when the pK_a of a hydrophobic probe
(i.e., one that is adsorbed at the surfactant/solvent interface) is
studied. If it is not negatively charged, the acid component of the
buffer, BH, will participate in the ion exchange processes. For the
case of AOT we have the following equilibria:

$$H_f + Na_b \rightleftharpoons H_{bl} + Na_{fl} \qquad \text{Equation 1}$$

$$BH_f + Na_{b2} \rightleftharpoons BH_{b2} + Na_{f2} \qquad \text{Equation 2}$$

$$BH_f + H_{b3} \rightleftharpoons BH_{b3} + H_f \qquad \text{Equation 3}$$

$$\text{and} \qquad BH_{b4} \rightleftharpoons B_{b4} + H_{b4} \qquad \text{Equation 4}$$

where the subscripts b1–b4, and f1, f2, refer to bound and free
species, respectively, involved in Equations 1–4. Note that $Na_{b2} =$

total Na_b - Na_{fl}; H_{b3} = H_{b1} + H_{b4}; BH_{b4} = BH_{b2} + BH_{b3}. If the buffer capacity is not exceeded upon its solubilization in the RM, then H_f, BH_f are, by definition, constants. In order to determine [H_b], one needs to know [BH_b] and this may not always be easy. A typical case is when the species (BH_b) and (BH_f) are not experimentally distinguishable. Recently we studied the pK_a of bromocresol green in AOT using such buffers (imidazole and Tris). An ion exchange resin was used as a model for the RM and (BH_b) was determined by elution from the resin bed[22]. This permits a calculation of (pH_b) and its values are lower than the pH_{st} counterparts. Moreover pH_b's increase much more slowly than the corresponding pH_{st} values. For example, the starting pH's of 0.2M imidazole solutions were increased from 6.44 to 7.87, but the corresponding pH_b values were found to increase only from 5.24 to 5.78 (AOT = 0.1M, R = 20)[22]. Additionally, due to the different ion selectivities for the different (BH) species (Equations 2, 3), the pH_b values depend on the buffer used. Compare, for example, the results for the imidazole with those for the Tris buffer (0.2M) when solubilized in AOT reversed micelles in heptane (AOT = 0.1M, R = 20). For the pH_{st} range of 7.5 to 8.0, the corresponding pH_b values were found to be 5.42-5.64, and 5.68-5.78, for imidazole and Tris, respectively.

The importance of using the appropriate pH scale can be appreciated from the following results. The pK_a of bromocresol green determined using the pH_{st} values showed a large dependence on the buffer, and on R. Thus when the latter was varied from 5.6 to 19.4 the ΔpK_a values were found to increase from -2.82 to -1.6, and from -3.72 to -2.73, for imidazole and Tris buffers, respectively. If the point at R = 5.6 was excluded (since at this low R the properties of the AOT-solubilized water deviate most from those of bulk water[3,13,16,21]), then the use of pH_b resulted in ΔpK_a of 0.45 ± 0.10 and 0.95 ± 0.10 units for imidazole and Tris buffers, respectively[22]. Not only was the micellar effect on the pK_a smaller, but also the anomalous dependence of the ΔpK_a on R disappeared.

A comment on the displacement mechanism suggested earlier[6] seems appropriate here. If this meachnism were a general one, then it would be expected that the same buffer would displace different probes of the same pK_a value (from the RM periphery to the center part of the pool) to the same extent. However, imidazole is completely ineffective in displacing bromothymol blue which has a pK_a value that is equal to that of p-nitrophenol. Although a difference in hydrophobicitiy between the two indicators can be invoked to explain the lack of displacement of the sulfophthalein indicator, then it is not clear why the same buffer can displace bromocresol green, which has a hydrophobicity similar to that of bromothymol blue. Additionally, the buffer displaces the indicator by preferen-

tial hydrogen-bonding to the $-SO_3^-$ group of the AOT[6]. Protonated Tris
has 6 protons to form hydrogen bonds, versus only two protons for
the imidazolium ion. Nevertheless, the latter buffer is more effective in producing the base indicating color of bromocresol green.
All these observations can readily be explained on the basis of ion
exchange processes taking place in the WP, and of the pH_b values
that the buffer can maintain at the surfactant/solvent interface[22].

CONCLUDING REMARKS

     Some points which are necessary for a correct determination of
pK_a and for a meaningful discussion of the obtained results will now
be considered[15]. At the outset, it should be emphasized that the indicator ratio (I) in the Henderson-Hasselbach equation refers to the
species in the micellar pseudo-phase. Thus the choice of indicators
is limited to those which are insoluble in the solvent, or else
those which bind strongly to the micelle. In this respect, nitro-
phenol indicators are unsuitable, since they are soluble in the organic solvents used[5-7], and because their association with the RM
depends on the experimental conditions, especially on the value of
R^{23}. Moreover, their λ_{max} and ε_{max} values in the UV-VIS region also
depend on R^7, making any Beer's law plot difficult. The second point
is that the appropriate pH values should be used in the calculation
of the pK_a's. In this respect, the use of very hydrophilic and/or
hydrophobic indicators eliminates the nagging worry that the indicator will change its solubilization site as the operational conditions (especially the value of R) are varied. In the case of ionic
micelles it is tempting to assume that the observed ΔpK_a value is
due to an electrostatic stabilization of one of the indicator forms
by the oppositely charged micelle. However, the possible effect of
other factors such as the lower polarity and dielectric constant of
the water in the pool[3,17] on the pK_a value should not be overlooked.

     This review has presented several points which are relevant to
the study of acid-base equilibria, and to reaction kinetics in RM's.
First the basis to calculate effective pH values in the different
regions of the RM was shown. The point was clearly demonstrated
that the above mentioned pH values are different from pH_{wp} (for un-
buffered systems) or (except for the center part of the pool, provided that the buffer capacity is not exceeded, vide supra) from
the pH_{st} values (for buffered water pools). A very significant point
is that different buffers having the same pH_{st} values and the same
initial concentrations, will not produce the same pH values at the
micelle solvent interface (except under the highly unlikely condition
that the ion selectivities for the different buffers are the same).
In practice, therefore, buffers cannot be used indiscriminately, and

one has to be careful in comparing the results of RM-mediated processes that occur at the detergent/solvent interface.

Most of the discussion has been limited to AOT aggregates. However, our results on positively charged RM's show that the above
considerations can also be extended to these systems[24]. It is clear
that very little work has been done so far and it is hoped that the
present review will result in an increase in the amount of work directed towards the solution of this very interesting, and fundamen-
tally important aspect of the chemistry of RM's.

ACKNOWLEDGEMENT

I thank the CNP_q and FAPESP Research Foundations for support.

REFERENCES

1. F.M. Menger and K. Yamada, J. Amer. Chem. Soc., 101 : 6731
   (1979).
2. S. Barbaric and P.L. Luisi, J. Amer. Chem. Soc., 103 : 4239
   (1981).
3. J.H. Fendler and E.J. Fendler, "Catalysis in Micellar and Macromolecular Systems", chapter 10, Academic Press, New York, 1975;
   J.H. Fendler, "Membrane Mimetic Chemistry"; chapters 3,12,
   Wiley-Interscience, New York, 1982.
4. F. Nome, S.A. Chang and J.H. Fendler, J. Colloid Interface Sci.,
   56 : 146 (1976).
5. F. Nome, S.A. Chang and J.H. Fendler, J. Chem. Soc. Faraday
   Trans.1, 72 : 296 (1976).
6. F.M. Menger and G. Saito, J. Amer. Chem. Soc., 100 : 4376
   (1978).
7. A. Levashov, V.I. Pantin and K. Martinek, Kolloid. Zh.,
   41 : 453 (1979).
8. H. Fujii, T. Kawai and H. Nishikawa, Bull. Chem. Soc. Japan,
   52 : 2051 (1979).
9. N. Miyoshi and G. Tomita, Z. Naturforsch., 35b : 736 (1980).
10. R.E. Smith and P.L. Luisi, Helv. Chim. Acta, 63 : 2302 (1980).
11. A.T. Terpko, R.J. Serafin and M.L. Buchholtz, J. Colloid Inter-
    face Sci., 84 : 202 (1981).
12. O.A. El Seoud, A. M. Chinelatto and M.R. Shimizu, J. Colloid
    Interface Sci., 88 : 420 (1982).
13. H. Fujii, T. Kawai, H. Nishikawa and G. Ebert, Colloid Polymer
    Sci., 260 : 697 (1982).
14. O.A. El Seoud and M.R. Shimizu, Colloid Polymer Sci., 260 : 794
    (1982).

15. O.A. El Seoud and R.C. Vieira, J. Colloid Interface Sci., in press.

16. S. Friberg and S.I. Ahmed, J. Phys. Chem., 75 : 2001 (1971); F.M. Menger, J.A. Donohue and R.F. Williams, J. Amer. Chem. Soc., 95 : 286 (1973); K. Martinek, A.V. Levashov, N.L. Klyachko and I.V. Berezin, Dokl. Akad. Nauk SSSR, 236 : 920 (1977); J. Sunamoto, H. Kondo and K. Akimaru, Chem. Letters, 821 (1978); H. Fujii, T. Kawai and H. Nishikawa, Bull. Chem. Soc., Japan, 52 : 1978 (1979).

17. A. Kitahara, Advan. Colloid Interface Sci., 12 : 109 (1980).

18. E. Banyai, in "The Indicators", E. Bishop, Ed., Pergamon Press, London, 1972.

19. R.B. Moon and J.H. Richards, J. Biol. Chem., 248 : 7276 (1973); J.M. Salhang, T. Yamane, R.G. Shulman and S. Ogawa, Proc. Natl. Acad. Sci. U.S.A., 72 : 4966 (1975); G. Navon, S. Ogawa, R.G. Shulman and T. Yamane, Proc. Natl. Acad. Sci. U.S.A., 74 : 87 (1977).

20. Y. Jean and H.J. Ache, J. Amer. Chem. Soc., 100 : 6320 (1978); K. Tamura and Z.A. Schelly, J. Amer. Chem. Soc., 101, 7643 (1979); 1013 (1981).

21. P. Ekwall, L. Mandell and K. Fontell, J. Colloid Interface Sci., 33 : 215 (1970); H.F. Eicke and J. Rehak, J. Colloid Interface Sci., 59 : 2883 (1976); M. Wong, J.K. Thomas and M. Grätzel, J. Amer. Chem. Soc., 98 : 2391 (1976); M. Wong and J.K. Thomas, J. Amer. Chem. Soc., 99 : 4730 (1977); E. Keh and B. Valeur, J. Colloid Interface Sci., 79 : 465 (1981).

22. O.A. El Seoud and A.M. Chinelatto, Sumitted to J. Colloid Interface Sci.

23. L.J. Magid, K. Kon-no and C.A. Martin, J. Phys. Chem., 85 : 1434 (1981).

24. O.A. El Seoud, A.M. Chinelatto and R.C. Vieira, unpublished results.

PROSPECTS FOR CHIRAL DISCRIMINATION IN REVERSED MICELLES

L. J. Magid

Department of Chemistry
University of Tennessee
Knoxville, Tennessee 37996 USA

Chiral discrimination in the aggregation properties of optically active surfactants and their use in
reaction catalysis are briefly reviewed. Suggestions
for enhancing the observed stereoselectivities in reversed micellar systems are presented.

Chiral Discrimination in Surfactant Aggregation and Solubilization

In normal micelle formation, critical micelle concentrations
(cmc's) of pure enantiomers are generally the same, within experimental error, as the cmc's of the racemates. Exceptions are the
sodium N-acylalanates and valinates (I), studied by Miyagishi and
Nishida[1], where the pure enantiomers have slightly lower

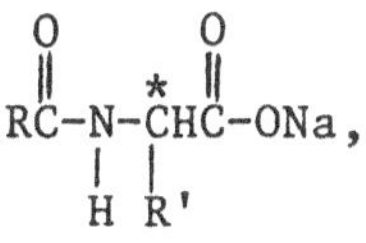

$$\begin{array}{cc} O & O \\ \| & * \| \\ RC-N-CHC-ONa, \\ | \ \ | \\ H \ R' \end{array} \quad \text{with } R' = CH_3 \text{ or } -CH(CH_3)_2$$

I

cmc's than the racemates. The maximum $\Delta\Delta G^O_m$ (difference in free
energy of micellization) observed was 74 cal/mol at 25°C for the
sodium N-dodecanoylalanate.

Arnett et al [2] have recently observed chiral discrimination in
monolayers of protonated N(α-methylbenzyl)stearamide, II, with the
pure enantiomers having a higher aggregation energy than the race-
mate. At 25°C on a subphase of 10 N H_2SO_4, the chiral discrimina-

95

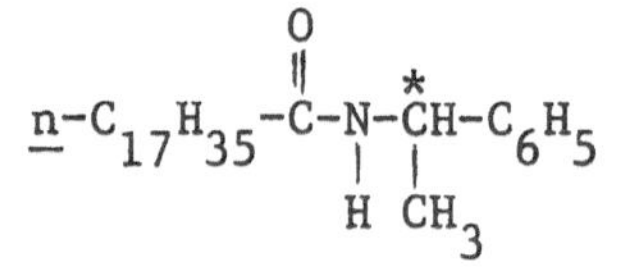

$$\underline{n}\text{-}C_{17}H_{35}\text{-}\overset{\overset{\textstyle O}{\|}}{C}\text{-}N\text{-}\overset{*}{C}H\text{-}C_6H_5$$
$$\qquad\qquad\quad |\ \ |$$
$$\qquad\qquad\quad H\ CH_3$$

II

tion factor in the surface free energies of the monolayers is 0.30
kcal/mol; the authors note that chiral discrimination in solution
is negligible by comparison.

In contrast to the stearamide case, where the chiral carbon
is adjacent to the surfactant's head group, chiral discrimination
was not observed by Arnett, et al [3] for monolayers of phosphatidylcholines. The enantiomers studied were sn-1 and sn-3 dipalmitoyl-
phosphatidylcholine (D-DPPC and L-DPPC respectively), III and IV.

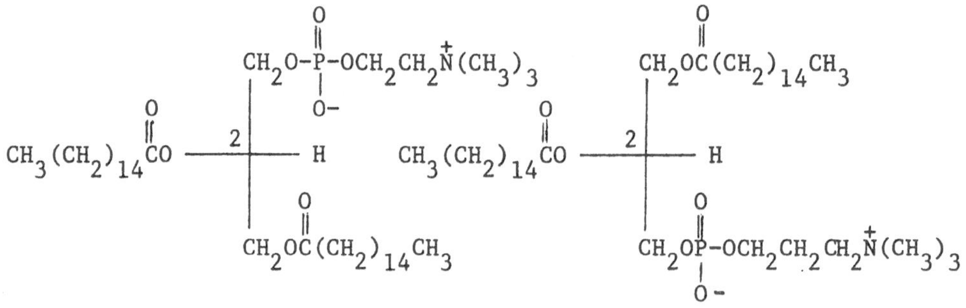

III   IV

Arnett proposes that C-2, the chiral carbon, being "buried" in the
monolayer may be the reason for the observed lack of chiral discrimination.

Although phosphatidylcholines form reversed micelles in hydrocarbon solvents[4], we are not aware of any comparison of pure enantiomers to racemates with respect to either cmc's (which are sometimes not appropriate) or cooperativity of aggregation. The optically active surfactants (+) and (−)-α-phenylethyldodecyldimethyl-
ammonium bromide (PDDAB), V, whose use in micellar catalysis[5] will

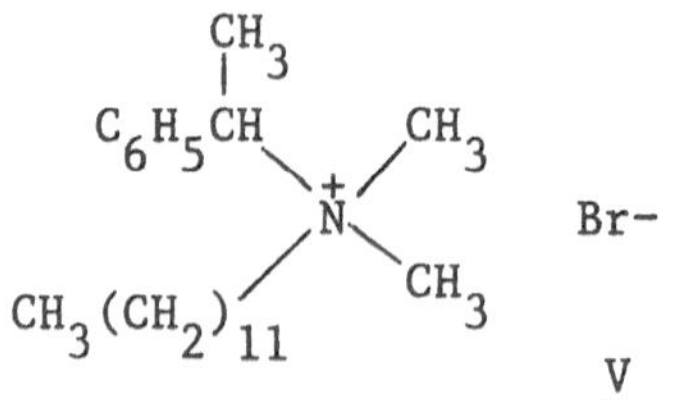

$$\begin{array}{c}
\qquad\qquad CH_3 \\
\qquad\qquad | \\
C_6H_5CH \qquad\quad CH_3 \\
\diagdown\ \ \ \diagup \\
\overset{+}{N} \qquad\qquad Br^- \\
\diagup\ \ \ \diagdown \\
CH_3(CH_2)_{11} \qquad CH_3
\end{array}$$

V

be discussed later, do have slightly different average aggregation
numbers ($\bar{n}$) in CCl_4. Thus $\bar{n}$ is 20.1 for (+) - PDDAB and 20.7 for (-)
- PDDAB.

Solubilization of optically active substrates by normal micelles
formed by optically active surfactants (or by comicelles where only
one of the surfactants is optically active) generally does not show
chiral discrimination. Thus in the cases of enantioselective micellar
catalysis discussed below, the binding constants to the micelles for
the two enantiomeric forms of the substrate are generally the same
within experimental error. An exception is found[6] for the interaction of D- and L-amino acids (alanine and valine) with micelles of
N-alkyl-N,N-dimethyl-(L or D)-alanine hydrobromide (alkyl betaines),
where differential refractive index measurements show that the
D-amino acids are more strongly adsorbed to the surface of normal
micelles made from D-surfactants.

Stereoselectivity in Micellar Catalysis: Normal Micelles and Vesicles

Much of the work on stereoselectivity in micellar catalysis
has employed ester derivatives of amino acids or di- and tripep-
tides. The general class of substrates represented by VI has been
investigated using chiral nucleophiles, the latter being incorporated either into a host normal micelle or vesicle (chiral or
achiral) or employed as functional micelle- or vesicle-forming
materials[7]. Nucleophiles of the VII class are common; Fn may be an
imidazole, cysteine, etc. side chain, while X is often a hydroxyl
group or a substituted amino group (this provides a molecular

$$
\begin{array}{cc}
\underset{\text{H}}{\underset{|}{\text{Y}}}\overset{\text{O}}{\overset{||}{-\text{C}}}-\underset{\text{H}}{\underset{|}{\text{N}}}-\underset{\text{R}}{\underset{|}{\overset{*}{\text{CH}}}}\overset{\text{O}}{\overset{||}{\text{C}}}-\text{O} & \\
\end{array}
$$

VI   VII

spacer which can be used to vary the depth of Fn in the aggregate).
The L/D stereoselectivity observed for substrates of type VI is
generally modest (with $\Delta\Delta G^{\ddagger}$ less than 1 kcal/mole); the record[8] is
12.2 for VI (R = $C_6H_5CH_2$- and Y = OCH_3) acylating N-(Z)-(L)-Leu-
(L)-His (VIII) in hexadecyltrimethylammonium bromide micelles. For
the deacylation step, the enantioselectivity is smaller: The L/D
ratio[9] is 3.18. These are far below the observed enantioselectivities in the acylation and deacylation steps (10^3 -10^4) for α-chymo-
trypsin - catalyzed hydrolysis[10] of peptide linkages.

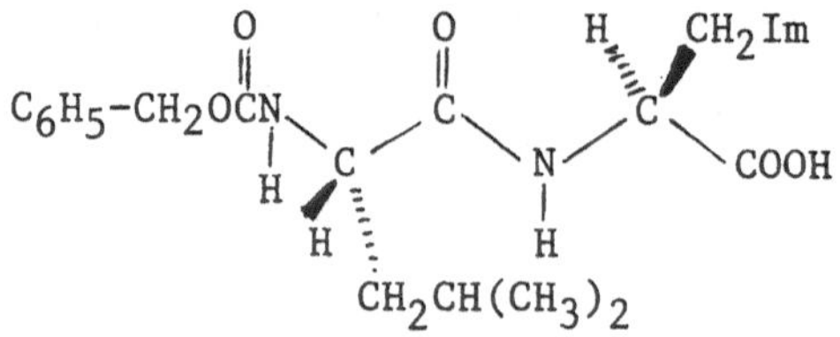

$$C_6H_5-CH_2OCN-\overset{H}{\underset{H}{\overset{|}{C}}}-\overset{O}{\overset{\|}{C}}-\overset{H}{\underset{H}{N}}-\overset{H}{\underset{}{\overset{|}{C}}}<\overset{CH_2 Im}{COOH}$$

$$\overset{|}{CH_2CH(CH_3)_2}$$

VIII

It has been recognized that strong and specific substrate-mi-
celle interactions are necessary requirements for substantial
chiral discrimination. Placing the chiral centers in hydrophobic
regions of the micelles rather than in the vicinity of the headgroups also leads to more pronounced discrimination[11]. These ideas
are nicely illustrated by the selectivities[7a] observed for substrates IX and X in normal micelles of XI.

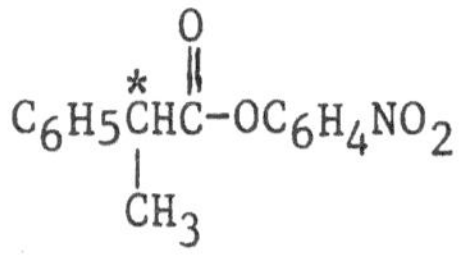

$$C_6H_5\overset{*}{C}HC\overset{O}{\overset{\|}{-}}OC_6H_4NO_2$$
$$\overset{|}{CH_3}$$

IX

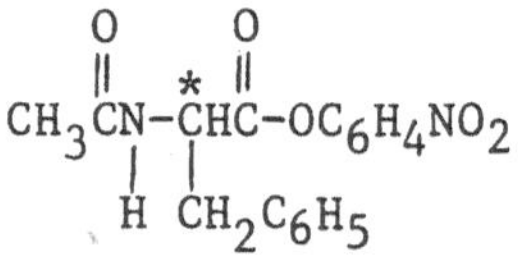

$$CH_3\overset{O}{\overset{\|}{C}}N-\overset{*}{C}HC\overset{O}{\overset{\|}{-}}OC_6H_4NO_2$$
$$\overset{|}{H}\ \overset{|}{CH_2C_6H_5}$$

X

$$CH_3(CH_2)_{11}\overset{|}{C}H(CH_2)_4\overset{+}{N}(CH_3)_3\quad Br-$$
$$\overset{*}{CONHCHCH_2}Im$$
$$\overset{|}{COOCH_3}$$

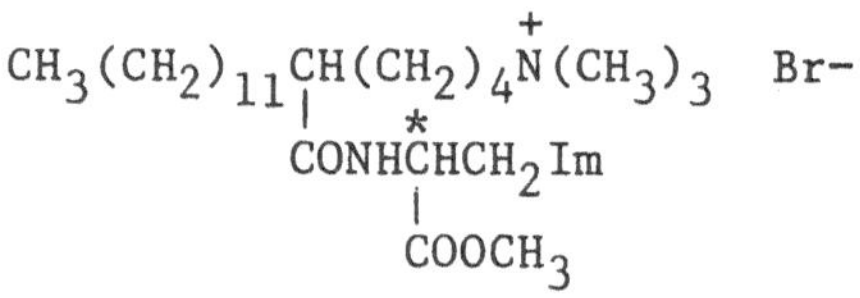

XI

The tetrahedral intermediate for the reaction of S-X with XI
is shown in XII; the S/R rate ratio is 3.0. For IX, which lacks

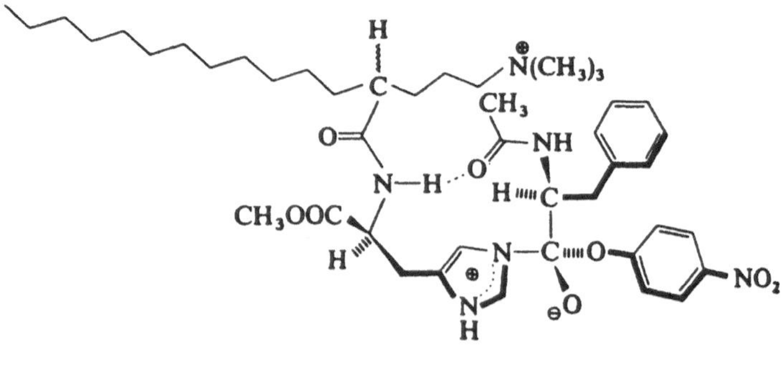

XII

the auxiliary H-bonding site of X, the selectivity is much lower:
the S/R rate ratio is 0.89.

Appropriately functionalized chiral crown ethers provide substantially higher enantioselectivities than do the micellar systems
discussed above. Thus the thiolysis[12] of dipeptide p-nitrophenyl
esters, XIII, by XIV in 95:5 CH_2Cl_2:CH_3CH_2OH (CF_3CO_2H, N-ethyl-
morpholine buffer) leads to an L/D rate ratio (R = $C_6H_5CH_2-$) of 90.
The complex of XIV with the dipeptide substrate glycyl-glycine
p-nitrophenyl ester (salt form) is shown in XV. Note the multiple

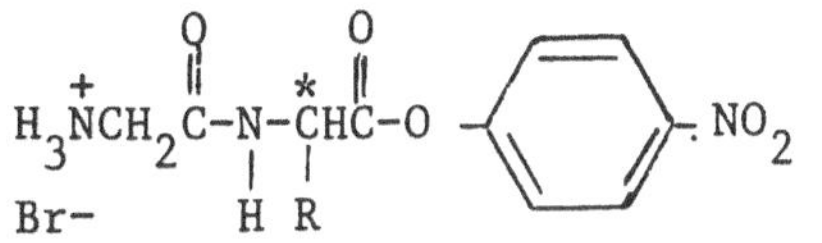

$$X = -CONH-(L)-CHCO_2CH_3$$
$$\underset{}{|}$$
$$CH_2SH$$

XIII

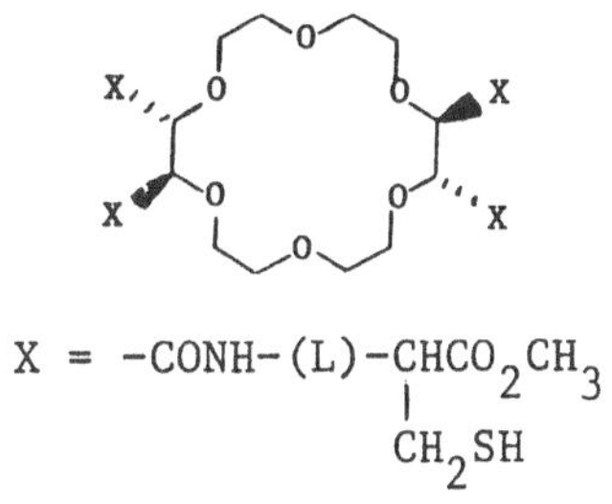

XIV

hydrogen bonding interactions available in the complex.

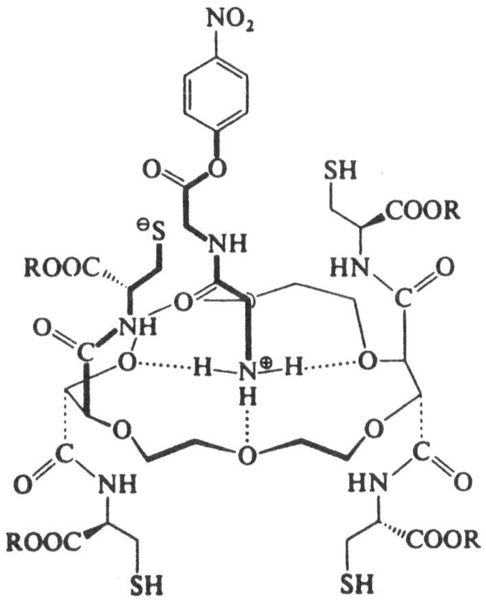

XV

The greater molecular ordering available in vesicles compared to normal micelles should increase the observed stereoselectivity[13]. In fact, Murakami found an L/D reactivity ratio of 4.4 for VI ($R = C_6H_5CH_2-$ and $Y = C_6H_5CH_2O-$) in vesicular XVI, compared to a value of 2.5 in micelles formed from the analog of XVI lacking one

$$(\underline{n}C_{12}H_{25})_2-N-\overset{\overset{O}{\|}}{C}\diagdown\overset{C}{\underset{H}{\underset{CH_2Im}{\overset{\bullet}{C}}}}\diagup\overset{\overset{H}{|}}{\underset{\overset{\|}{O}}{N}}\diagdown\overset{C}{\underset{}{}}(CH_2)_5\overset{+}{N}(CH_3)_3 \ Br^-$$

XVI

dodecyl chain[14]. However, the occurrence of enhanced stereoselectivity in vesicular systems is mechanism-dependent. Moss[7e] has recently shown that the greater molecular ordering in vesicular systems can in fact also decrease the observed stereoselectivity. Thus esterolytic cleavage of diastereomeric peptide esters (XVII shows a dipeptide example) by pure vesicles of XVIII shows less LL/DL diastereoselectivity than mixed micelles or vesicles of XVIII with

$$C_6H_5CH_2O-\overset{\overset{O}{\|}}{C}-N-\overset{*}{\underset{\underset{R}{|}}{C}H}-\overset{\overset{O}{\|}}{C}-N\diagup\diagdown \\ \underset{H}{|} \quad \overset{*}{\underset{H}{|}}\diagdown\overset{C-O}{\underset{\overset{\|}{O}}{}}-\text{—}NO_2$$

XVII

$$(\underline{n}-C_{16}H_{33})_2\overset{+}{\underset{\underset{CH_3}{|}}{N}}CH_2CH_2SH \ \ Cl^-$$

XVIII

a second surfactant. The LL diastereomers of XVII have clefts into which a hexadecyl chain of XVIII fits well, positioning XVIII's thiolate moiety in an optimum arrangement for attack at the ester's carbonyl carbon. Moss proposes that this interaction is inhibited in pure vesicular XVIII because imposed molecular ordering of the two alkyl chains inhibits the flexibility needed for a single chain to bind. More flexibility is possible when XVIII is co-aggregated.

Stereoselectivity in Micellar Catalysis: Reversed Micelles

There has been speculation in the literature[15,16] that micelle-solubilizate interactions should be more specific (in the sense of (a) one solubilizate orientation being markedly favored over others or (b) multiple contact points being possible) in reversed than in normal micelles. Moss and Sunshine[17] looked at the hydrolysis of

XIX in reversed micelles formed from XX in 1-hexanol; essentially
no stereoselectivity was observed. Kon-no et al [5,18] have investigated the aminolysis of XIX and the related hydroxy compound using
(+) and (-) -α-phenethylamine as the nucleophile. Enantioselectivities were modest (up to 2.5), and the use of an optically active

$$C_6H_5\overset{*}{\underset{OCH_3}{CH}}\overset{O}{\overset{\parallel}{C}}-O-\text{⟨benzene ring⟩}-NO_2$$

XIX

$$CH_3(CH_2)_{15}\overset{CH_3}{\underset{CH_3}{\overset{+}{N}}}-R \quad Br^-$$

$$R = -\underset{CH_3}{CH}C_6H_5$$

XX

(but nonfunctional) surfactant made relatively little difference.

     From the examples of stereoselective catalysis in normal mi-
cellar and vesicular systems, it seems evident that providing multiple hydrogen bonding and/or hydrophobic interactions which significantly stabilize one diastereomeric transition state relative to
the other will be needed to realize stereoselectivities of significant magnitude in reversed micellar systems. To date these strata-
gems have not been applied in the reversed micelle cases. An obvious
first effort might be a study of the reaction of X and XI (the
latter co-micellized with a surfactant which forms reversed micelles). It is also of particular interest (for one of these opti-
mized systems) to determine whether the use of an optically active
surfactant confers an additional kinetic advantage on the reaction
of optically active substrates with optically active nucleophiles,
since this effect has been observed in normal micellar systems[7c].

     The surfactant Aerosol OT (XXI with R equal to $CH_3(CH_2)_3CH$
$(CH_2CH_3)CH_2-$) has been used extensively to form reversed micelles and
water-in-oil microemulsions[19]. It is structurally quite similar to
IV, but one of its chiral carbons (C_1) is adjacent to the headgroup, which is not the case in IV. The conformational properties
of Aerosol OT about the C_1-C_1' bond at hydrocarbon/water interfaces
are analogous to those of the sn-3 phosphatidylcholines[19]. Analogs
of Aerosol OT can be prepared in which C_1 is the only chiral carbon
(R equal to 4-heptyl, for example); by analogy to Arnett's work
with the stearamides, substantial chiral discrimination might be
expected in the aggregation properties of these analogs. The related amide surfactants, XXII, offer the additional attractive
feature of subheadgroup hydrogen bonding where the surfactant can

$$
\begin{array}{cc}
\underset{\text{XXI}}{
\begin{array}{c}
\overset{\displaystyle O}{\underset{\displaystyle}{}} \\
\text{CH}_2\overset{\text{O}}{\overset{\|}{\text{C}}}\text{-OR} \\
\text{RO-}\overset{\text{O}}{\overset{\|}{\text{C}}}\text{----H} \\
\text{SO}_3\text{Na}
\end{array}}
&
\underset{\text{XXII}}{
\begin{array}{c}
\text{CH}_2\overset{\text{O}}{\overset{\|}{\text{C}}}\text{-N-R} \\
\text{R-N-}\overset{\text{O}}{\overset{\|}{\text{C}}}\text{----H} \\
\text{SO}_3\text{Na}
\end{array}}
\end{array}
$$

XXI

XXII

XXIII

XXIV

act as both hydrogen-bond acceptor and donor. This is reminiscent of the so-called hydrogen belt[20] in bilayer membranes. Additional double-tailed surfactants which may be prepared easily in optically active form are types XXIII and XXIV, which are derivatives of L-aspartic acid. These are expected to have aggregation properties similar to the Aerosol OT analogs.

REFERENCES

1. S. Miyagishi and M. Nishida, J.Colloid Interface Sci., 65 : 380 (1978).
2. E.M. Arnett, B.J. Kinzig, M.V. Stewart, O. Thompson, J. Chao and R.J. Verbiar, J. Am. Chem. Soc., 104 : 389 (1982).
3. a. M.V. Stewart and E.M. Arnett, Top. Stereochem., 13 : 195 (1982)
   b. E.M. Arnett and J.M. Gold, J. Am. Chem. Soc., 104 : 636 (1982).
4. See S.-T. Chen and C.S. Springer, Chem. Phys. Lipids, 23 : 23 and references therein (1979).
5. K. Kon-no, M. Tosaka and A. Kitahara, J. Colloid Interface Sci., 86 : 288 (1982).
6. A.H. Beckett, G. Kirk and A.S. Virji, J. Pharm. Pharmac. 19 : 827 (1967).
7. a. J.M. Brown and C.A. Bunton, J. Chem. Soc., Chem. Commun. 969 (1974).
   b. Y. Ihara, J. Chem. Soc., Perkin Trans, 2 : 1483 (1980).

7. c. K. Ohkubo, K. Sugahara, K. Yoshinaga and R. Ueoka, J. Chem. Soc., Chem. Commun., 637 (1980).

   d. R.A. Moss, Y.-S. Lee and K.W. Alwis, J. Am. Chem. Soc., 102 : 6646; Tetrahedron Lett., 22 : 283 (1981).

   e. R.A. Moss, T. Taguchi and G.O. Bizzigotti, ibid. 23 : 1985 (1982).

8. Y. Ihara, N. Kunikiyo, T. Kunimasa, M. Nango and N. Kuroki, Chem. Lett., 667 (1981).

9. Y. Ihara, Y. Kimura, M. Nango and N. Kuroki, Makromol. Chem., Rapid Commun., 3 : 521 (1982).

10. D.W. Ingles, J.R. Knowles and J.A. Tomlinson, Biochem. Biophys. Res. Commun., 23 : 619 (1966).

11. J.H. Fendler in "Membrane Mimetic Chemistry", John Wiley & Sons, New York, p. 322 (1982).

12. J.-M. Lehn and C. Sirlin, J. Chem. Soc. Chem. Commun., 949 (1978).

13. T. Kunitake, N. Nakashima, S. Hayashida and K. Yonemori, Chem. Lett., 1413 (1979).

14. Y. Murakami, A. Nakano, A. Yoshimatsu and K. Fukuya, J. Am. Chem. Soc., 103 : 728 (1981).

15. J.H. Fendler and E.J. Fendler, "Catalysis in Micellar and Macromolecular Systems", Academic Press, New York (1975).

16. J. Sunamoto, K. Iwamoto, M. Akutagawa, M. Nagase and H. Kondo, J. Am. Chem. Soc., 104 : 4904 (1982).

17. R.A. Moss and W.L. Sunshine, J. Org. Chem., 39 : 1083 (1974).

18. K. Kon-no, M. Tosaka and A. Kitahara, J. Colloid Interface Sci., 79 : 581 (1981).

19. See for example the references in L.J. Magid and C.A. Martin, this volume.

20. H. Brockerhoff, "Bioorganic Chemistry", van Tamlen E.E., Ed. Academic Press, New York, Vol. 3, Chapter 1 (1977).

HYDROCARBON AROMATICITY AND W/O MICROEMULSIONS STABILIZED BY NON-IONIC SURFACTANTS

S.E.Friberg, H.Christenson and G. Bertrand

Chemistry Department
University of Missouri-Rolla
Rolla, MO 65401 USA

and

D.W.Larsen
Chemistry Department
University of Missouri-St.Louis
St.Louis, MO 65401 USA

The aromatic character of hydrocarbons has a pronounced influence on the temperature-dependent behavior
of microemulsions stabilized by non-ionic surfactants that
have a polyethylene gylol alkyl ether structure.

Calorimetric and spectroscopic investigations
indicated that the aromatic hydrocarbons compete to some
extent with the water molecules for the binding sites at
the ether bridges of the compounds. It is suggested that
this competition leads to a reduced hydration of the oxyethylene groups similar to the one observed with an
increase in temperature.

INTRODUCTION

The polyethylene glycol alkyl ether surfactants give a solubilization of hydrocarbons in water and vice versa which are extremely
temperature sensitive, as has been amply discussed in the literature
[1-11]. A general description of the results can be given as follows.

In a limited temperature range ($\Delta T \simeq 30^\circ C$) the micellar
association changes from the "normal" structures found in aqueous
solutions to the inverse structures found in hydrocarbon solutions

over a surfactant phase that is probably a disordered bicontinuous
structure[6,12,13]. The surfactant phase appears in a narrow temperature region which has been named the HLB-temperature by Shinoda
who developed an emulsifier selection system based on this approach[14].

An important feature of this situation is the fact that the
structure of the hydrocarbon has a pronounced effect on the value
of the HLB-temperature; it will be reduced by approximately $50^{\circ}C$ by
changing from an aliphatic to an aromatic hydrocarbon.

The reason for this strong influence is an intriguing scientific
problem and it was judged that an investigation into the location
and molecular interaction of the solubilized hydrocarbon molecules
would contribute to the solution of the problem. A combination of
spectroscopic and calorimetric techniques was found useful[15,16].
In the present article the results will be combined with some
recent information to give a comprehensive description of the phenomenon.

EXPERIMENTAL

Materials

Penta- and tetraethylene glycol dodecyl ether from Nikkol Co.,
Japan were $\geq$ 99% pure, according to gas chromatograms from the
manufacturer. Benzene, c-hexane (Fisher certified), n-decane and
2-(2-n-butoxyethoxyethanol) (Aldrich, Gold Label) were used without
further purification.

Nuclear Magnetic Resonance Spectra

Proton-NMR measurements were performed on a Varian EM-360
spectrometer with TMS as an internal standard with a check that the
minute amounts of TMS did not change the relative position of the
lines. Some of the spectra were also recorded on a Jeol FX100
Fourier transform spectrometer at 100 MHz with a probe temperature
of $30^{\circ}C$ and with deuterium as an internal lock.

Calorimetry Measurements

All measurements were performed on a Tronac 550 Isothermal
Titration Calorimeter operating at 25.00 ± $0.01^{\circ}C$, equipped with
a 2.0 ml Gilmont buret capable of measuring titrant volumes to
0.0002 ml. Solutes were added to 25.0 ml of the solvent mixture
prepared by weight, in 4 to 10 increments of 0.1 to 0.5 ml for benzene, and 0.02 to 0.1 ml for water. Further details have been given
elsewhere[15].

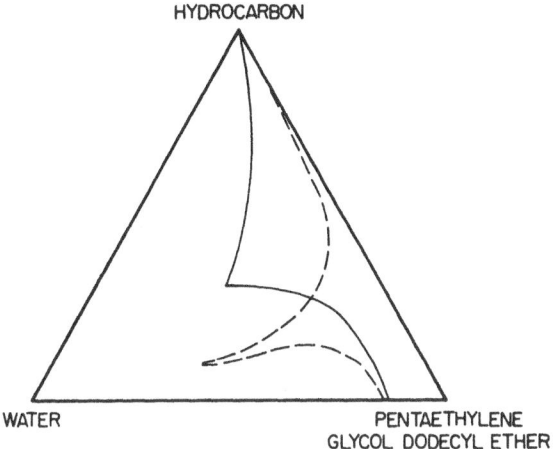

Figure 1. Solubility areas in water,pentaethylene glycol dodecyl ether and a) benzene (----); b) cyclohexane (——).

RESULTS AND DISCUSSION

    The dependence on the nature of the hydrocarbon for the solubilization of water is shown in Figure 1.The solubility area for benzene shows that the maximum solubility of water takes place at considerably higher ratios than is the case for cylcohexane. In addition, the benzene solution gives a smaller initial solubilization of water compared with the cyclohexane solution; the water/surfactant molecular ratio at high hydrocarbon content was 5 for the cylcohexane versus 0.5 for benzene. This result,indicating a diminished hydration in the presence of benzene relative to cyclohexane may,with advantage, be compared with the earlier results from NMR spectra[16].These showed a separation of resonance frequencies for the different oxyethylene groups,indicating a concentration gradient of benzene along the polar part of the surfactant chain for dilute solutions of surfactant in benzene. Later NMR results for cylcohexane showed a minimum resolution of the resonance frequencies; as a matter of fact,the spectrum was rather similar to that obtained for the native surfactant[16].

    Addition of water also gave a different response for solutions with benzene or cyclohexane. In the benzene system the addition of small amounts of water gave a further split of some of the frequencies, showing a gradual translation of the water from the terminal OH group towards the alkyl chain. In the cyclohexane system the addition of water caused a small down-field shift and no further separation of the resonance frequencies.

    A reasonable conclusion from these results was that there might be a competition between the benzene and water molecules for the polar part of the surfactant chain. Such a competition would be

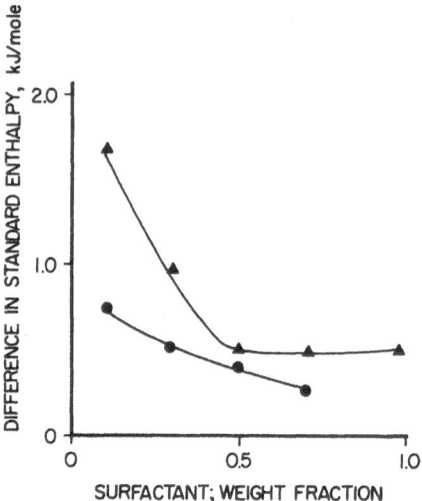

Figure 2. The increment in standard heat of solution versus weight fraction of tetraethylene glycol dodecylether in decane for benzene (●) when water is added and for water (▲) when benzene is added.

expected to manifest itself in partial molar enthalpies of solution of benzene and water in surfactant/hydrocarbon solutions.

The results of calorimetric investigations supported this hypothesis[15]. Figure 2 shows the difference in standard heat of solution of benzene with no water present and with a water concentration of 0.9 M and the corresponding difference for water with and without benzene in decane/tetraethylene glycol dodecyl ether solutions. In both cases the increment is positive, e.g. making the standard heat of solution of one compound more endothermic in the presence of the other.

A competition between water and benzene should also be quantified in terms of enthalpy values for benzene interacting with the ether groups and the hydroxyl group of the surfactant molecule. An investigation using 2-(2-n-butoxyethoxy) ethanol and tetraethylene glycol dodecyl ether gave the results in Figure 3. The reduction of the endothermic heat of solution in decane (≃ 3 kJ/mole) was a linear function of the molar concentration of oxygen. The 20 M concentration represents a pure ether and the difference ≃ 4 kJ/mole can be used to estimate the enthalpy of binding of benzene to ether oxygens. The underlying assumption is that in the absence of an aliphatic chain all of the added benzene molecules are bound to ether oxygens. This is a reasonable postulate, considering the standard heat of solution is at infinite dilution.

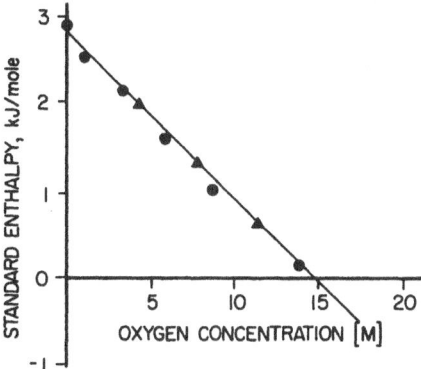

Figure 3. The standard heat of solution for benzene versus the molar concentration of oxygen atoms for solutions of decane and 2-(2-n-butoxyethyoxy) ethanol (▲) or tetraethylene glycol dodecyl ether (●).

Applying the Boltzman distribution law and ignoring entropy to estimate benzene partition between polar and non-polar parts of a surfactant molecule, we obtain a partition ratio of 5, which is of the same magnitude as data from UV investigations[16].

The results also gave some insight into the consequences of this competition for the association to reverse miscelles (W/O microemulsion droplets). The following facts form the basis for the suggested mechanisms.

A rapid increase of water solubilization in benzene is initiated (Figure 1) at such a high surfactant concentration that the surfactant molecules are associated even in the absence of water, as shown by the results of NMR and IR spectra[16]. The former show the disappearance of the gradient of benzene concentration along the oxyethylene chain that is characteristic for low surfactant concentrations, while the IR spectra show that all the surfactant molecules are hydrogen bonded. In the cyclohexane, on the other hand, the formation of reverse micelles already took place at such low concentrations that IR spectra showed no indication of hydrogen bonds.

The following mechanism offers a reasonable interpretation of these results. The weak competition from benzene with water for the oxyethylene groups prevents the water from spreading along the whole oxyethylene chain in the dilute surfactant solution in benzene. Hence, the association to reverse micelles is prevented and the water molecules are mainly attached to the terminal hydroxyl groups of the surfactant. At the high concentrations at which the solubilization increases to a significant degree, the benzene molecules are all hydrogen-bonded, as shown by the IR pattern and in addition, the NMR spectra[16] showed the environment to be identical along the chain. The signal position is now similar to that found for the pure surfactant (Figure 3, Reference 16).

These facts indicate a high degree of self-association for the surfactant molecules, e.g. the conditions in the non-aqueous benzene solution are similar to those in the pure surfactant. This association enables interattachment of the water molecules at a distance from the terminal OH group that is sufficient to give rise to continued association to reverse micelles as water is added. The point to be emphasized is the fact that the association to reverse micelles in benzene can take place only after the initial association in water-free solutions has proceeded to a significant degree.

On the other hand, in the aliphatic hydrocarbon the addition of water leads to reverse micellization already at such dilute concentrations that a substantial part of the surfactant molecules do not even take part in the intermolecular hydrogen bonds. In this case, even the primary association of surfactant molecules prior to the subsequent association to reverse micelles is caused by bound water molecules.

An attractive, but so far not sufficiently substantiated explanation for this difference in behavior lies in the fact that benzene and the aliphatic hydrocarbon differ in regard to their capability to bind to ether groups. The benzene enthalpy of the magnitude of 1 kcal/mole appears sufficient to prevent the water molecules from attaching to the ether oxygen in dilute surfactant solutions in benzene. In contrast, the aliphatic hydrocarbons have not such an energy hindrance and the water molecules are free to interact with the entire polar chain, with micellization as a consequence.

The evidence so far supports this hypothesis, at least for the example with W/O systems discussed here. However, further and more detailed evidence is needed before a final conclusion can be made, especially before an extension can be made to the O/W and to bicontinuous systems.

REFERENCES

1. K.Shinoda, "Colloidal Surfactants" Academic Press (1963)
2. K.Shinoda, J.Coll.Int.Sci., 24: 4 (1967)
3. K.Shinoda, Ibid. 34: 278 (1970)
4. K.Shinoda and T. Ogawa, Ibid. 24: 56 (1967)
5. S.Friberg and I. Buraczewska, Progr.Coll.Polymer Sci.,63:1 (1978)
6. S.Friberg, I.Lapczynska and G. Fillberg, J.Coll.Int.Sci.,56:19 (1976)

7. S.E.Friberg, I.Buraczewska and J.C.Ravery in "Micellization Solubilization and Microemulsions" (K.L.Mittal, Editor) Plenum Press (1977) 901.

8. T.A.Bostock, M.P.McDonald, G.J.T. Tiddy and L. Warring, Surface Active Agents, SCJ Symp. at Univ.of Nottingham, Sept. 1979.

9. T.A.Bostock, M.H.Boyle, M.P.McDonald and R.M. Wood, J.Coll. Int.Sci., 73: 368 (1980)

10. J.C.Lang and R.D.Morgan, J.Chem.Phys., 73: 5849 (1980)

11. P.G.Nilsson and B. Lindman, J.Phys.Chem., 86: 271 (1982)

12. L.E.Scriven, Nature, 263: 123 (1976)

13. Y. Talmon and S. Prager, J.Chem.Phys.,69: 2984 (1978)

14. K.Shinoda and H. Arai, J.Phys.Chem.,68: 3485 (1964)

15. M.Nakamura, G.L.Bertrand and S.E.Friberg, J.Coll.Int.Sci., (in Press)

16. H. Christenson and S.E.Friberg, J.Coll.Int.Sci., 75: 276 (1980)

AGGREGATION STATES AND DYNAMICS OF NONIONIC POLYOXYETHYLENE

SURFACTANTS

Anthony Ribeiro

Stanford Magnetic Resonance Laboratory
Stanford University
Stanford, California 94305

INTRODUCTION

The nonionic polyoxyethylene ether (POE) surfactants exhibit solvent dependent aggregation states[1,2]. Solvents with two or more potential hydrogen bonding centers like ethylene glycol[3] and form-amide[4-6] promote micelle formation similar to that in water. In contrast, in aliphatic hydrocarbon solvents, such as cyclohexane or decane, reverse (or inverted) micelles may form[2,7]. However, polar solvents with a single hydrogen bonding center do not appear to support micelle formation[4], and several reports describe the destruction of micelles by the lower alcohols, ethanol and methanol[8,11].

We have been interested in the structure of the nonionic detergents for a number of years and have previously used ^{1}H and ^{13}C NMR techniques to study isooctylphenylpolyoxyethylene (OPE)[12,13] and alkylpolyoxyethylene (APE)[14] ether micelles in aqueous media. Here we report the use of ^{1}H NMR chemical shifts and spin-lattice relaxation times (T_1) to monitor the aggregation and dynamics of the isooctylphenylpolyoxyethylenes (OPE or Triton X series) in a variety of solution environments.

EXPERIMENTAL

Triton X-100 (Rohm and Haas) is a polydisperse preparation of isooctylphenylpolyoxyethylene ethers with an average chain length of 9.5 units. Concentrations are expressed in terms of an average molecular weight of 624. Monodisperse preparations of OPE are ob-

tained through chromatography and synthesis [15,17]. The sample of
OPE_{10} used here was a gift of Dr. C.F. Allen, Pomona College, CA.

^{1}H NMR spectra were recorded as previously described[12,14].
Spin-lattice relaxation times (T_1) were determined by the inversion
recovery Fourier transform (IRFT) method of Vold et al.[18]. For the
T_1 measurements, samples in 5 mm tubes with small pieces of 8 mm
tubing fused to the top of the tubes were degassed to 1 x 10^{-5} torr
with at least 5-6 freeze-pump-thaw cycles. The samples in D_2O were
frozen in dry ice-acetone ($-80^{\circ}C$) and those in organic solvents
were frozen in liquid N_2 ($-196^{\circ}C$). In general some 10-15 valid
points were obtained for each line in a relaxation experiment as
shown in Figure 1 of Ref. 12. The T_1 relaxation for each line analyzed was found to follow a single exponential decay.

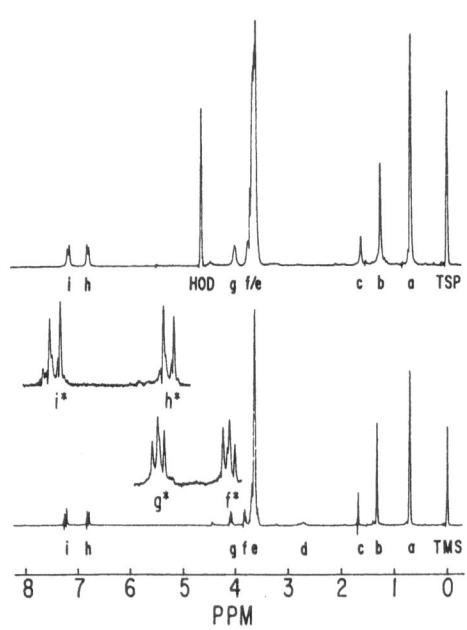

Figure 1. *220 MHz* [1]*H NMR spectra at 37^oC of OPE$_{9.5}$ (Triton X-100)*
in (Top) D$_2$O referred to sodium 3-trimethylsilylpropionate-
2,2,3,3,-d$_4$ (TSP) and (Bottom) CDCl$_3$ referenced to tetramethyl-
silane (TMS).

RESULTS AND DISCUSSION

Solvent Dependence

The ^{1}H NMR spectrum of polydisperse $OPE_{9.5}$ (Triton X-100) as micelles in D_2O (Top) and in $CDCl_3$ solution (Bottom) is shown with the assignments in Figure 1. The ^{1}H NMR spectra and T_1 relaxation times (Table II) for monodisperse OPE_9 and OPE_{10} preparations are not significantly different from that of polydisperse Triton X-100. Integration of relative areas is consistent with the oxyethylene chain lengths given. The OPE_9 prepared as in references 15-17 is more than 99% pure; minor contaminants could be eliminated by preparative thin-layer chromatography.

Tables I and II summarize the ^{1}H chemical shifts and T_1 relaxation times for the OPE's in different solvent environments at similar spectrometer conditions. The chemical shift data for the OPE's in most environments are not significantly varying from their values as neat detergents. Exceptions are seen in (1) the wider range of chemical shifts for the oxyethylene protons as micelles in D_2O, suggesting they are in a more polar environment than in the bulk state[19] and (2) the large upfield shifts of the surfactant in benzene[19], suggesting interactions of different detergent segments with benzene rings. These shifts are considerably smaller in deuterobenzene. In fact, we see no significant shift changes in the alkyl and phenol regions, and only some small shifts of the oxyethylene protons. The interaction of various detergent segments with neighboring benzene rings of the solvent appears to be substantially reduced in deuterobenzene.

While the chemical shifts are not significantly different for most environments, the T_1 spin-lattice relaxation times differ substantially among the solvents considered. The relaxation times for both the hydrophobic and hydrophilic portions of the OPE molecule are long and very similar in deuteromethanol, deuterochloroform and deuterobenzene. Considerable decreases in the T_1 for the alkyl, phenyl and oxyethylene groups are seen for the detergent in micellar form in D_2O. Decreases in T_1 are also observed for the alkyl and phenyl groups in DMSO, while both a short and long T_1 (main peak) are seen in the oxyethylene.

T_1 values of surfactant molecules are known to decrease upon formation of molecular aggregates[20]. The decrease in T_1 with aggregation is not well understood. Internal chain motions (both translational and rotational) could be reduced in the aggregated species relative to the monomeric form. Another factor to consider is the change in environment that a molecule undergoes during aggregation

Table I. ^1H NMR chemical shifts of $OPE_{9.5}$ in various solvents[a]

$$CH_3 - \underset{\underset{CH_3}{|}}{\overset{\overset{CH_3}{|}}{C}} - CH_2 - \underset{\underset{CH_3}{|}}{\overset{\overset{CH_3}{|}}{C}} - C \underset{\underset{h}{\overset{H}{C}}}{\overset{\overset{H}{C}}{\underset{}{}}} C - OCH_2 - CH_2 - (OCH_2CH_2)_{n-1}OH$$

Solvent	Alkyl Region				Oxyethylene Region			Phenol Region	
	a	b	c	d	e	f	g	h	i
Bulk[b]	0.71	1.31	1.69		3.55	3.76	4.04	6.84	7.23
D_2O	0.700	1.26	1.63		3.56-3.72[c]	3.79	4.00	6.80	7.18
D_2O[b]	0.70	1.28	1.65		3.65-3.76[c]	3.8	4.03	6.84	7.22
Ethylene-d_4-glycol[b]	0.72	1.35	1.74		3.53-3.55				
Dioxane[b]	0.70	1.32	1.73		overlaps solvent			6.82	7.29
$CDCl_3$	0.705	1.34	1.69	2.06	3.63[d]	3.82	4.09	6.80	7.23
Benzene[b]	0.31	0.83	1.19		2.89-3.02[d]			overlaps solvent	
Deuterobenzene	0.705	1.29	1.67	2.96	3.44,3.47[d]		4.11	6.77	7.14

[a]Data obtained at 220 MHz and 40°C, except as noted in footnote b. Shifts on the δ scale in ppm with positive numbers downfield from TMS.
[b]Data from reference 19 at 220 MHz and 18°C, corrected from external $CHCl_3$ to the TMS scale by assuming a shift of 7.35 ppm for the external $CHCl_3$ signal. The discrepancy in temperature is not important, as the surfactant in D_2O shows identical shifts over a wide range of temperatures.
[c]Range of chemical shifts for several overlapping lines observed in D_2O.
[d]Several components can be observed.

Table II. ^1H spin-lattice relaxation times of isooctylphenylpolyoxyethylene ethers in various solvents[a]

Chemical structure: $CH_3-C(CH_3)(CH_3)-CH_2-C(CH_3)(CH_3)-C_6H_4-OCH_2-CH_2-(OCH_2CH_2)_{n-1}OH$ with position labels a, c, b (alkyl) and h, i, g, f, e, d (aromatic/oxyethylene).

Compound	Conc. (mM)	Solvent	Alkyl Region				Oxyethylene Region			Phenol Region[b]	
			a	b	c	d	e	f	g	h	i
TX-100	110	$CDCl_3$	1.1	0.59	0.73	0.87	1.3,1.5[c]	1.2	1.1	2.1,1.8	1.1,1.1
TX-100	330	CD_3OD	1.1	0.50	0.63	-	1.4,1.5[c]	1.3	1.0	1.8,1.7	1.1,0.95
TX-100	117	C_6D_6	1.1	0.54	0.67	2.37	1.9,1.6[c] and 1.4	1.6	1.1	1.8,1.6	1.2,1.3
TX-100	100	D_2O (25°C)	0.18	0.073	0.075	-	0.25-0.60[d]			0.21,0.21	0.19,0.19
TX-100	100	D_2O	0.23	0.098	0.096	-	0.37-0.89[d]	0.38	0.17	0.26,0.24	0.19,0.18
TX-100	147	DMSO	0.56	0.23	0.23	-	0.83,1.8[e]			0.83,0.75	0.50,0.43
TX-100	147	DMSO(56°C)	0.84	0.39	0.39	-	1.2,2.4[e]			1.3,1.1	0.71,0.67
OPE10	100	D_2O	0.23	0.094	0.092	-				0.24,0.23	0.20,0.19

[a] T_1 values (sec) at 34°C with peaks listed in the downfield direction from TMS.
[b] T_1 value for each component of each doublet with the upfield component listed first.
[c] T_1 value for large main signal listed first and then smaller shoulders at the lower field position.
[d] T_1 range for several overlapping components.
[e] T_1 value for a smaller upfield signal and then a larger lower field component.

or micelle formation. For a monomer in a deuterated solvent, the intermolecular dipolar relaxation should be mainly caused by interaction between protons of the surfactant and deuterons of the solvent. On the other hand, in an aggregated micellar species in a deuterated solvent, the intermolecular relaxation should occur mainly by proton-proton (surfactant-surfactant) interactions, and this is expected to lead to a relative decrease in the observed T_1.

Evidence that the dipolar relaxation mechanism plays a substantial role in the relaxation of the OPE detergents is provided by their temperature dependence. The T_1 values of the alkyl singlet and the oxyethylene peaks in D_2O and DMSO-d_6 (Table II) increase with temperature, a behavior very characteristic of the dipolar relaxation mechanism. In addition, the phenyl doublets may feature a contribution from spin-coupling relaxation. It is clear that the relaxation times observed for the OPE's in $CDCl_3$, CD_3OD and C_6D_6 favor an interpretation in terms of monomeric detergent molecules interacting with deuterated solvents. In the case of benzene, this is further substantiated by the different shifts observed in benzene and deuterobenzene. The relaxation times observed in D_2O and DMSO-d_6 on the other hand favor an interpretation in terms of intermolecular proton-proton (surfactant-surfactant) interactions in an aggregated species. These interpretations appear consistent with the ideas that the nonionic detergents are monomeric in apolar solvents with a single hydrogen bonding center (e.g. methanol), while extensive micellar aggregation occurs in solvents which possess two or more hydrogen bonding centers (e.g. water)[3-6].

Aqueous Micelles

Here we discuss the more interesting NMR relaxation characteristics of nonionic detergent micelles in aqueous media, aspects of which are presented elsewhere[12-14,19]. In general, the ^{1}H resonances of OPE$_{9-10}$ in micelles in D_2O are broader than in $CDCl_3$ (Figure 1). The methyl and methylene singlets in the hydrophobic part relax as single exponentials, indicating that they are single Lorentzian lines. Similar exponential decay is seen for the phenyl doublets and the triplets f and g corresponding to the POE unit adjacent to the phenyl ring. However, the main POE peak relaxes with the high field components nulling first and then a progressive relaxing of the various components located at lower fields. The main POE peak is thus a series of closely-spaced chemically-shifted lines representative of various oxyethylene groups in the micellar environment. The sharp peak between f and e at ∿3.72 ppm in Figure 1 has been assigned to the last oxymethylene group (next to the terminal hydroxyl) on the outside of the micelle in a fully hydrated environment[19]. In addition, an increase of the POE chain length from $OPE_{9.5}$ to OPE_{13} to OPE_{16} to OPE_{30} gives a preferential increase in the intensity of the low field components, suggesting that they reflect POE groups on the outer parts of the nonionic micelle[19,21]. The combination of these assignments with the observed relaxation suggests that a gradient in mobility occurs from the hydrophobic/ hydrophilic interface (short T_1 for g and f in first POE group near phenyl ring) to the outermost POE groups of the nonionic micelle (long T_1 for low field components of main POE peak). A careful analysis at 220 MHz allowed the resolution of reliable T_1 values for 11 components of the main POE peak (Table II, reference 12). The lack of water penetration to the hydrophobic core of these nonionic micelles has been explored with relaxation time measurements in H_2O/D_2O mixtures[12,19] and in the presence of paramagnetic ions[12,14].

ACKNOWLEDGEMENTS

This work was supported by NIH Grants RR00708 and RR00711 and NSF Grants GP32829 and GP23633.

REFERENCES

1. G.C. Kresheck, "Water: A Comprehensive Treatise Vol. 4. Aqueous Solutions of Amphiphiles and Macromolecules", F. Franks, ed., Plenum Press, New York (1975).
2. J.H. Fendler and E.J. Fendler, "Catalysis in Micellar and Macromolecular Systems", Academic Press, New York (1975).
3. A. Ray and G. Némethy, J. Phys. Chem. 75 : 809 (1971).
4. A. Ray, Nature (London) 231 : 313 (1971).
5. C. McDonald, J. Pharm. Pharmacol. 22 : 148 (1970).
6. C. McDonald, J. Pharm. Pharmacol. 22 : 774 (1970).
7. P.S. Sheih and J.H. Fendler, J. Chem. Soc. Far. Trans. I 73 : 1480 (1977).
8. M.J. Sasaki and N. Sata, Koll. Z. 199 : 49 (1964).
9. M.J. Schick and A.H. Gilbert, J. Colloid Sci. 20 : 464 (1965).
10. P. Becher, J. Collid Sci. 20 : 728 (1965).
11. P. Becher and S.E. Trifiletti, J. Colloid Interface Sci. 43 : 485 (1973).
12. A.A. Ribeiro and E.A. Dennis, Biochemistry 14 : 3746 (1975).
13. A.A. Ribeiro and E.A. Dennis, J. Phys. Chem. 80 : 1746 (1976).
14. A.A. Ribeiro and E.A. Dennis, J. Phys. Chem. 81 : 957 (1977).
15. C.F. Allen and L.I. Rice, J. Chromatogr. 110 : 151 (1975).
16. R.J. Robson and E.A. Dennis, Biochem. Biophys. Acta 508 : 513 (1978).

17. R.C. Mansfield and J.E. Locke, J. Amer. Oil Chem. Soc.
    41 : 267 (1964).
18. R.L. Vold, J.S. Waugh, M.P. Klein and D.E. Phelps, J. Chem.
    Phys. 48 : 3831 (1968).
19. F. Podo, A. Ray and G. Némethy, J. Amer. Chem. Soc. 95 : 6164
    (1973).
20. T. Nakagawa and F. Tokiwa, "Surface and Colloid Science. Vol.9",
    E. Matijevic, ed., John Wiley and Sons, New York (1976).
21. A. Ribeiro, Ph.D. Thesis, University of California San Diego
    (1975).

DYNAMIC BEHAVIOR OF MICROEMULSIONS

A.M. Cazabat, D. Chatenay, P. Guering, D. Langevin,
J. Meunier and O. Sorba

Laboratoire de Spectroscopie Hertzienne de l'E.N.S.
24, rue Lhomond
75231 Paris Cedex 05 - France

The dynamic behavior of water-in-oil microemulsions
has been studied by using different techniques:
light scattering, electrical conductivity, viscosity
and electrical birefringence measurements. A tentative picture of the inversion process from water-in-
oil to oil-in-water structures is presented. When
the interfacial layer is relatively rigid,droplets
can be concentrated to large volume fractions and
inversion is relatively sharp. Where the interfacial
layer is more fluid the structure of the droplets
disappears progressively above the percolation
threshold,and in the very large inversion region the
structure is probably bicontinuous.

INTRODUCTION

Microemulsions are transparent dispersions of oil and water
made with surfactant molecules. When the oil (or water) content is
low, it is generally accepted that the structure is that of small
oil (resp.water) droplets surrounded by surfactant molecules and
dispersed into the water (resp.the oil)[1]. When the volume fraction
ϕ of the droplets increases the structure can invert without any
apparent discontinuity in the physical properties from an oil-in-
water (o/w) to a water-in-oil (w/o) structure. Up to now it is not
clear whether the droplet structure exists in the whole concentration range and if the inversion is sharp, or, if the inversion
is progressive accompanied by structural changes, such as the
appearance of bicontinuous structures[2,3].

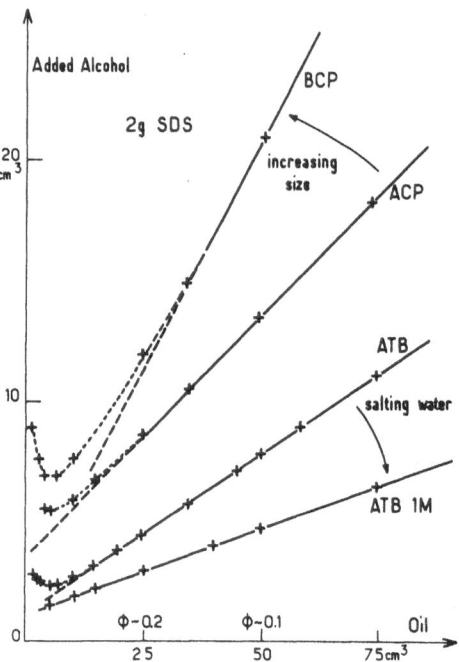

Figure 1. Alcohol versus oil content in several microemulsion series (for 2 g of SDS). The series is referenced by three letters designating the water/SDS ratio, oil and alcohol (see text), followed by salt molarity.

In order to clarify these points we will present in this paper the results of different types of experiments: static and dynamic light scattering, viscosity, electrical conductivity and transient electrical birefringence measurements.

MICROEMULSION COMPOSITIONS

Inversion was approached in two different ways: 1) by varying the droplets' concentration, keeping their size constant, and 2) by varying the surfactant partitioning between oil and water by adding salt.

1)The type 1 systems are mixtures of water, sodium chloride in some cases (salt molarity is then indicated), toluene (T), or cyclohexane (C), sodium dodecyl sulfate (SDS) and butanol (B), or pentanol (P). The size of the droplets is fixed by keeping the water-to-soap ratio constant: 1(wt/wt) for the series α, 1.2 for the series A and 2.4 for the series B. The microemulsions of a given series are those which are in equilibrium with excess water and their composition in the pseudo ternary phase diagram is represented by a curve called

the "dilution" curve. Several examples of dilution curves are given
in Figure 1. When the oil content is great enough (the volume fraction ϕ of the droplets < 10%) the dependence is linear, indicating
that the alcohol partitioning between the water and the oil continuous phase is constant. This suggests that the size of the droplets remains constant, as confirmed by neutron[4] and light-scattering[5]
experiments. For smaller oil contents the alcohol partitioning
varies about a ϕ value, depending on the series' composition, indicating that more alcohol is incorporated in the interfacial region,
thus allowing the change of sign in curvature between w/o and o/w
structures.

2) Here, the Type 2 systems were composed of brine (47% wt),
toluene (47% wt), SDS (2%) and butanol (4%). The brine contains
water and sodium chloride, and according to the salinity S the mixture separates into :

   S < 5.4 wt%   o/w microemulsion in equilibrium with excess oil

   S > 7.5   w/o microemulsion in equilibrium with excess water

   5.4 < S < 7.5 middle phase microemulsion in equilibrium with
                 excess oil and water

   For each salinity in the two-phase systems the microemulsions
were diluted and the dilution curves were found to be linear (volume
fraction of droplets remains below 20%). In the three-phase system,
dilution lines were not found, indicating possible bicontinuous
structures[6].

LIGHT-SCATTERING EXPERIMENTS

   Static and dynamic light-scattering experiments have been performed[5,6]. The general theories lead to the following expressions
for the scattered intensity I and for the diffusion coefficient D
obtained from the correlation function of the scattered light :

   $$I(q) \propto P(q) \, S(q) \qquad\qquad\qquad Equation \ 1$$

   $$D(q) \propto \frac{H(q)}{S(q)} \qquad\qquad\qquad Equation \ 2$$

q being the scattering wave vector, P the form factor, S the structure factor and H the hydrodynamic interactions term. Here the
characteristic structure size is small compared with the light
wavelength : $P(q) \sim 1$.

   In dilute systems: $0.01 < \phi < 0.1$, I and D are independent of
q and the above expressions become very simple :

$$I = \frac{\phi}{\frac{\partial \Pi}{\partial \phi}} \; ; \quad \frac{\partial \Pi}{\partial \phi} = \frac{kT}{\nu}(1 + B\phi); \quad \nu = \frac{4}{3}\pi R^3 \qquad \text{Equation} \quad 3$$

$$D = \frac{kT}{6\pi\eta_c R_H}(1 + \alpha\phi) \qquad\qquad\qquad \text{Equation} \quad 4$$

where Π is the osmotic pressure of the droplets, R their radius, R_H their hydrodynamic radius and η_c the continuous phase viscosity; B and α are virial coefficients: for hard spheres, B = 8 and α = 1.5.

Results of the measurements of droplet size and virial coefficients can be found in references[5] and[6]. The general conclusions are as follows :

1) Sizes: these mainly depend on the water/SDS ratio in the w/o domain, the area per polar head is then roughly constant: $A \sim 50$ Å^2/mol; similarly, the sizes in the o/w domain depend on the oil/SDS ratio : $A \sim 65$ Å^2/mol.

2) Interactions: in all these systems, interactions are attractive. The attraction strength increases with oil penetration into the droplet interfaces for w/o systems ($R_H > R$), and with droplet elongation in the o/w systems. This happens when salt is added to the o/w systems, and when salt is removed from the w/o systems. Attractive forces are also stronger in w/o systems when short-chain alcohols are used, when replacing toluene by cyclohexane, or when increasing the water/SDS ratio (i.e. increasing the size).

Theory[5] allows us to relate B and α through $\alpha \sim 1.5 + (B-8)/2$. Small systematic discrepancies are observed in all cases, indicating the possible influence of the transient character of the droplets[7].

Examples of the variation of osmotic compressibility with concentration are shown in Figure 2. Hard sphere-like behavior is observed for ATP microemulsions, whereas in the series ATB, where pentanol has been replaced by butanol, strong attraction is observed; its strength decreases when salt is added (ATB1M) and with smaller particle size (αTB). Attraction is still greater when toluene is replaced by cyclohexane (ACB), in which case phase separation occurs around $\phi \sim 0.05$, above which two microemulsions are in equilibrium. Figure 2 clearly shows that ATB microemulsions approach a critical point around $\phi \sim 8\%$. This fact is confirmed by the appearance of angular variations of I and D which, according to theory, should follow :

$$I(q) \propto \frac{1}{1 + q^2 \xi^2} \qquad\qquad\qquad \text{Equation} \quad 5$$

$$D(q) = \frac{kT}{6\pi\eta\xi} K(q\xi) \qquad\qquad\qquad \text{Equation} \quad 6$$

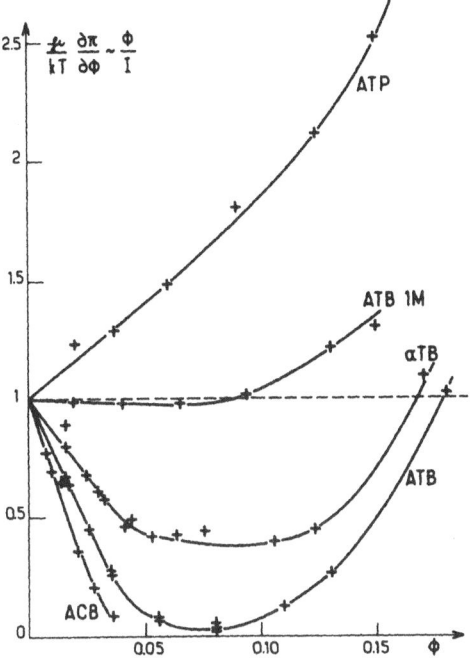

Figure 2. Reduced osmotic compressibility versus volume fraction of droplets for several microemulsion series.

where ξ is the correlation length, η the microemulsion viscosity and K the Kawasaki function. The three determinations of ξ are all in good agreement.

Light-scattering data for Type 2 systems are given in Figure 3. The intensity of light scattering increases when the phase separation boundaries are approached. Particle size also increases (from 100 to 200 Å) and there are more attractive interactions.

Angular asymmetry also appears close to the phase boundaries. This can be explained by the distortion of the oil droplets into ellipsoidal shapes in the o/w microemulsion side (S < S_1), but in the w/o side (S > S_2), the only origin of the asymmetry is the vicinity of a critical point in the phase diagram (critical end point)[6]. Again, the three determinations of ξ are in good agreement.

For the more concentrated systems, angular asymmetries are no longer observed, but theoretical predictions for the variation with concentration of structure factors and hydrodynamic interactions are not yet available.

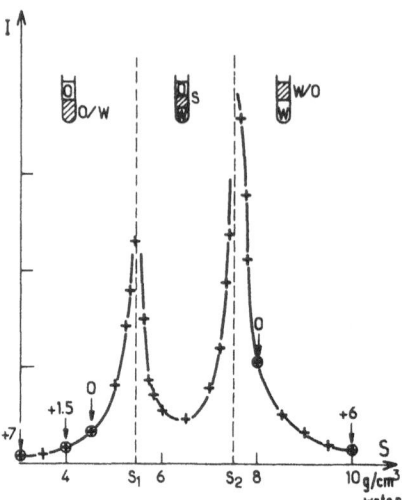

Figure 3. Backward scattered light intensity versus salinity for Type 2 systems. The numbers above the arrows are the virial coefficients B.

ELECTRICAL CONDUCTIVITY

Examples of electrical conductivity variations versus the volume fraction of droplets are shown in Figure 4a. For ATB systems a steep increase is observed around $\phi \sim 0.1$, characteristic of percolation phenomena, i.e. transient interconnection of droplets over macroscopic distances[6]. The "geometrical" percolation threshold for hard spheres is $\phi \sim 0.14$, which is the volume fraction above which an infinite number of spheres can be found to be in contact with each other. The geometrical threshold for attracting spheres has never been calculated, but it might be expected to be smaller than 0.14. Although geometrical percolation exists also for systems ATB1M and ATP around, let's say $0.1 \lesssim \phi \lesssim 0.14$, "electrical" percolation phenomena are not observed, probably because, although droplets are in contact, they do not exchange their water content, thereby forming transient opened structures. This is confirmed by the fact that the interface is more rigid than for ATB systems. There is no oil penetration and they are "hard-sphere-like" systems. The formation of opened structures at the percolation threshold can be viewed as the onset of a bicontinuous structure.

Similar observations can be made on systems where electrical percolation is observed in the w/o microemulsions close to S_2, where the attractive potential becomes large enough. Figure 4b represents the ratio of the microemulsion and brine conductivities versus the volume fraction of <u>brine</u>. Theory[9] predicts that they should be proportional by analogy with porous media. This is indeed the case in

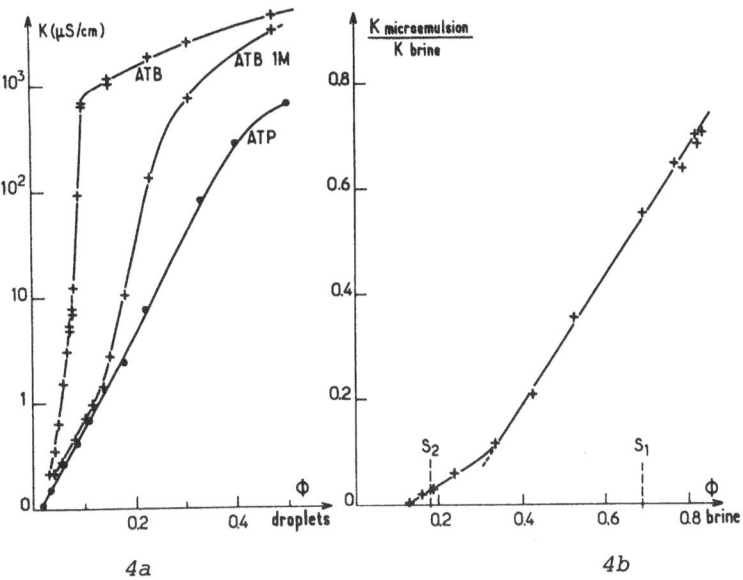

4a   4b

*Figure 4.   (a) Electrical conductivity variation with volume frac-
tion of droplets for several microemulsion series; (b) Ratio of
microemulsion and brine conductivities versus the microemulsion's
volume fraction of brine.*

Figure 4b, except at low brine fractions, where the amount of
"closed" structures might not yet be negligible.

DISCUSSION

     A first observation is that electrical percolation threshold
and critical points are systematically close in w/o microemulsions.
The preceding results allow us to understand why electrical percolation is found in "attractive" systems; opened systems have to
exist. On the other hand, critical concentrations for solutions
of spheres are around $\phi \sim 0.15^{10}$, i.e. very close to the percolation threshold $\phi \sim 0.1$. This coincidence explains why these two
phenomena are observed in the same volume fraction range, whereas
they can be very far apart in other systems (phase separation in
gels)[11].

     A correct theoretical treatment of the microemulsions would
then involve the elaboration of a site-bond percolation theory[11],
taking into account the transient character of the bonds (stirred
percolation). This theory is not yet available. Let us note however
that light scattering is a probe of the concentration fluctuations
that diverge at the critical point, whereas electrical conductivity
probes the connection fluctuations that diverge at the percolation
threshold. The interpretation of other types of experiments can be

much more difficult, since either kind of fluctuation can give a contribution.

Let us take for instance the case of shear viscosity[12]. ATB systems show an anomaly around $\phi \sim 0.1$ that can be due to critical effects or to hydrodynamic contributions to the flow of large transient aggregates. Similarly, ultrasonic absorption is anomalous for ATB systems in the same range. Critical contribution can reasonably be excluded and the anomaly explained by the modification of exchange processes due to the appearance of opened structures[13]. Finally, let us mention briefly the preliminary transient Kerr effect measurements made on ATB and ATP systems. The ATP "hard-sphere-like"system shows a small electric birefringence and short relaxation time ($\sim$1 µs), whereas the "attractive" ATB system shows a giant birefringence and a combination of short ($\sim$ 1 µs) and long ($\sim$ 1 ms) relaxation times. We are presently trying to interpret these results within the framework of the above ideas.

CONCLUSIONS

A general picture of the inversion process in w/o microemulsions emerges from these results :

1) When the interfacial layer is rigid, droplets can be concentrated to large volume fractions (up to about 0.3) and, as evidenced from interfacial compositions (Figure 1) and electrical conductivity data (Figure 4), inversion is relatively sharp, around $\phi \sim$ 0.5.

2) When the interfacial layer is fluid the structure of the droplets disappears progressively above the percolation threshold $\phi \sim 0.1$ and in the very large inversion region, the structure is probably bicontinuous. It seems possible that the transition to o/w structure implies the existence of ellipsoidal microemulsion droplets, as observed in the single particular case investigated here.

ACKNOWLEDGEMENTS

We acknowledge numerous fruitful discussions with R.Kjellander and R. Zana.

REFERENCES

1. "Microemulsions" ed.L.M.Prince, Acad.Press, (1977)
2. L.E.Scriven, Nature 263: 123, (1976)
3. Y.Talmon, S. Prager, J.Chem.Phys.,69: 2984 (1978)
4. M.Dvolaitzky, M.Guyot, M.Lagües, J.P.Lepesant, R.Ober, C.Sauterey, C. Taupin, J.Chem.Phys. 69: 3279 (1978)

5. A.M.Cazabat, D.Langevin, J.Chem.Phys.,74: 3148 (1981)
6. A.M.Cazabat, D.Langevin, J. Meunier, A.Pouchelon, Adv.Coll.Int.
   Sci., 16: 175, (1982) and references therein.
7. G.D.Phillies, J.Coll.Int.Sci., 86: 226 (1982)
8. M. Lagües, J.Phys.Lett. 40: L-331, (1979)
9. K.E.Bennett, J.C.Hatfield, H.T. Davies, C.W.Macosko, I.E.Scriven,
   in "Microemulsions", ed. I.D.Robb, Plenum Press (1982)
10. R. Kjellander, Far.Trans. (in press)
11. T.Tanaka, G. Swislow, I. Ohmine, Phys.Rev.Lett. 42: 1556 (1979)
12. A.M. Cazabat, D. Langevin, O. Sorba, J.Phys.Lett., 43: L-505
   (1979)
13. R.Zana, J.Lang, O. Sorba, A.M. Cazabat, D. Langevin, J.Phys.Lett.,
   (in press).

MICROEMULSIONS STABILIZED BY OLEATE/PENTANOL

Eva Sjöblom[1], Ulf Henriksson[2], and Peter Stilbs[3]

[1]The Institute for Surface Chemistry
Box 5607,   11486 Stockholm, Sweden

[2]Department of Physical Chemistry
The Royal Institute of Technology
10044   Stockholm, Sweden

[3]Institute of Physical Chemistry
Uppsala University
75121   Uppsala, Sweden

INTRODUCTION

In various applications of microemulsions it is of interest to control the distribution and transport of different species. For example, in cleaning processes it is important that the uptake and transport to the appropriate solubilization site is rapid, both for water-soluble and oil-soluble material, while in other applications the opposite is desired. In order to better understand the transport mechanism in microemulsion systems we have studied the self-diffusion of the different components by means of pulsed gradient spin echo FT-NMR. In addition, ^{23}Na NMR has been used to monitor changes at the molecular level in the hydrophobic/hydrophilic interfacial region. The investigated systems contain water, sodium or potassium oleate, pentanol and benzene or decane (Figure 1).

EXPERIMENTAL

Chemicals: The following chemicals were used without further purification: cis-9, -10-octadecenoic acid (Fluka A.G. 99.5%), pentanol (Fluka A.G. 99%), benzene (Merck 99.5%), decane (Merck 95%) and twice distilled water. The sodium oleate and potassium oleate were prepared by titration of an ethanolic ethoxide solution with octadecenoic acid.

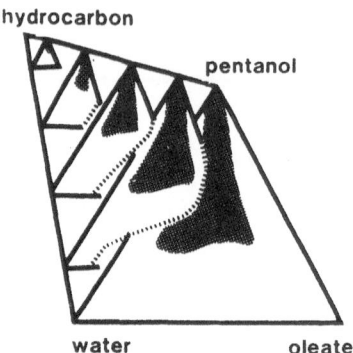

Figure 1. Schematic phase diagram showing the W/O microemulsion region in the quaternary system water-sodium or potassium oleate-pentanol-hydrocarbon.

Vapor Pressure Measurements : The vapor pressures above equilibrated binary solutions of pentanol/hydrocarbon were deter-mined at 20ºC on a Carlo Erba 4200 gas chromatograph equipped with a head space sampler, HS 250.

NMR Measurements:The ^{23}Na NMR spectra were recorded at 27ºC on a Bruker CXP-100 spectrometer. The field was locked using the ^{2}H signal from external CD$_3$CN. The ^{23}Na line shapes were found to be Lorentzian and the spin-spin relaxation times (T$_2$) were determined from the experimental line widths corrected for contributions from magnetic field inhomogenities. The self-diffusion measurements were performed on protons using an improved Fourier transform pulsed-gradient spin-echo technique[1]. A Jeol FX-100 spectrometer operating at 99.6 MHz and 23ºC was used.

RESULTS AND DISCUSSION

The self-diffusion is measured over macroscopic distances and gives information on whether a component is confined inside a closed aggregate or not. For a reversed micellar system with a well-defined interface, the following sequence for the self-diffusion coefficients is expected[2]:

$$D_{HC} > D_{ROH} > D_{H_2O}$$

The diffusion of pentanol in the ternary microemulsions is always rapid (Figure 2), showing that pentanol is located mainly in the continuous phase. As the water content is increased the water diffusion increases and exceeds the value for pentanol. Although

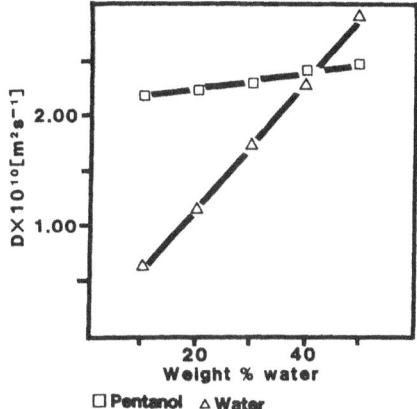

Figure 2. Self-diffusion coefficients for water (△) and pentanol (□) in ternary microemulsions with a constant weight ratio pentanol: oleate = 68:32.

the fairly high solubilility of water in pentanol is taken into account the results indicate that no typical reversed micelles with a well-defined interface are present in the water-rich ternary micro-emulsions[3]. The self-diffusion coefficients for pentanol, benzene and decane, shown in Figure 3, clearly show that hydrocarbon/pentanol con-stitutes the continuous phase in the quaternary microemulsions. As the concentration of hydrocarbon is increased the water diffusion is reduced, and in excess of ~25% hydrocarbon the values approach those obtained for typical reversed micellar systems[2]. This is in-terpreted in terms of a more organized hydrophobic/hydrophilic inter-face which offers an effective barrier to the translational motion of the water molecules.

Microemulsions are thermodynamically stable single-phase systems[4]. However, in many cases a microemulsion can be considered to consist of three pseudo-phases in equilibrium: an aqueous, a non-aqueous and an interfacial region containing the surfactant and variable amounts of the alcohol[5]. Since the system is in internal equilibrium the chemical potential of a given component is the same in the different pseudo-phases. This means that added hydrocarbon which is dissolved in the non-aqueous region lowers the alcohol chemical potential. As a result, alcohol is transferred from the aqueous and interfacial regions to the non-aqueous region. As demonstrated in Figure 4, this results in a closer packing of the oleate ions. This is accompanied by an increased [23]Na relaxation rate, indicating an increased counter-ion binding (Figure 5)[6].

The effect of added hydrocarbon on the self-diffusion and the [23]Na relaxation rate is most pronounced for microemulsions containing benzene. This can be understood by studying the alcohol chemical potential in alcohol/hydrocarbon mixtures which, at a given mole fraction, is considerably lower in solutions of benzene (Figure 6).

This reduction in the chemical potential is a consequence of the interaction between the alcohol hydroxyl group and the π-electrons of the benzene molecules, having an association enthalpy of ~ 5 kJ/mol[7].

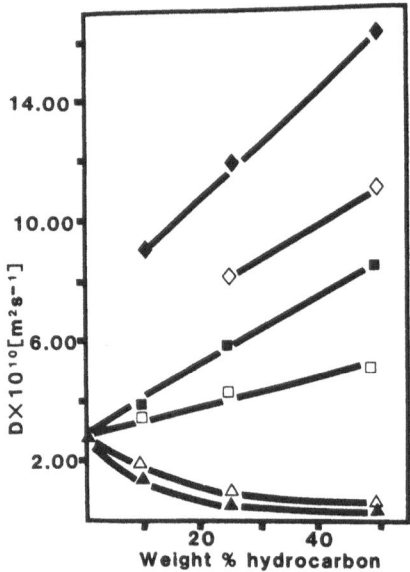

Figure 3. Self-diffusion coefficients for water (Δ), pentanol (□) and hydrocarbon (◊) in quaternary microemulsions with constant weight ratio pentanol:water:oleate = 35:49:16. Filled symbols represent microemulsions containing benzene and open symbols microemulsions containing decane.

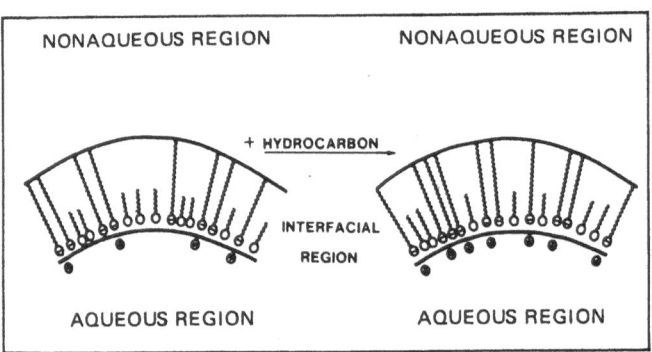

Figure 4. Schematic illustration showing the division of a microemulsion into an aqueous, a non-aqueous and an interfacial region

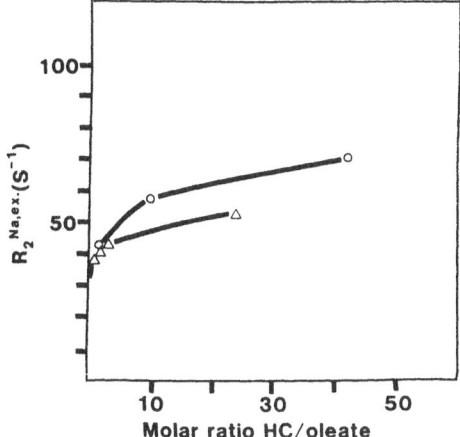

Figure 5. ^{23}Na excess relaxation rates $(R_2{}^{Na}, ex)$ for quaternary microemulsions. Molar ratio water:pentanol:sodium oleate = 44.5: 10.4:1. The hydrocarbon is benzene (o) and decane (Δ).

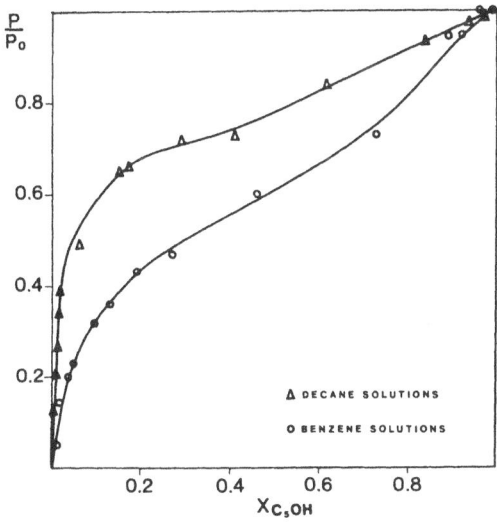

Figure 6. Relative vapor pressures for pentanol in binary solutions of pentanol/hydrocarbon as function of mole fraction pentanol (X_{C_5OH}). Solutions containing benzene are marked (o) and those containing decane (Δ).

SUMMARY

     At high water concentrations and in the absence of hydrocarbon,
the aqueous and non-aqueous regions in the studied microemulsions
are separated by a fairly flexible interfacial region containing
the oleate and a large fraction of the pentanol. The addition of
hydrocarbon redistributes alcohol from the interfacial to the non-
aqueous region, resulting in a closer packing of the oleate ions.
Thus, the barrier to translational diffusion of the water molecules,
as well as the counter ion binding, is increased.

REFERENCES

1. P. Stilbs and M.E. Mosley, Chem. Scripta, 15 : 175 (1980).
2. B. Lindman, N. Kamenka, T.M. Kathopoulis, B. Brun and
   P.-G. Nilsson, J. Phys. Chem, 84 : 2485 (1980).
3. E. Sjöblom, U. Henriksson and P. Stilbs, Proc. VII Scand. Symp.
   Surface Chem., Lyngby, Denmark 1982 : p. 233, K.S. Birdi, Ed.
4. I. Danielsson and B. Lindman, Colloids and Surfaces, 3 : 391
   (1981).
5. J. Biais, P. Bothorel, B. Clin and P. Lalanne, J. Disp. Sci.
   Tech. 4, 2 : 67 (1981).
6. E. Sjöblom and U. Henriksson, J. Phys. Chem., 86 : 4451)1982).
7. E. Sjöblom and U. Henriksson, "Surfactants in Solution",
   K.L. Mittal Ed., Plenum Press, New York (submitted).

EXISTENCE OF UNSTABLE TRANSPARENT "SOLUTIONS" IN THE AOT-DECANE-WATER SYSTEM

P. Delord and F.C. Larché

LA 233   Université des Sciences et Techniques du Languedoc

34060 Montpellier Cedex, France

INTRODUCTION

The sodium bis-(2 ethylhexyl)sulfosuccinate (AOT) is a widely used product and its behavior in mixtures with oil and water has been extensively studied. The domain of existence of isotropic transparent "solutions" often covers a large region on the oil-rich side of these systems. It is believed that inverse micelles are formed up to unusually large w/o ratios[1,3]. Erratic results with repeated measurements of self-diffusion on some mixtures of the AOT-decane-water system[4] and the results of Rosano et al.[5] prompted us to investigate the possible existence of transparent, unstable, but slowly separating emulsions in the region hitherto described as homogeneous. Preliminary observations have indeed indicated that some of the mixtures slowly separate into two phases upon standing[6]. The latest results of this study are presented here.

EXPERIMENTAL

The AOT used is a Fluka product of "purum" quality. It was further purified following the method of Rogers and Windsor[7]. The components were weighed and mixed in either sealed or screw-cap glass tubes in the order AOT, decane, water. The resulting "solution" was kept at 25 ±0.1°C for three to six months. Visual observations were made at approximately two-week intervals. The light microscopy observations were made with dark field and Zernike phase contrast equipment (Leitz).

RESULTS

After a period of one to three months some solutions that had
appeared upon preparation to be monophasic were distinctly diphasic.
This phenomenon occurs in the zone denoted $L_2 + L_2'$ in Figure 1. The
process that takes place is somewhat similar to the separation of an
ordinary emulsion. After 24 to 48 hours at 25°C most of the tubes
of this zone presented a gradient in refractive index in a region
near the air/liquid interface. A slight mechanical perturbation
makes this change of optical properties readily visible to the naked
eye. As time evolves the thickness of the solution that is affected
by this phenomenon increases until an interface becomes visible.
At this stage it is impossible to reverse this situation to a transparent solution by mechanical agitation. The emulsion obtained
separates in a matter of minutes. In the early stages however it is
sufficient to shake the tube to reconstitute what appears to be a
transparent, uniform solution. Assuming that the observed phenomena correspond to a slow phase separation, one would expect a
three-phase zone to the left of $L_2 + L_2'$. This has been found after
storage of what appeared to be mixtures of a homogeneous liquid and
a liquid crystal. The upper phase is almost pure decane, the middle
phase contains 20.1% water and the liquid crystal 50.8%. Under a
polarizing microscope the texture of this last phase is that of a
lamellar phase, a not unexpected result if one considers the published diagram of the AOT-water system[8]. The tie lines between L_2
and L_2' rotate, following almost a line at constant H_2O/AOT ratio,
the water and the AOT content of the upper phase remaining less than
0.5%, varying from (H_2O)/(AOT) = 30 to (H_2O)/(AOT) = 15, these ratios
being 1.21 and 0.61 by weight respectively. Finally, the solution
containing 5.0% water and 12.3% AOT has a blueish appearance of a
near critical phase and is very sensitive to temperature. Although
this has not been confirmed, a critical point around that composition
would coincide with the maximum in viscosity found by Rouvière et al[9].

It is interesting to note that the limit between $L_2 + L_2'$ and the
three-phase zone corresponds well to the limit obtained by Huang and
Kim at 23°C[10] as the line separating a one-phase from a two-phase
region.

The various observations made on solutions of the $(L_2 + L_2')$ zone
seem to indicate that the phase separation starts relatively early,
but is not easily detected by visual inspection. However, the observation of the inhomogeneous zone on top of the solution with light
microscopy and phase contrast indicates the present of droplets of
an isotropic phase floating in another isotropic phase. Because this
sometimes appears less than 24 hours after the preparation of the
solution, it seems to rule out the possibility of a bacterial degradation as the cause of the observed phenomena. The reproducibility
of the results, and the agreement with the thermodynamic rules in
phase diagrams are further indications that this is not the case.

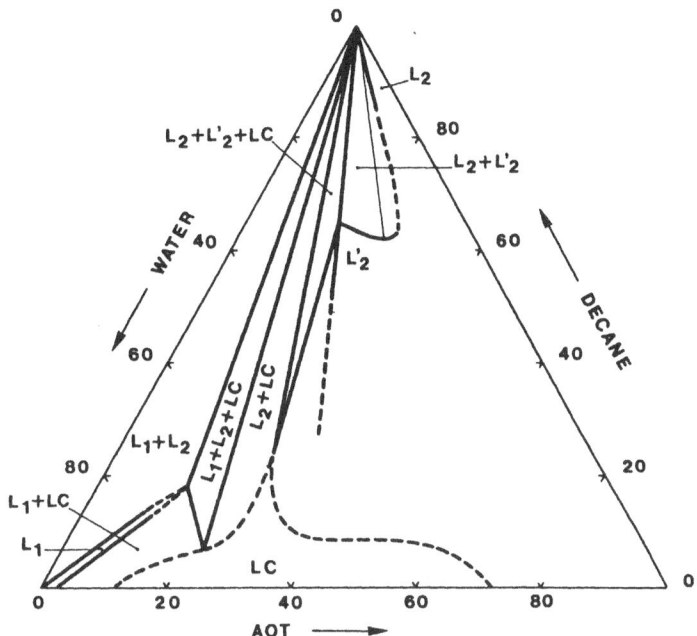

Figure 1. Tentative phase diagram of the AOT-decane-water system at 25°C.

There are three optically asymmetric carbons in the AOT molecule. This implies the existence of 8 optical isomers. It is known that the phase rule is different for the different diasterometric systems[11], and one would expect some effects due to the separation of these isomers into non-racemic proportions among the phases. We have measured the rotation of the polarization on fresh and separated solutions. None of the tested solutions is optically active and we concluded that the racemic mixture can be considered in this particular system as a unique component. With other oils, this effect may not be negligible[12].

DISCUSSION AND CONCLUSIONS

The presence of a slowly separating two-phase zone in a region of the AOT-decane-water system, previously assumed to be homogeneous, now seems to be well established. If the presence of a critical point in the $(L_2 + L_2')$ zone is confirmed it would have some interesting consequences. In particular a zone will exist where the solution is thermodynamically unstable(rather than metastable) and is expected to separate by spinodal decomposition[13].This mechanism is usually very fast, but the scale of the wave compositions, or particles produced, can be extremely small. Outside this zone the decomposition mechanism is different, and therefore differences might be expected in the kinetics of phase separation between these two regions.

Previous low-angle X-ray results[14] are in agreement with a
micelle structure of the low water concentration phase (L_2).
Earlier results have also indicated that micelles exist at high concentration[1]. This would imply that we have a system where a gas-
like and a liquid-like micellar solution exists, a result that would
be quite compatible with a critical point. This has already been
proposed and discussed as a possibility in microemulsions[15,16].

The existence of a zone where two isotropic liquids are in
equilibrium could be expected in ternary AOT-water-oil systems
with an oil other than decane. With normal alkanes we would even
expect similar kinetics. This could be the reason for the presence
of large particles together with the more standard micelles seen by
quasi-elastic light scattering in AOT-water-heptane[17]. It is also
interesting to point out that the measurement of self-diffusion seems
to be one of the most sensitive techniques for the early detection
of such a phenomenon. The usually good reproducibility of such experiments is however drastically reduced, sometimes up to the point
of giving quite erratic results.

AOT micelles are used as confining media for several chemical
processes; if the micellar solution undergoes a phase separation
there should be some deviations from the results that are expected
for a simple micellar solution. Some of these deviations in
chemical reactions may be of interest for a further investigation
of the phenomena that have been described here.

ACKNOWLEDGEMENTS

The authors would like to thank J.Rouvière for the Karl-Fischer
determinations of water in some phase-separated preparations.
Fruitful discussions with J.W.Cahn are gratefully acknowledged.

REFERENCES

1. P.Ekwall, L.Mandell and K. Fontell, J.Coll.Int.Sci.,33: 215 (1970)
2. H.Kunieda and K. Shinoda, J.Coll.Int.Sci.,70: 577 (1979)
3. B.Tamamushi and N. Watanabe, Coll.Polymer Sci., 258: 174 (1980)
4. N. Kamenka, private communication
5. H.L. Rosano, T. Lan, A. Weiss, J.H.Whittam and W.E.F.Gerbacia,
   J.Phys.Chem., 85: 468 (1981)
6. T.Assih, P.Delord and F.C. Larché, Symp.on Surfactants in
   Solution, Lund, (1982) to be published.
7. J. Rogers and P.A.Windsor, J.Coll.Int.Sci., 30: 247 (1980).
8. K. Fontell, J.Coll.Int.Sci., 44: 318 (1973)
9. J. Rouvière, J.M.Couret, M. Lindheimer, J.L. Dejardin and R.
   Marrony, J.Chim.Phys. 76: 289 (1979)

10. J.S.Huang and M.W.Kim, NATO Advanced Study Institutes, B73: 809 (1981)
11. R.L. Scott, J.Chem.Soc., Faraday Trans. II 75: 356 (1977)
12. A.N.Maitra and M.F.Eicke, J.Phys.Chem., 85: 2687 (1981)
13. J.W.Cahn, Trans.Met.Soc. AIME 242: 166 (1968)
14. T. Assih, F.C.Larché and P. Delord, J.Coll.Int.Sci.,89: 35(1982)
15. C.Miller, R.Hwan, W. Benton, T. Furt,Jr., J.Coll.Int.Sci., 61: 554 (1977)
16. P.G. de Gennes and C. Taupin, J.Phys.Chem., 86: 2294 (1982)
17. E. Gulari, B. Bedwell and S.Alkhafaji, J.Coll.Int.Sci., 77: 202 (1980).

FLUORESCENCE: A METHOD TO OBTAIN INFORMATION ABOUT REVERSE MICELLAR SYSTEMS

E. Geladé and F. C. De Schryver

Department of Chemistry
Katholieke Universiteit Leuven
3030 Leuven (Heverlee) Belgium

Using the fluorescence properties of several naphthalene derivatives in the presence and absence of quenchers, it was possible to characterize DAP- and AOT-micelles in cyclohexane. Based on simple fluorescence lifetime measurements, it is proposed that the aggregation mechanisms for both systems are identical. The meaning of the CMC is also discussed. The use of fluorescence quenching to obtain information on e.g. the mean aggregation number and on the solubilizate exchange between colliding micelles is discussed.

In all cases, the total quencher concentration, and not the quencher concentration in the waterpool, must be used. Exciplex formation with triethylamine and energy transfer to Tb^{3+}, proved to be very good methods to localize the solubilized probes in the reverse micellar system.

INTRODUCTION

Although reverse micellar systems (RMs) have been under investigation for some time now[1], the use of luminescence methods to obtain information on these aggregates has only more recently been introduced[2]. The most common procedure is to solubilize a fluorescence probe in RMs. From measurements of the luminescence intensities, lifetimes (both in the absence and the presence of quenchers)[2,3], fluorescence depolarization[2a,2e], and the energy transfer[2a,2b] of this probe, information about the RMs is obtained.

In this paper, several aspects of the fluorescence emission
are treated. Depending on the method used, information can be
obtained about the localization of the probe, the polarity of the
interface, the aggregation behavior of the RMs and on the
processes where solubilized molecules are exchanged between
micelles by intermicellar collisions.

EXPERIMENTAL SECTION

Dodecylammonium propionate (DAP) was prepared by the method
of Kitahara[4], followed by several recrystallizations from n-hexane
and vacuum drying over P_2O_5. Aerosol OT (AOT) was purified by the
method of Magid et al.[5]. Both surfactants were kept in a desiccator
over P_2O_5.

The synthesis and/or purification of the probes 1-methyl-
naphthalene (1-MeN), 2-(1-naphthyl)acetic acid (NAA), 4-(1-naphthyl)
butyric acid (NBA), 6-(1-naphthyl)hexanoic acid (NHA) and
12-(1-naphthyl)dodecanoic acid (NDA), of the quenchers triethylamine (TEA), potassium iodide (KI) and sodium nitrate ($NaNO_3$), and
of the solvent cyclohexane ($c-C_6H_{12}$) has previously been published[3d,6]. Terbium chloride ($TbCl_3 \cdot 6H_2O$, Alfa Ventron) was used
as received.

Absorption spectra were taken on a Perkin-Elmer 124 spectrophotometer. Corrected fluorescence spectra were measured on a
"FICA spectrofluorimètre absolu et différentiel" or on a SPEX
fluorolog apparatus.

Fluorescence lifetimes were measured on a single photon
counting apparatus and the decay curves were analyzed with decon-
volution, using a PDP-11/computer.

RESULTS AND DISCUSSION

In this work, the main system investigated is DAP solubilized
in cyclohexane, unless otherwise stated. Using the fluorescence
methods, information on these aggregates and the influence of
added water, is obtained.

A. Exciplex Formation: Information on Probe Localization and on Polarity Changes in the Environment of the Waterpool

Before the fluorescence property of molecules can be used to probe a certain system, it is necessary to know the mean localization of this probe. Since exciplex formation is dependent on the polarity of the environment, it is a good method to obtain this information[7].

First, the properties of the probes were measured in pure cyclohexane. As fluorescence probes, 1-MeN and several naphthaléne derivatives with a detergent-like apolar chain (NAA, NBA, NHA and NDA) were used. In the excited state they form exciplexes with the donor triethylamine. The emission of this complex is at lower energy than that of the naphthalene fluorescence[6]. From the emission maximum (λ_{max}) and the quantum yield (ϕ_E) of this exciplex

Table I. *Values of the maximum of the exciplex emission (λ_{max}), the ratio of the quantum yield of exciplex emission (ϕ_E) over the quantum yield of monomer emission (ϕ_M), and the Stern-Vomer constant (K_{SV}) for homogeneous cyclohexane (H) compared with those for the reverse micellar system (RMs), DAP (0.12M)/cyclo-hexane/R = 1.375, for several naphthalene derivatives, quenched by TEA. ϕ_E is determined at [TEA] = 2.155 x 10^{-2}M. NAA did not show any exciplex fluorescence in the RMs.*

Probe	λ_{max} (nm)		ϕ_E/ϕ_M (10^{-2})		K_{SV} (M^{-1})	
	H	RMs	H	RMs	H	RMs
1MeN	408	408	34	33	19	55
NDA	407	413	25	16	16.2	65
NHA	408	423	28	10	17.8	71.1
NBA	410	427	43	4	35	56.8
NAA	429	–	136.8	–	135.7	26.8

fluorescence (Table I) it can be concluded that the influence of the chain cannot be neglected. The probes 1-MeN, NDA and NHA behave almost identically, but further shortening the chain results in increased λ_{max} and ϕ_E. Thus, in homogeneous cyclohexane, NAA forms the most stabilized exciplex.

Taking these differences into account, the fluorescence properties of the probes were measured in the DAP-RMs. NAA, although quenched by TEA, did not show any exciplex emission at all. The results for the different probes are listed in Table I. Comparing the results for both media (Table I) it appears that the properties of the exciplex emission of 1-MeN in both media are almost identical. However, for the probes with a detergent-like chain, the values of λ_{max} and ϕ_E/ϕ_M (ϕ_M is the quantum yield of naphthalene fluorescence) for both media differ more as the chain length is shortened. These results can only be explained by a different localization of the probes in RMs. The apolar probe, 1-MeN, resides mainly in the bulk apolar phase, while the other probes are bound with their carboxylic acid group to the water-pool. The number of methylene units between this headgroup and the naphthalene chromophore therefore determines the average distance between the chromophore and the waterpool.

Besides the possibility of exciplex formation to localize probes in RMs, this fluorescence method also enables us to obtain information on the polarity in the environment of the probe. Equation (1) relates ν_{max} to the dielectric constant of the medium[8].

$$\nu_{max} = \nu_{vac} - \frac{\mu^2}{4\pi\varepsilon_o h c \rho^3} \left[2\left(\frac{\varepsilon - 1}{2\varepsilon_o + 1}\right) - \frac{n^2 - 1}{2n^2 + 1}\right] \quad (1)$$

which can be put in the form

$$\nu_{max} = \nu_{vac} - S \cdot f\ (\varepsilon, n) \quad\quad\quad\quad\quad (2)$$

For the case of naphthalene/TEA, Hammond et al.[8] were able to determine the slope (S) and the intercept (ν_{vac}).

Assuming that these values are the same for our naphthalene derivatives, it is possible to calculate ε (Table II).

The values of the refractive indices (n) in the table are the most extreme values of the solvents, in which Hammond et al. observed exciplex formation.

Table II. *Calculated ε-values [eq.(1)] on the basis of the*
emission maximum (λ_{max} or ν_{max}) of the exciplex band, using the
refractive indices of diethylether (n_1) and cyclohexane (n_2).

Probe	$\lambda_{max}(nm)$	$\nu_{max}(cm^{-1})$	ε	
			$n_1 = 1.35243$	$n_2 = 1.42623$
1MeN	408	24510	-	2.0058
NDA	413	24213	2.10	2.20
NHA	423	23640	2.55	2.65
NBA	427	23419	2.80	2.90
NAA	n.e.[a]	n.e.[a]	n.e.[a]	n.e.[a]

n.e.[a] : *no exciplex emission*

B. Energy Transfer: Information on Probe Localization and Mobility of Solubilized Ions

Micelles can organize donor and acceptor molecules leading to
more efficient energy transfer processes in these systems[2a,2b,9,10].

The energy transfer of the triplet of the several naphthalene
derivatives (NAA, NBA, NHA, NDA and 1-MeN) to Tb^{3+}, solubilized in
the waterpool, was investigated. Since this energy transfer occurs
at short distances[10a], the study of this process as a function of
the chain length must also give an indication of the probe localization. Indeed, the energy transfer is, under the experimental
conditions used, only efficient for the system NAA/Tb^{3+} (Figure 1).
No energy transfer from NDA or 1 MeN could be observed. The very
weak emission at 546 nm is only due to direct excitation of Tb^{3+}
at 283 nm, the wavelength used to excite the naphthalene chromophore.

For further investigation, the system NAA/Tb^{3+} was chosen.
The influence of increasing the water concentration on the sensitized Tb^{3+}-emission (Figure 2) indicates a changing environment of
Tb^{3+} up to R ∼ 1. Analogous results were obtained by Eicke et al.
with AOT micelles for the energy transfer of hydroxyphenyl acid
to Tb^{3+} [2b] (Figure 2).

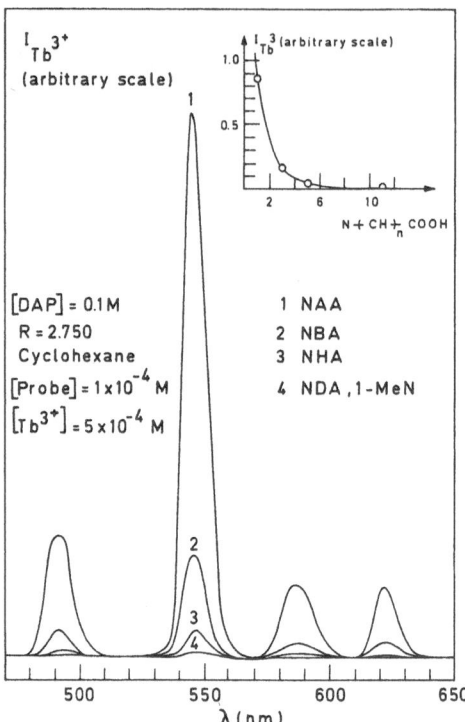

Figure 1. *Emission spectrum of Tb^{3+}, solubilized in the waterpool*
of DAP-micelles in cyclohexane. The emission is due to energy
transfer from the triplet state of several naphthalene derivatives,
which are excited at 283 nm. The insert represents the variation
of the emission intensity at λ_{max} (546 nm), as a function of the
number of methylene units between the naphthalene chromophore and
the carboxylic acid group.

However, instead of an increasing Tb^{3+}-luminescence with
increasing R, they found a decrease, which tends to reach a plateau
at a higher R-value. The variations of the intensity can in both
cases be explained by a ligand change around Tb^{3+}. At first, Tb^{3+}
is surrounded by detergent molecules. When water is added, the
polar headgroups are partially hydrated but also Tb^{3+}, since water
is a good ligand. Depending on the quenching capacity of the
ligands, the luminescence intensity of Tb^{3+} will increase or decrease with increasing R-value.

The fact that the Tb^{3+}-emission with AOT reaches a plateau
at a higher R-value than with DAP, finds its origin in the higher
interaction of water with the sulfonate groups in AOT, than with

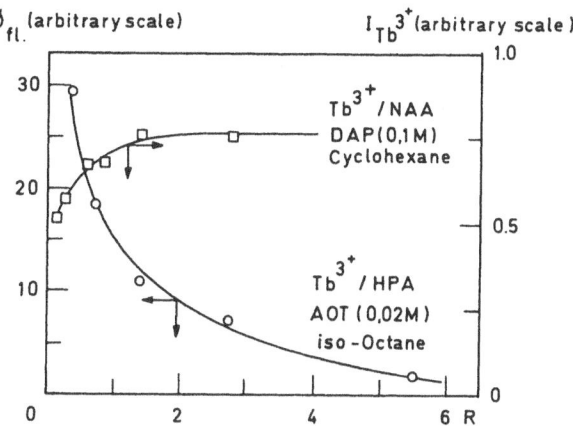

Figure 2.   *Influence of the amount of added water (R) on the energy
transfer between NAA and Tb^{3+}, solubilized in DAP-micelles in
cyclohexane, compared with the influence on the energy transfer
between hydroxyphenyl acetic acid (HPA) and Tb^{3+}, solubilized in
AOT-micelles in iso-octane. $I_{Tb}3+$ is the intensity of the Tb^{3+}-
emission at 546 nm and ϕ_{fl} is the quantum yield of Tb^{3+}-emission.*

the headgroups in DAP[11] and also in the higher maximal hydration
number of the AOT headgroups $(R \simeq 9)^{[12]}$ as compared with DAP $(R \simeq 4)^{[2a]}$

A mechanism describing energy transfer in DAP-reverse micelles
was proposed by Fendler et al. for the system pyrene butyric acid
(PBA)/Tb^{3+} [2a]. This mechanism explains why the emission intensity
of Tb^{3+} as a function of the Tb^{3+} concentration reaches almost a
plateau.

Our results indicate that energy transfer also can be used to
localize probes in RMs, and that the binding of Tb^{3+} to the polar
headgroups decreases with increasing water concentration.

C. Fluorescence Quenching: Information on the Size of Reverse
Micelles and on Solubilizate Exchange between Micelles

The results in the previous two parts are mainly based on
stationary measurements. In this part we focus our attention only
on non-stationary measurements, although stationary measurements
can also give information on RMs[3a],[3b]. The latter method is only
useful if certain conditions, such as very rapid (static) quenching

and the absence of redistribution of solubilizates within the time-scale of reaction, are fulfilled. This is usually not the case; therefore the non-stationary measurements are more generally applicable.

In the non-stationary conditions of light excitation, the time dependence of the fluorescence intensity (decay curve) of solubilized probes in the absence and presence of quenchers is measured. However, the decay curves are dependent on the localization (bound or not-bound) of both the probe and the quencher. Four different cases can be considered.

Case 1: Probe and quencher are not bound. As concluded in part A, the quenching of 1-MeN by TEA must be regarded as an example of this case. At room temperature, the decay curves of 1-MeN in homogeneous and heterogeneous are mono-exponential at all quencher concentrations. The quenching rate constant can be calculated from the well-known Stern-Volmer relation; since the probe and the quencher are both in the apolar phase the overall quencher concentrations must be used. However, the value of this rate constant in RMs is almost three times higher than in pure cyclohexane (Table I), indicating that the influence of the detergent is not negligible. This is probably due in part to a higher quencher concentration (smaller available volume). In this case, no information on the RM itself can be obtained.

Case 2: The probe is bound (i.e. spatially restricted within the micelle), the quencher is not. Also from the exciplex study, it followed that all the naphthalene derivatives with a chain (NAA, NBA, NHA and NDA) are bound probes, while the quencher TEA is not. The results are discussed under case 3.

Case 3: The quencher is bound (i.e. spatially restricted within the micelle), the probe is not. An example of case 3 is the quenching of 1-MeN, solubilized in the apolar phase by inorganic ionic quenchers (e.g. KI), which can only be solubilized in the waterpool of the micelles.

Hence, in case 2 or 3, either the probe or the quencher is bound to the micelle. In both cases, the decay curves of the probe are always mono-exponential, independent of the quencher concentration. The reason heretofore is that the probe "sees" a time-averaged quencher concentration. The quenching of the excited probe (solubilized in the micelle (case 2) or in the bulk cyclohexane phase (case 3)) is dependent on diffusion-controlled collisions of the micelles (with solubilized probe or quenchers) with the non-solubilized entities in the apolar bulk phase. In cases 2 and 3,

the use of the total quencher concentration (in the bulk phase) is
therefore a better estimate than the use of the local quencher concentration (in the waterpool). A plot of the inverse of the lifetime as a function of the total quencher concentration also obeys
the Stern-Volmer relation. The quenching rate constant (k_q) is
only dependent on the ability of the quencher to reach the micelle-
bound probe or vice versa, and is therefore dependent on the
density of the interface.

This explains the variation of k_q with the amount of added
water (R = [H_2O]/[DAP], Figure 3). Indeed, it is known (vide infra),
that the first amounts of added water hydrate the polar headgroups,
resulting in a decrease of the surface area per headgroup and in
an increase of the aggregation number (N_{agg}). At the moment of
maximal hydration (R = 4, in the case of DAP[2a]), the surface area
per headgroup becomes almost constant[13], although N_{agg} keeps increasing, since the "free" water now increases the volume of the
waterpool. Thus, the micelle density (distance between the apolar
chains) increases as long as the surface area per headgroup de-

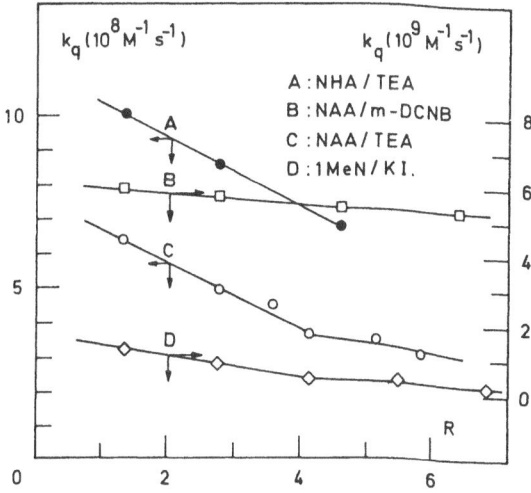

Figure 3.   *Influence of the water concentration (R = [H_2O]/[DAP])
on the quenching rate constant (k_q) if one solubilizate (the probe
or the quencher) is not bound to the micelle (cases 2 and 3). The
RM system is DAP (0.12 M)/cyclohexane/water. The quenchers are
triethylamine (TEA), m-dicyanobenzence (m-DCNB) and potassium
iodide (KI).*

creases (i.e. until R = 4). The ability of the quencher to reach
the micelle-bound probe (or vice versa) decreases up to this
R-value, and becomes almost constant at higher R-values. This is
very clearly reflected in the variation of k_q with R. The small
variations at higher R-values are probably due to the higher mobility of the quencher (see part B).

Case 4: Probe and quencher are both bound. If we first assume that
neither the probe nor the quencher can leave the micelle during the
lifetime of the excited probe, it is obvious that the decay curves
are then determined by the distribution of the quenchers and
probes. At probe or quencher concentrations that are low enough to
ensure that the probability of solubilization of a probe or
quencher in a certain micelle is not dependent on the presence of
(an)other probe(s) or quencher(s) in that micelle, the distribution of probes and quenchers is best described by a Poisson-dis-
tribution. We think that this condition is certainly fulfilled
experimentally for the probe distribution (at the concentrations
used there is a maximum of one probe per micelle), and also for
the quencher distribution if the mean number of quenchers per
micelle (A_3 =[Quencher]/[Micelle]) is not much greater than one.
The latter point is probably dependent on the size of the quencher
(e.g. N,N'-dimethylviologen compared with KI) and on the R-value
(at higher R-values, the solubilization of e.g. a second quencher
in a particular micelle will be more independent of the presence
of the first quencher than at low R-values). Under these circum-
stances the decay curves are multi-exponential and can be described
by:

$$I(t) = A_1 \, exp \, \left[\, -A_2 t - A_3 \, [1-exp(-A_4 t)] \, \right] \qquad (3)$$

where $A_1 = I_0$ = the fluorescence intensity at t = 0; $A_2 = k_0$ =
$1/\tau_0$; τ_0 is the lifetime of the probe in the absence of quenchers;
$A_3 = [Q_t]/[M]$ = the mean number of quenchers (Q) per micelle (M);
$A_4 = k_{qm}$.

From the kinetic scheme, which is analogous to that for aqueous micelles[14], it follows that it is the total quencher concentration that must be used. However, the assumption that neither
the probe not the quencher can leave the micelle during the lifetime of the excited probe is in practice not a common situation.
Indeed, it has been proved[2b,3a,3b] that solubilizate exchange between waterpools by intermicellar collisions is possible within the
time scale of quenching.

When we take into account only the exchange of the quencher
(vide infra), the decay curves can be described by the same overall
equation (Equation (3)). Parameter A_2, however, now becomes dependent on the total quencher concentration ($[Q_t]$):

$$A_2 = k_o + k_e[Q_t] \qquad\qquad (4)$$

where k_e is the rate constant of quencher exchange.

The quenching of bound NAA by the rather small ionic quenchers,
KI and $NaNO_3$ (solubilized in the waterpool), is an example of case
4. If the quencher concentrations are not too low[3d], the decay
curves of NAA are multi-exponential and can very well be fitted by
convoluting the observed lamp curve with Equation (3). Parameters
A_2 and A_3 both change with $[Q_t]$, indicating that ion exchange between waterpools by intermicellar collisions is possible during the
lifetime of NAA ($\sim$45 ns). One can calculate k_e from parameter A_2
and the micellar concentration [M] from parameter A_3 and therefore
also the mean aggregation number, since $N_{agg} \cdot [M] = [DAP] - "CMC"$.
The results for both quenchers at different water concentrations
are listed in Table III.

Table III.   The rate constants (k_e and k_{qm}) and N_{agg}, calculated
using parameters A_2, A_3 and A_4 [Equations (3) and (4)] for systems
based on NAA-cyclohexane where $[Q_t]$, [DAP] and $[H_2O]$ are variable.

R	Q	N_{agg}	$k_e (\times 10^8 M^{-1}s^{-1})$	$k_{qm} (\times 10^7 s^{-1})$
1.375	KI	28	8.0	5-7
	$NaNO_3$	30	11.1	3-5
2.750	KI	40	13	3-5
	$NaNO_3$	37	21.4	2-4
4.125	KI	67	-	2-4

From the values of N_{agg}, it is clear that small amounts of water increase the micellar size, since N_{agg} in the absence of water was measured to be about 5, using vapor pressure osmometry[2a]. It is also obvious that further increasing the water concentration increases N_{agg}, which is in agreement with results for other systems[13,15]. However, it must be pointed out that the values of N_{agg} are mean values, because the aggregation mechanism of DAP-micelles is described by a multiple self-association (part D[1c,16]) and because micellar collissions, with probably also exchange of detergent molecules, occur within the lifetime of the excited probe.

This increase of N_{agg} explains in part the decrease of k_{qm} with increasing R-value[17]. A second aspect to be considered is the variation of the localization of NAA within the interface[3d].

The main interest is in the value of k_e. In comparison with results for AOT[3a,3b], the observed k_e-value is rather high, indicating that about 10% of the intermicellar collisions are efficient ($k_{diff} \sim 10^{10}M^{-1}s^{-1}$). How can such an efficient exchange be explained? There are probably two main reasons. First, there is a structural effect. For a collision to be efficient, it is necessary that the waterpools of the colliding micelles come in contact. This process must be dependent on the rigidity or density of the interface, where the apolar tails are located. AOT has two branched chains per polar headgroup, while DAP has alternating one short and one long unbranched apolar tail per headgroup. From this, one can imagine that exchange processes must be more efficient between DAP-micelles then between AOT-micelles. That the structure of the interface is really important is also indicated by Atik and Thomas[3a]. They studied the effect of additives on k_e in AOT-micelles and found that e.g. 0.3 M benzyl alcohol, which is assumed to be at the interface, increases the rate of exchange by 30-fold, because the order of surfactant headgroups is disrupted (Table IV). Thus, we may conclude that the "open" structure of the interface of DAP is one reason that k_e is high.

The second reason must lie in the fact that in the derivation of the kinetic scheme, only the exchange of quenchers has been taken into account. This assumption is certainly the weakest point in the scheme, since there is no reason that exchange of probes should not occur. However, if the concentrations of the probes and quencher are so chosen that the mean number per micelle is always below one, possible exchange of the probe will have almost no influence on the kinetics, and the derived values of N_{agg}, k_e and k_{qm} are meaningful. k_e is then a mean value for the rate constants of the probe and quencher exchange. In our work,

Table IV. *Effect of the structure of the interface and the localization of the solubilizates on the exchange rate constant (k_e).*

Detergent	Solvent	R	Probe[a]	Quencher[b]	Additive	$k_e (10^7 M^{-1} s^{-1})$	Ref.
AOT	n-heptane	11	$Ru(bipy)_3^{2+}$	$K_3Fe(CN)_6$	none	1.3	3a
	n-heptane	11	$Ru(bipy)_3^{2+}$	$K_3Fe(CN)_6$	benzylalcohol (0.3M)	33.0	3a
	n-hexane	15	MgTPP	$C_8MV_2^{2+}$	none	19.0	18
	n-hexane	5	MgTPP	$C_8MV_2^{2+}$	none	12.0	18
DAP	cyclohexane	1.375	NAA	KI	none	80.0	3d

(a) $Ru(bipy)_3^{2+}$: ruthenium(II)tris(bipyridyl)

  MgTPP : magnesium tetraphenyl porphyrin

  NAA : naphthalene acetic acid

(b) C_8MV^{2+} : N,N'-dioctylviologen

the probe NAA is not in the waterpool but bound at the hydrocarbon-water interface (see part A). The exchange of such a probe is on one hand determined by the chance that the collisions occur in the environment of the probe and on the other hand by the strength of the binding of the probe to the interface.

If a solubilizate is localized at the interface, we think that the possibility of exchange by collision is always higher than if the solubilizate is localized in the waterpool. Even if at the moment of collision the waterpools do not come in contact, exchange is possible. This reasoning explains partly the high k_e-value in our case and probably also the higher k_e-values reported by Pileni et al.[18] (Table IV), who used a quencher with two apolar tails. In their case, however, at low water concentrations, the large size of the porphyrin probe would certainly also cause structural changes of the micelle.

Thus the high exchange efficiency in DAP-micelles can be explained on the one hand by the "open" structure of the micellar interface and on the other hand by the higher exchange possibility of the probe localized in the interface.

As we already mentioned, the exchange of such a probe is dependent on the strength of the binding to the interface. As discussed in part D, NAA becomes less bound if R increases. This partly explains why k_e increases with increasing R-value.

If R > 4.125, the kinetic scheme cannot be used for this system anymore[3d]. Indeed, as pointed out above, the kinetic scheme is only valid if the mean number of quenchers (probes) per micelle is below one. In that case, however, only minor deviations from mono-exponentiality are observed, which makes the uncertainty of the results too high.

As a last point, we will indicate here a possibility to obtain k_e-values in a rather easy way. When $k_e \gg k_{qm}$ or $[Q_t]$ is very low, Equation (3) reduces to

$$I(t) = A_1 \exp [-A_2 t] \qquad (5)$$

The decay curves become mono-exponential. In the case of NAA, quenched by KI, this occurs if $[KI_t] < 10^{-4}M$.

The plot of A_2 as a function of $[Q_t]$ is linear with a slope equal to k_e and an intercept equal to k_o. Thus, the quenching is determined by the transport of quenchers (and probes) between the micelles. The apparent quenching rate constant is in fact equal to the exchange rate constant. Working at these low quencher concentrations with the DAP-RMs, it appeared that only if R > 4, is k_e large enough (almost diffusion controlled) so that a quenching effect can be seen.

D. Fluorescence Lifetime Measurements in Absence of Quencher: Information about Aggregation Behavior

One of the unresolved problems of reverse micelles is their aggregation behavior. A few years ago several papers were published, where these micelles were divided in two classes showing strikingly different aggregation behaviors[16,19]. DAP is the prototype of class I, characterized by a stepwise sequential multiple-equilibrium association without any CMC. AOT is the prototype of class II, the aggregation of which can be described by a monomer $\rightleftharpoons$ n-mer model with a CMC. However, ^{1}H-NMR[20], positron annihilation[16c], and dielectric increment measurements[11,21], the results of which are not dependent on a solubilized probe, all indicate for both prototypes a well-defined discontinuity in the curve describing the followed property plotted against the detergent concentration. Also in the case of DAP, there must be a cooperative effect. Conductometric measurements on AOT in cyclohexane[22] and ultrasonic absorption measurements on DAP and AOT[19c] do not show any evidence of a CMC, which is in contradiction with the existence of a CMC for these systems. In the case of AOT, Eicke proposed a model to explain the results[21a]. He postulated that pre-micellar linear aggregates are formed by adding monomers to trimeric nuclei (no real CMC). At a certain concentration, the "operational CMC", these linear aggregates close to form reverse micelles. A conformational change would be responsible for the apparent CMC.

Another explanation is that the apparent CMC is due to association of several pre-micellar aggregates (PA) to form a reverse micelle (RM) (e.g. $PA_n + PA_m \rightleftharpoons RM_{n+m}$)[19c]. Concerning the aggregation of AOT, the latter mechanism has been proved to be less probable[11].

In this respect, we tried to obtain information on the aggregation behavior of both DAP and AOT, using the variation of the fluorescence lifetime (τ^o) of NAA as a function of the detergent concentration in the absence and the presence of additional water (Figures 3 and 4).

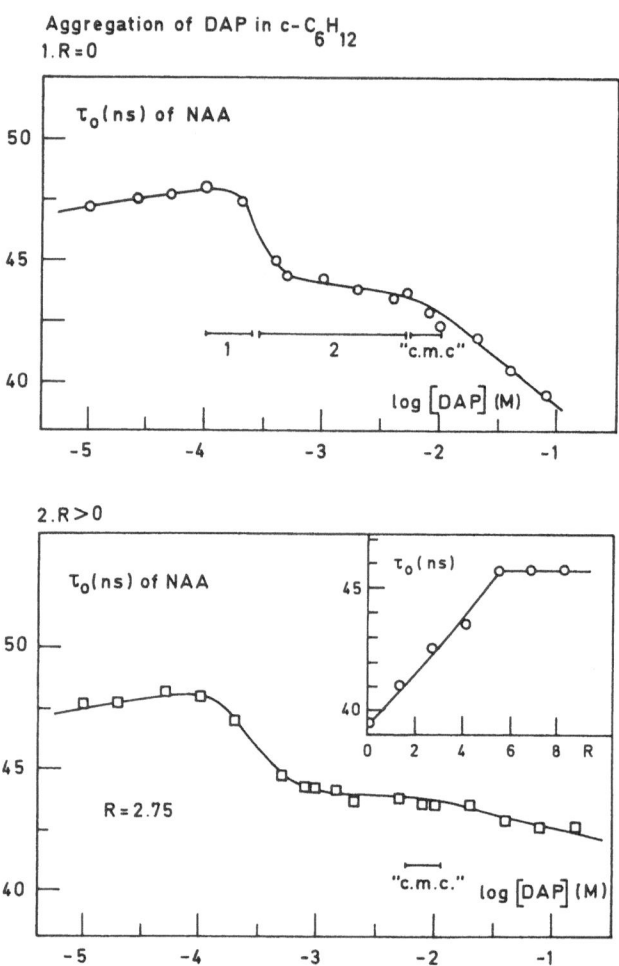

Figure 4. *The fluorescence lifetime (τ_o) of NAA in the absence*
of quenchers as a function of the logarithm of the DAP-concentra-
tion when R = 0 (upper curve) and when R = 2.75 (lower curve).
The insert shows the variation of τ_o of NAA with the amount of
water solubilized in the RMs DAP (0.08 M)/cyclohexane.

1. R = 0 : no added water. In the case of DAP (Figure 4), solubilized in cyclohexane, the curve shows two discontinuities,
region 1 and "CMC", and an intermediate, region 2. The region "CMC"
corresponds to the CMC-value obtained by the positron annihilation
method[16c]. The other values in the literature are in region 2[11,16e23].

The shape of the curve is explained by the model of Eicke for
AOT, although the alternative model, proposed by Zana[19c] cannot be
excluded for DAP by this method. Region 1 represents the association of NAA (<10^{-5}M) with one or already a few associated DAP-mole-
cules to form a linear pre-micellar aggregate. Increasing the detergent concentration (region 2) only increases the length of this
pre-micellar aggregate; the position of NAA almost remains constant
(constant τ_0).

The next discontinuity, region "CMC", reflects the structural
change from a linear to a spherical aggregate or reverse micelle.
Above this concentration range, the change in τ_0 of NAA is probably
due to the increase of the size of the aggregates, which is in
agreement with a stepwise sequential multiple equilibrium association model, as evidenced by e.g. vapor pressure osmometric measurements[16a].

Since we assume reverse micelles to be spherical aggregates,
region "CMC" ((6-10)x10^{-3}M), must be regarded as the operational
CMC. Thus, although the first discontinuity is most conspicuous,
this has nothing to do with a CMC. The CMC-values for DAP in the
literature, determined by using probes, probably all reflect this
association of the probe with the pre-micellar aggregate[16e,23].
The concentration at which this happens is of course dependent on
the nature of the probe. The CMC-value can in those cases also be
influenced by a probe-induced micelle stabilization[1a]. For this
system, NAA seems to be a very good probe to follow the aggregation of DAP, probably because of the analogy with the propionate
counterion.

This is not the case for AOT. From solubilization studies of
polar substances in reverse micelles, Kitahara and Kon-no[24] found
that the association of acetic acid with AOT is almost 15 times
less than with DAP. This phenomenon explains the rather small lifetime variations at R = 0 (Figure 5). The general aspect of this
curve, however, seems to be in agreement with the model proposed
by Eicke. The initial increase of the NAA-lifetime indicates the
formation of linear pre-micellar aggregates, without NAA-associa-
tion, resulting in a decrease of the polarity of the environment
of NAA. At the moment of cyclization, region "CMC", NAA is solubilized and its lifetime decreases. The variations at higher concen-

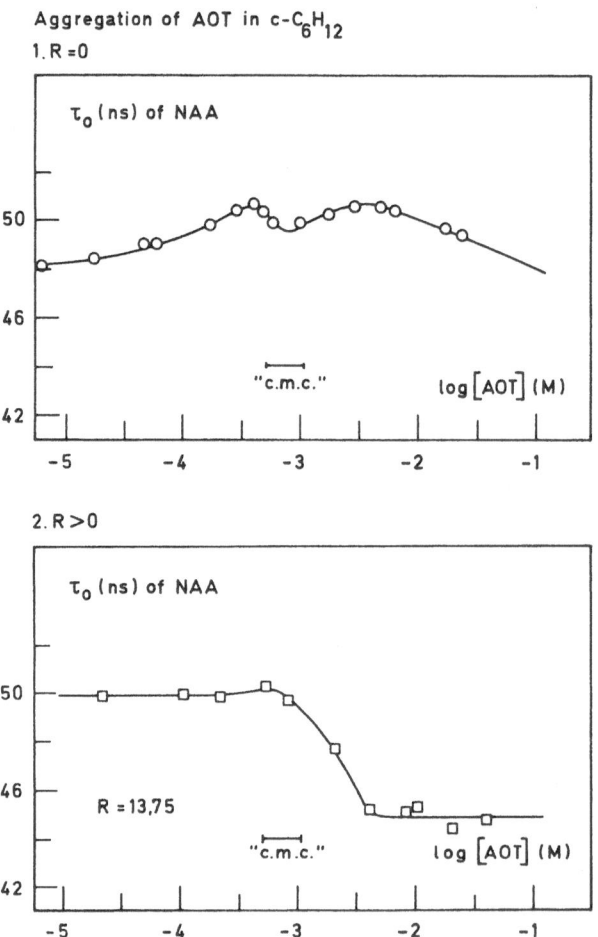

Figure 5.   *The fluorescence lifetime (τ$_o$) of NAA in the absence of quenchers as a function of the logarithm of the AOT-concentration when R = 0 (upper curve) and when R = 13.75 (lower curve).*

trations could be explained by an increase of the aggregation number, continuously changing the localization of NAA between the headgroups and the ester moieties. The value of the "CMC"-region $((6-10) \times 10^{-4}M)$ is in good agreement with values in the literature.

From these curves it seems that the aggregation behavior of DAP- and AOT-micelles can be described by the same mechanism, in contradiction with the earlier proposed totally different aggregation mechanisms. Concerning the meaning of the "CMC", we would better define it as the "(spherical micelle)-concentration".

2. $R > 0$. For both systems, the "CMC" is independent of the amount of water. The general aspect of the curves, however, changes. In the case of DAP, $R = 2.75$ (Figure 4), we still have two discontinuities, although the second ("CMC"-region) is less obvious. The reason therefore is the change in localization of NAA in the spherical micelle with increasing R-value. The variation of the NAA-lifetime with R, if $[DAP] = 0.08$ M, is given in the insert of Figure 4. An explanation is presented in our previous work[3d]. Thus, the higher the R-value, the higher τ_0 of NAA, and therefore the less is the difference in the lifetime before and after the "CMC"-region.

In the case of AOT, $R = 13.75$ (Figure 5), NAA seems to be better solubilized at the moment of becoming spherical. The most interesting result is maybe the fact that the initial increase of τ_0 almost disappeared, a phenomenon which is already there at lower R-values (not shown). This probably indicates that the water, hydrating the polar headgroups, causes the detergent molecules to form pre-micellar aggregates already at those low detergent concentrations. That this occurs with AOT and not with DAP is due to the strong interaction of water with the AOT-headgroups as compared with the DAP-headgroups[1a,11]. From our results at other R-values it is also obvious that for AOT, at concentrations above the "CMC"-region, τ_0, and therefore the localization of NAA, only becomes constant if $R > 10$, the maximal hydration number.

CONCLUDING REMARKS

The study of the fluorescence properties of probes solubilized in RMs, greatly contributes to the characterization of the RMs. From simple lifetime measurements, it can be concluded that the aggregation of DAP- or AOT-detergent molecules in apolar media occurs already at very low detergent concentrations, resulting in (linear) pre-micellar aggregates, which keep growing until, in a certain concentration region, they form closed aggregates or reverse

micelles. The size of these aggregates is dependent on the amount
of water solubilized in the system, as evidenced by quenching experiments. For better understanding and interpreting the results obtained
by fluorescence quenching, the localization of the probe and the
quencher must be known. Since exciplex formation is dependent on the
polarity of the environment, it is a method of choice to obtain this
information. From energy transfer processes and mainly from the
fluorescence quenching of micelle bound probes by bound quenchers,
it can be concluded that the mobility of solubilizates in the system
is very high. An exchange of solubilizates between two colliding
micelles is possible within the fluorescence lifetime of the probe.
The rate of this process is determined by the structure of the interface of the micelle, by the localization of the probe and the
quencher in the micelle, and by the water concentration. The mobility inside a micelle also increases with increasing water amount.

ACKNOWLEDGEMENTS

    Support of the FKFO and the University Research Fund is grate-
fully acknowledged. The NFWO is thanked for a predoctoral fellow-
ship to E.G. Financial support by IWONL - Agfa-Gevaert is gratefully acknowledged.

REFERENCES

1. a. H.F. Eicke, Topics in Current Chemistry, Vol. 87 86 (1979).
   b. J.H. Fendler, Acc. Chem. Res. Vol. 9 : 153 (1976).
   c. A.S. Kertes and H. Gutman, Surface and Colloid Sci., Ed. by
      Matijevic, John Wiley & Sons, N.Y., 8 : 193-295 (1976).
   d. A. Kitahara, "Cationic Surfactants", Ed. by E. Jungerman,
      N.Y., 289-310 (1970).
2. a. G.D. Correll, R.N. Cheser, III, F. Nome, and J.H. Fendler,
      J. Am. Chem. Soc., 100 : 1254 (1978).
   b. H.F. Eicke, J.C.W. Shepherd and A. Steinemann, J. Colloid
      Interf. Sci., 56 : 168 (1976).
   c. U.K.A. Klein, D.J. Miller and M. Hauser, Spectrochim. Acta,
      32A, 379 (1976); J. Chem. Soc. Faraday Trans. I, 73 : 1654
      (1977).
   d. M. Wong and J.K. Thomas, "Micellization, Solubilization and
      Microemulsions", Ed. by K.L. Mittal, Plenum Press N.Y.,
      Vol. 2 : 647 (1977).
   e. M. Wong, J.K. Thomas and M. Grätzel, J. Am. Chem. Soc.
      98 : 2391 (1976).

3. a. S.S. Atik and J.K. Thomas, J. Am. Chem. Soc., 103 : 3543 (1981); J. Am. Chem. Soc., 103 : 4367 (1981); J. Phys. Chem., 85 : 3921 (1981); J.Am. Chem. Soc., 103 : 7403 (1981).
    b. P.D.I. Fletcher and B.H. Robinson; Ber Bunsenges. Phys. Chem., 85 : 863 (1981).
    c. M.A.J. Rodgers and J.C. Becker, J. Phys. Chem., 84 : 2762 (1980).
    d. E. Geladé and F.C. De Schryver, J. Photochemistry, 18 : 223 (1982).
4. A. Kitahara, Bull. Chem. Soc. Japan, 28 : 234 (1955).
5. C.A. Martin and L.J. Magid, J. Phys. Chem., 85 : 3938 (1981).
6. E. Geladé, N. Boens and F.C. De Schryver, J. Am. Chem. Soc., will be published in the issue of November 17.
7. ref. 2-4 in ref. 6.
8. S.P. Van and G.S. Hammond, J. Am. Soc., 100 : 3895 (1978).
9. J.K. Thomas, Acc. Chem. Res., 10 : 133 (1977).
10. a. M. Almgren, F. Grieser, J.K. Thomas, J. Am. Chem. Soc., 101 : 2021 (1979).
    b. J.R. Escabi-Perez, F. Nome and J.H. Fendler, J. Am. Chem. Soc., 99 : 7749 (1977).
11. H.F. Eicke and H. Christen, Helv. Chim. Acta, 61 : 2258 (1978).
12. M. Wong, J.K. Thomas and T. Nowak, J. Am. Chem. Soc., 99 : 4730 (1977).
13. H.F. Eicke and J. Rehake, Helv. Chim. Acta, 59 : 2883 (1976).
14. a. J.C. Dederen, M. Van der Auweraer and F.C. De Schryver, J. Phys. Chem., 85 : 1198 (1981).
    b. P.P. Infelta, M. Grätzel and J.K. Thomas, J. Phys. Chem., 78 : 190 (1974).
    c. M. Tachiya, Chem. Phys. Lett., 33 : 289 (1975).
15. R.A. Day, B.H. Robinson, J.H.R. Clarke and T.V. Doherty, J. Chem. Soc., Faraday Trans. I : 132 (1979).
16. a. F.Y.F. Lo, B.M. Escott, E.J. Fendler, E.T. Adams,Jr., R.D. Larsen and P.W. Smith, J. Phys. Chem., 79 : 2609 (1975).
    b. N. Muller, J. Phys. Chem., 79 : 287 (1975).
    c. L.A. Fucugauchi, B. Djermouni, E.D. Handel and H.J. Ache, J. Am. Chem. Soc., 101 : 2841 (1979).
    d. H.F. Eicke and A. Denns, J. Colloid Interface Sci., 64 : 386 (1978).
    e. U. Hermann and Z.A. Schelley, J. Am. Chem. Soc., 101 : 2665 (1979).
    f. K. Tsujii, J. Sunamoto, F. Nome and J.H. Fendler, J. Am. Chem. Soc., 82 : 423 (1978).
17. M. Van der Auweraer, J.C. Dederen, E. Geladé and F.C. De Schryver, J. Chem. Phys., 74 : 1140 (1981).

18. M.P. Pileni, J.M. Furois, B. Hickel, C. Ferradini and
    J. Richault, poster presentation at the "International
    Symposium on Surfactants in Solution", June–July 1982,
    Lund (Sweden).
19. a. N. Muller, J. Colloid Interface Sci., 63 : 383 (1978).
    b. L. Magid, "Solution Chemistry of Surfactants", Ed. by
       K.L. Mittal, Plenum Press N.Y., Vol. 1 : 427 (1979).
    c. R. Zana, "Solution Chemistry of Surfactants", Ed. by
       K.L. Mittal, Plenum Press N.Y., Vol. 1 : 455 (1979).
20. J.H. Fendler, F. Nome and H.C. Woert, J. Am. Chem. Soc.,
    96 : 6745 (1974).
21. a. H.F. Eicke, "Micellization, Solubilization and Micro-
       emulsions", Ed. by K.L. Mittal, Plenum Press N.Y.,
       Vol. 1 : 429 (1977).
    b. H.F. Eicke, R. Hopmann and H. Christen, Ber. Bunsenges.
       Phys. Chem., 79 : 667 (1975).
22. A. Denat, B. Gosse and J.P. Gosse, J. Electrostatics, 12 : 197
    (1982).
23. S. Muto, Y. Shimazaki and K. Meguro, J. Colloid Interface Sci.,
    49 : 173 (1974).
24. A. Kitahara and K. Kon-no, "Micellization, Solubilization and
    Microemulsions", Ed. by K.L. Mittal, Plenum Press N.Y.,
    Vol. 2 : 1675 (1977).

PICOSECOND STUDIES OF ROSE BENGAL FLUORESCENCE IN REVERSE MICELLAR

SYSTEMS

Michael A.J. Rodgers

Center for Fast Kinetics Research
University of Texas at Austin
Austin, Texas 78712 USA

INTRODUCTION

The nature and properties of microheterogeneous systems have
been increasingly investigated in recent years. Several reasons
exist for this interest: one concerns the similarity with biological systems where compartmentalization is necessary to proper
functioning and microheterogeneity is a requisite for this. Other
reasons are of technological nature: surfactant systems are being
used in the oil industry for recovery of oil from spent wells and
in research into solar energy storage, where compartmentalization
is envisaged as a way in which separated charges can be prevented
from undergoing energy-degrading back reactions.

One aspect of supra-molecular assemblies deals with dispersions
of surfactant in non-polar organic liquids[1,2]. These dispersions
at an appropriate concentration contain reverse micelles (RM) that
are capable of solubilizing large amounts of water into the polar
core. Such species attract the curiosity of molecular scientists
since several questions are immediately posed :

What are the physical characteristics of the core and
interfacial regions?

What can be ascertained about the fluidity, polarity,
dielectric nature and hydrogen-bonding characteristics
in those regions?

Are any restrictions imposed on the diffusion of molecules
from one compartment into another?

Where are solubilized materials held?

> How are the rates and energetics of chemical and
> physical processes altered by compartmentalization?

For this work the fluorescence properties of xanthene dyes were used
in an attempt to probe the interior phase of RM's.

Xanthene dyes such as Rose Bengal (RB) and Erythrosin B (EB)
absorb in the visible spectrum and give out strong fluorescence,
the intensity of which is markedly medium-dependent. The photophysical scheme is simple :

$$X + h\nu \rightarrow X^*(S_1)$$

$$X^*(S_1) \xrightarrow{k_F} X(S_0) + h\nu_F$$

$$X^*(S_1) \xrightarrow{k_G} X(S_0)$$

$$X^*(S_1) \xrightarrow{k_T} X^*(T_1)$$

whence, the fluorescence lifetime, τ_M, is given by

$$\tau_M = 1/k_M = 1/(k_F + k_G + k_T)$$

and the quantum yield of fluorescence, ϕ_F, is

$$\phi_F = k_F/(k_F + k_G + k_T) \qquad \text{and}$$

$$k_F = \phi_F/\tau_M$$

Evidence has accumulated that K_T is strongly influenced by the
hydrogen-bonding power of the environment of the dye[3,5]. This leads
to large differences in τ_M and ϕ_F between media of high hydrogen-
bonding power (water) and aprotic liquids such as acetone[1].Similar
effects are seen when xanthene dyes are dissolved in aqueous micelles
of ionic surfactants[6,7]. This has been interpreted as indicating
that water molecules at the water-micelle interface have a much
different hydrogen-bonding power than that of bulk water.

In this paper the application of the xanthene dye probe to RMs
in various aprotic solvents is described. Since τ_M is typically
less than 1 ns, equipment with time resolution in the 10^{-11}s range
was necessary for the study.

EXPERIMENTAL

Fluorescence Lifetimes

Excitation was by a ca 35 ps pulse of 532 nm radiation from a frequency-doubled Quantel NG50 mode-locked Nd:YAG laser. Fluorescence was detected at right angles by a computerized streak camera system based on a Hamamatsu temporal photometer. The incident beam was polarized vertically and magic-angle detection was employed. For each sample ten replicate determinations were made and averaged by computer to improve signal-to-noise. A detailed description and a schematic layout have been published[6].

RESULTS

A. Aerosol OT (AOT) in Iso-octane

1. Effect of Water on k_M for RB. A typical fluorescence time profile is displayed in Figure 1. Throughout this work all decays followed a single exponential. Figure 2 shows how k_M (= $1/\tau_M$) for RB fluorescence varied with H_2O (expressed as R_W = H_2O / AOT) in dispersions of AOT (0.25 M) in iso-octane. The RB concentration was near 20 μM. RB was insoluble in micellar systems containing no added water.

2. Effect of Water on k_M for Hexadecyl RB. The hexadecylester of RB ($C_{16}RB$, a gift from Professor J.D.Spikes, University of Utah) was used in a series similar to the above. $C_{16}RB$ was soluble in the R_W = 0 dispersion. The concentration effect is shown in Figure 3.

3. Effect of AOT on k_M. Values of k_M = 1.3 ns^{-1} were obtained for AOT in iso-octane RM's containing H_2O at R_W = 1.0 in range 0.02 M to 0.25 M AOT.

B. AOT in Different Solvents

RM's of AOT (0.1 M) in several solvents (heptane, toluene and chlorobenzene) were examined for k_M changes as a function of R_W. $C_{16}RB$ was used as probe. The data are in Figure 4.

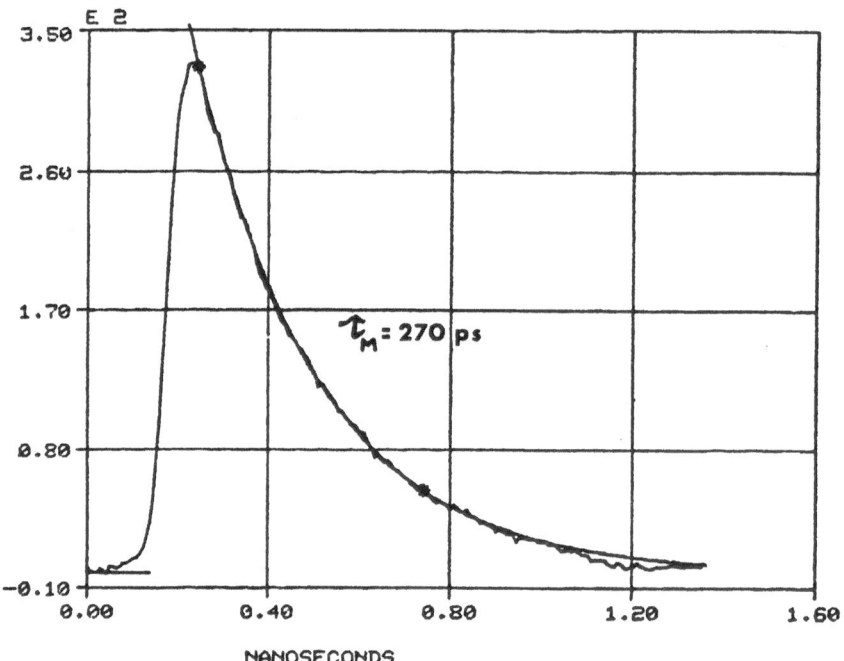

Figure 1. *Typical decay profile of $C_{16}RB$ fluorescence (250 mM AOT/ iso-octane at R_W = 40)*

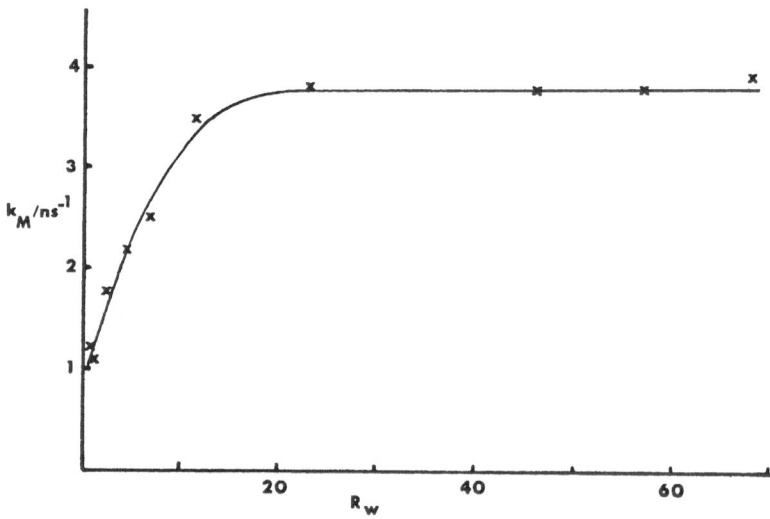

Figure 2. *R_W effect on k_M for RB (20µM) in AOT (0.25 M) in iso-octane.*

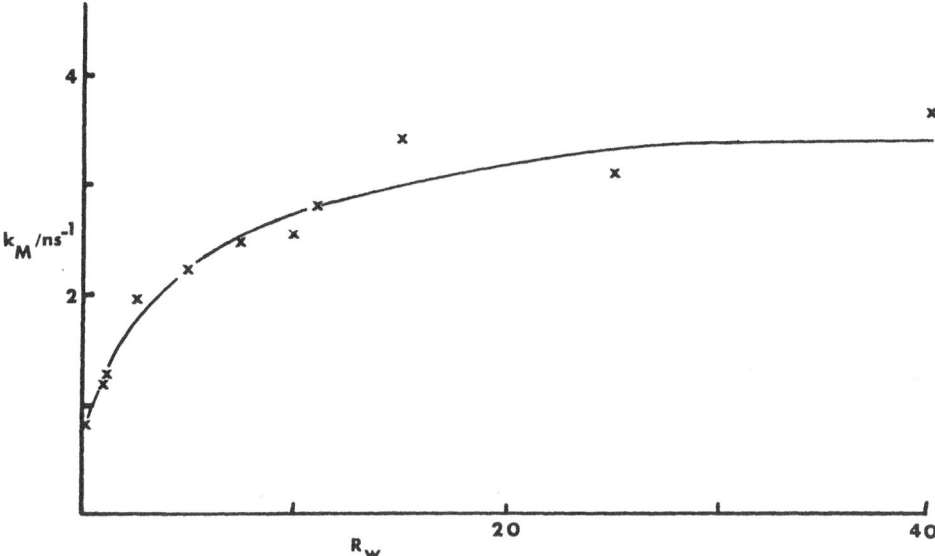

Figure 3. R_W effect on K_M for $C_{16}RB$ (20 μM) in AOT (0.25 M) in iso-octane.

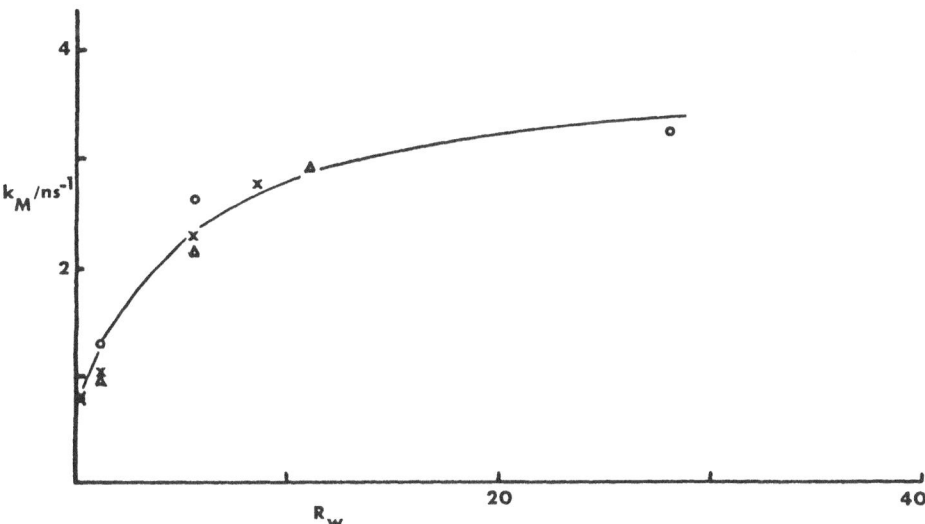

Figure 4. R_W effect on K_M for $C_{16}RB$ (20 M) in AOT (0.1 M) in heptane (0), chlorobenzene (x) and Toluene (Δ).

C. Alcohol effects

    1. n-octanol. Experiments were conducted with $C_{16}RB$ (20 μM) in AOT (0.25 M)/iso-octane in the presence and absence of added n-octanol and added H_2O. The k_M values are in Table I.

    2. Methanol. Table I also shows some k_M values for RM dispersions containing added methanol at $R_W = 0$.

Table I : Effect on k_M of adding n-octanol (OCT) or methanol
           (MET) to AOT (0.1 M) in iso-octane

R_W	R_{OCT}	R_{MET}	k_M/ns^{-1}	k_M^o/ns^{-1} (a)
0	1.3	0	0.79	0.77
0	2.6	0	1.13	0.77
5	2.6	0	2.64	2.40
11	2.6	0	2.93	3.00
0	0	4.9	1.38	0.77
0	0	9.8	1.64	0.77

(a) k_M^o is the observed rate constant in the absence of
       any added alcohol.

D. Benzyl hexadecyl dimethyl ammonium chloride (BHDC) in benzene

    The fluorescence decay constant of RB (20 μM) in RM dispersion of BHDC (0.1 M) in benzene was measured as a function of R_W. The values are collected in Table II.

Table II : Effect on k_M of water concentration added to
         BHDC (0.1 M) in benzene

R_W	k_M/ns^{-1}	τ_M/ns
0	0.48	2.1
5.6	0.74	1.3
16.7	0.84	1.2
27.8	0.91	1.1

Table III :   Solvent effects on k_M at $R_W = 0$ (dry RM's)

Surfactant (M)	Solvent	k_M/ns^{-1}	τ_M/ns
AOT (0.1)	heptane	0.83	1.20
AOT (0.1)	chlorobenzene	0.77	1.30
AOT (0.1)	toluene	0.71	1.41
AOT (0.25)	iso-octane	0.77	1.30
BHDC (0.1)	benzene	0.48	2.10
none	tert-butanol[a]	0.81	1.24
none	acetonitrile	0.56	1.80

(a)   Reference 5.

DISCUSSION

The underlying assumption herein is that the environmental hydrogen-bonding power governs the fluorescence lifetime (via k_T changes) in RM systems as in homogeneous media. Table III assembles all the k_M values for $C_{16}RB$ in AOT and BHDC RM's in the various solvents used at $R_W = 0$. Clearly for AOT there is no solvent effect and the mean value of k_M is 0.77 ns^{-1} ($\tau_M = 1.3$ ns). Highly aprotic solvents, such as acetonitrile, show τ_M values near 2 ns, whereas the weakest H-bonding alcohol studied by Cramer and Spears[5] was tert-butanol, which had $\tau_M = 1240$ ps, very close to the 'dry' RM values here. Zulauf and Eicke[8] maintain that even with 'dry' materials, RM's always contain low concentrations of H_2O which serves to cement the headgroups together with extra-strong H-bonds. The present work indicates that such H_2O molecules provide an environment for the probe that has a hydrogen-bonding power similar to that of tert-butanol.

A scrutiny of Figures 1 and 2 allows the conclusion that the k_M values for RB and its hexadecyl ester are not significantly different at all R_W values employed. Hence the xanthene moiéties on the two probes find equivalent environments in RM dispersions. Thus the hexadecyl residue confers no special characteristics in terms of a solubilization site. One clear advantage of the C_{16} derivative however, is that it is soluble at $R_W = 0$ and virtually insoluble in water. Thus as R_W increases from zero it is a safe assumption that the C_{16} probe remains firmly anchored in the interface, near to the hydrophobic regions. Hence the increase in k_M from 0.77 ns^{-1} at $R_W = 0$ to the limiting value of 3.6 ns^{-1} at $R_W > 20$ is a result

of changes in the hydrogen-bonding power of the interfacial region
as more H_2O molecules are included into the swelling RM's. Since
RB and $C_{16}RB$ behave so similarly it therefore follows that RB itself
prefers the headgroup region(through hydrophobic interactions) even
though it is readily soluble in water. Additional evidence arises
out of the observation that the limiting value of k_M = 3.6 ns^{-1} at
R_W > 20 is far lower than the k_M value in neat H_2O (12.7 ns^{-1}).
Nmr studies have shown[9] that solubilized H_2O has the relaxation
properties of bulk water at R_W > 8. Thus if the fluorescent probe
were dissociated from the interfacial region at R > 8 it would be
expected to exhibit a k_M value more in line with that in bulk water.

Figure 4 shows that the fluorescence lifetime of $C_{16}RB$ is independent of the nature of the dispersion medium. Day et al.[10] have
shown that for RM's composed of AOT, at low R_W there are pronounced
differences in the density of the water pool, surfactant aggregation
number and water pool radius from one solvent to another. This work
shows that such solvent-derived dependences are not reflected in the
H-bonding character of the interfacial region.

Table I compares the incremental effects of added alcohols compared to added water alone. With n-octanol as coadditive, additions
of R_{OCT} = 2.6 have only a small effect on k_M at R_W = 0 and even
smaller effects at higher R_W values. Thus n-octanol added to RM's
does not cause any appreciable changes in the H-bonding environment
near the probe location. Methanol,however, is probably more hydrophilic than n-octanol and will seek out polar regions. At R_W = 0,
added CH_3OH causes increases in k_M(Table I), but it has a smaller
effect on k_M than H_2O at the same R value (Figure 2). However, this
may be due to the fact that CH_3OH itself has a weaker H-bonding power
than H_2O. In fact, k_M for RB in neat CH_3OH is 2.1 ns^{-1}, not very
much larger than the 1.64 ns^{-1} observed here at R_{MET} = 9.8. It
therefore appears possible that of the two alcohols, at R_W = 0,
CH_3OH resides preferentially within the AOT aggregate, whereas
n-$C_8H_{17}OH$ resides in the exterior phase. On this point the current
evidence is sketchy and more work is needed.

BHDC is a cationic surfactant and the anionic xanthene dye will
have an electrostatic reason in addition to a hydrophobic reason for
being preferentially associated with the benzene-BHDC-water interfacial regions. Table II shows that in RM dispersions of BHDC (0.1 M)
in benzene, the value of k_M for RB was only ca. 2x higher at R_W = 28
than at R_W = 0 (for AOT dispersions the incremental factor was near
5x). The k_M value at R_W = 0 was 0.48 ns^{-1} (τ_M = 2.1 ns), which
closely approaches the aprotic solvent values (e.g. acetonitrile)
of 0.56 ns^{-1} (τ_M = 1.8 ns).

The data suggest that RB in BHDC systems find a more aprotic
environment than it does in AOT dispersions. Probably the electrostatic effects allow the probe to bury itself deeper into the surfactant layer, behind the headgroups, where it finds only aprotic

aryl and hydrocarbon residues. Under these conditions the addition
of H_2O molecules into the polar interior would conceivably have less
influence on the probe than if it were located closer to the aqueous
region as, for example, in AOT systems where electrostatic repulsions
disfavor deep penetration.

ACKNOWLEDGEMENTS

These experiments and the data analyses were performed at the
Center for Fast Kinetics Research which is supported jointly by
the Biotechnology Branch of the Division of Research Resources of
NIH (RR00886) and by the University of Texas at Austin.

REFERENCES

1. H.F.Eicke, Topics in Current Chemistry, Ed. by M.J.S. Dewar
    et al., Springer-Verlag, Berlin, Vol.87; p.6 (1980)
2. J.H. Fendler, Membrane Mimetic Chemistry, Chapter 3,
    Wiley-Interscience, New York, (1982)
3. W. Yu, F.Pellegrino, M. Grant and R.R.Alfano,
    J.Chem.Phys. 67: 1766 (1977)
4. G.R. Fleming, A.E.W.Knight, J.M. Morris,R.J.S.Morrison and
    G.W. Robinson, J.Am.Chem.Soc. 99: 4306 (1977)
5. L.E. Cramer and K.G. Spears,   J.Am.Chem.Soc., 100, 221 (1978)
6. M.A.J. Rodgers, Chem.Phys.Lett. 78; 509 (1981)
7. M.A.J. Rodgers, J.Phys.Chem.   85:3372 (1981)
8. M. Zulauf and H.F.Eicke, J.Phys.Chem. 83: 480 (1979)
9. M. Wong, J.K. Thomas and T. Nowak, J.Am.Chem.Soc. 99: 4370,
    (1977)
10. R.A.Day, B.H.Robinson, J.H.R. Clarke and J.V. Doherty, J.Chem.
    Soc. Faraday I, 75; 132 (1979).

MAGNESIUM PORPHYRIN SENSITIZED REDUCTION OF VIOLOGEN SULFONATE

IN REVERSE MICELLES

M.P. Pileni and J.M. Furois

Université P. et M. Curie
Laboratoire de Chimie-Physique
75005 Paris, France

C.E.N. Saclay D.P.C./S.C.M.
91191 Gif sur Yvette, France

The photoelectron transfer from magnesium tetraphenyl porphyrin to viologen sulfonate in reverse micelles has been observed. The electron transfer rate
constant and the intermicellar exchange rate constant
have been determined. The latter depends on the water
content in the water pool.

INTRODUCTION

Water can readily be dispersed in an organic medium such as
isooctane using sodium bis(2-ethylhexyl) sulfosuccinate (Aerosol
OT, AOT) as a dispersant. AOT possesses a hydrophylic head which, in
non-polar solvents in the presence of water, forms aggregates known
as reverse micelles. A relatively large amount of water can easily
be incorporated into the micellar core and the state and the properties of such water pools have previously been investigated[1]. The
number of water molecules in a pool can vary considerably and depends principally on w_o, the ratio $[H_2O]/[AOT]$ in the system.

In this paper we report the photoelectron transfer from magnesium tetraphenylporphyrin (MgTPP) to viologen sulfonate (PVS) in
reverse micelles. Such a photoelectron transfer provides new possibilities for the light energy conversion studies. From a kinetic
study the photoelectron rate constant and the bi-molecular rate

constants for the exchange processes involving the water pool collisions are deduced at various w_o values.

EXPERIMENTAL

Isooctane and methanol were obtained from Fluka and were of
"puriss" quality. They were used without further purification. The
incorporated water was deionized and twice distilled. Aerosol OT
from Merck was purified by dissolving it in methanol and adding active charcoal. This solution was stirred for several hours, concentrated by evaporation, and the AOT was extracted with isooctane.
Methanol and water were eliminated from the AOT isooctane solution
by distillation, taking into account the azeotropic point for iso-
octane-water.

The structure of the viologen sulfonate is:

$$^-O_3S-(CH_2)_3-\overset{+}{N}\underset{CH=CH}{\overset{CH-CH}{\diagup\diagdown}}C-C\underset{CH=CH}{\overset{CH-CH}{\diagup\diagdown}}\overset{+}{N}-(CH_2)_3-SO_3^-$$

Magnesium tetraphenylporphyrin was a Sigma product.

Laser photolysis experiments were performed using Quantel frequency doubled Nd laser with a pulse of 20 ns duration and an energy of 50-100 mJ.

RESULTS AND DISCUSSION

The absorption spectrum of magnesium tetraphenylporphyrin
(MgTPP) in micellar solution of AOT (0.1 M) is characterized by two
bands centered at 610 nm and 560 nm as well as the Soret band at
420 nm.

Figure 1A shows the absorption spectrum of the MgTPP triplet
which exhibits a maximum at 460 nm. The triplet is relatively stable
in the absence of oxygen. Its decay follows first order kinetics
with a specific rate of $3.5 \times 10^4 s^{-1}$.

We shall now examine the triplet behavior in the presence of
an electron acceptor such as viologen sulfonate. Figure 1B shows
transient spectra obtained with an AOT solution containing both
MgTPP and viologen sulfonate (PVS). The absorption curve recorded at
the end of the laser pulse closely resembles the spectrum for the
triplet. After 1.5 ps one notices a decrease in these triplet peaks

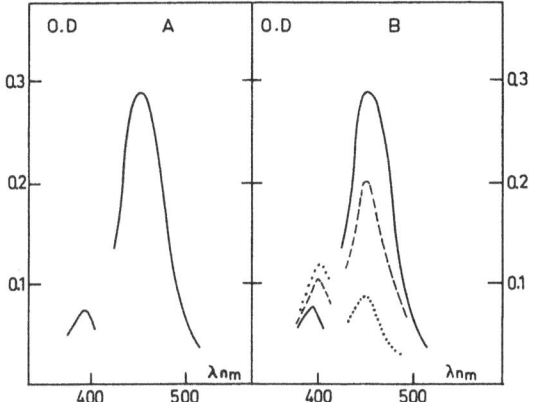

Figure 1. *A) Triplet-triplet absorption spectrum of MgTPP in 0.1M AOT reverse micelles.*
            B) Absorption spectra obtained (———) at the end of the laser pulse, (— — —) 7 ps after the laser pulse and (-------) 15 ps after the laser pulse of 2x10⁻⁵M MgTPP in 0.1M AOT in iso-octane and 0.25 mM PVS.

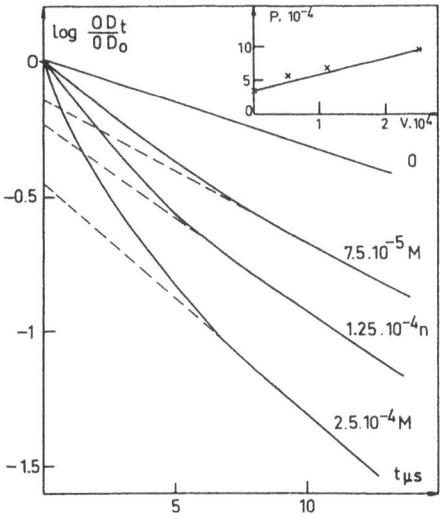

Figure 2. *Decay of MgTPP triplet at 460 nm at different PVS con-centrations. [MgTPP] = 2x10⁻⁵M.*

with the simultaneous formation of new bands centered around 400 nm. By analogy with the spectral data for MgTPP$^+$ and for PVS$^-$, this peak can be assigned to the MgTPP$^+$ cation and the reduced viologen, which was produced via quenching of the triplet state:

$$MgTPP(T_1) + PVS \rightarrow MgTPP^+ + PVS^- \qquad Equation\ 1$$

The kinetics of reaction 1 will now be analyzed in terms of various water contents, w_o. MgTPP is insoluble in water and sparsely soluble in isooctane. Addition of AOT up to $3 \times 10^{-2}M$ considerably increases the MgTPP solubility, indicating that the MgTPP is probably located in the hydrocarbon tail or at the interface of the reverse micelle. The viologen sulfonate is water soluble and is expected to be located in the water pool of the reverse micelle. If the distribution of PVS over the aggregates is governed by Poisson's law, then the time dependence of the MgTPP triplet concentration is given by[2]:

$$[^3MgTPP] = [^3MgTPP]_o\ exp\left\{-(k_o+k_e[PVS])t \right.$$
$$\left. -\bar{n}[1-exp(-k_q t)]\right\} \qquad Equation\ 2$$

where $[^3MgTPP]$ and $[^3MgTPP]_o$ are the MgTPP triplet concentrations at time t and time zero respectively, k_o is the first order rate constant governing the MgTPP triplet decay in the absence of viologens, k_e is the bimolecular rate constant for the exchange process involving water pool collisions and $\bar{n}$ is the average number of viologen molecules per water pool, $\bar{n} = [PVS]/[WP]$ and $[WP]$ is the water pool concentration. For long time intervals Equation 2 is reduced to:

$$[^3MgTPP]/[^3MgTPP]_o = -(k_o+k_e \cdot [PVS]) \cdot t-\bar{n} = -p \cdot t$$

Figure 2 shows the variation of the logarithm of the OD_t/OD_o versus the time, (the optical densities at time t and time zero are OD_t and OD_o, respectively). The slope of the long time decay gives $(k_o+k_e \cdot [PVS])=p$. By plotting the slope versus viologen concentration, the exchange rate constant can be deduced. From Table I we deduce that the intermicellar exchange rate constant depends on the water content in the pool. Using these k_e values in Equation 2 to fit the short time, experimental data enables us to determine k_q (Table II).

From these kinetic studies we have determined the exchange rate constant of the water pool. The values obtained show an increase with the water content in the water pool. However, the values are higher by a factor of 10 than those obtained by B. Robinson[3] but are similar to those given by de Schryver[4]. The difference could probably be due to the fact that the porphyrin is located in the

Table I. *Exchange rate constants obtained from Equation 2*
          using PVS as a quencher of MgTPP triplet state

w_o	5	15	25
$k_e \cdot 10^{-8} (M^{-1} \cdot s^{-1})$	0.7	2.3	2.8

micellar core in our case and in Robinson's case the chromophores
are located in the water pool.

CONCLUSIONS

From this study we have pointed out that photoelectron transfer occurs in reverse micelles. In the case of MgTPP–PVS, the photoelectron rate constant is independent of the water content, w_o, and
the bimolecular exchange rate constant increases with the increase
in water.

Table II. *Electron transfer rate constant obtained from*
          Equation 1 using PVS as a quencher of MgTPP state

w_o	5	15	25
$k_q \cdot 10^{-5} (s^{-1})$	7	7	7

REFERENCES

1. J.H. Fendler, Account Chem. Res. 9 : 1953 (1976).
   H.F. Eicke, "Micellization, Solubilization and Microemulsion"
   K.L. Mittal, Editor, Vol.2, p.493, Plenum Press, New York
   M.P. Pileni, B. Hickel, C. Ferradini and J. Puchault, Chem.
   Phys. Letters, 92 : 308 (1982).
2. S.S. Atik and J.K. Thomas, Chem. Phys. Letters, 79 : 351 (1981).
3. B.H. Robinson and D.C. Steytler, J.C.S. Faraday1, 75 : 481 (1979).
4. F.C. De Schryver, personal communication.

^{13}C NMR STUDIES OF MOLECULAR CONFORMATIONS AND INTERACTIONS IN THE CURVED SURFACTANT MONOLAYERS OF AEROSOL OT WATER-IN-OIL MICROEMULSIONS

L. J. Magid and C.A. Martin

Department of Chemistry
University of Tennessee
Knoxville, Tennessee 37996 USA

Carbon-13 NMR data for Aerosol OT reversed micelles and water-in-oil microemulsions are used to elucidate hydrocarbon chain dynamics and rotational isomerism. Evidence is presented that benzene and carbon tetrachloride penetrate the surfactant interphase to a greater extent than cyclohexane.

INTRODUCTION

Aerosol OT [AOT, sodium bis(2-ethylhexyl)sulfosuccinate] forms reversed micelles and w/o microemulsions in apolar solvents: these aggregates have been studied extensively by Eicke[1] and others[2]. When AOT is carefully purified to remove trace amounts of inorganic impurities[3], it is capable of solubilizing large amounts of water. There are good experimental and theoretical reasons[4] for dividing the L_2 region of an AOT/H_2O/apolar solvent system into reversed micellar and w/o microemulsion domains based on the molar ratio, R, of added water to AOT. When R is small (up to ca. 10), all the water present is involved in hydrating the sulfonate head groups (and perhaps the esters' carbonyl oxygens); a highly structured hydrogen-bonded head group, counterion and water network exists in the aggregates' cores. The appearance of free water (R>10-15), which coincides with the aggregates' acquiring a water core encased by a curved surfactant monolayer having a constant area per head group, marks the transition to the w/o microemulsion domain.

In many aromatic or chlorinated hydrocarbon solvents, AOT's water solubilizing capacity is not high enough for an extended w/o

microemulsion region to exist. However, in certain aliphatic hydrocarbons, such as isooctane and cyclohexane, R values as high as 70
are possible. It is of interest to ascertain whether these differences in solubilizing capacity have consequences on a molecular
level, such as characteristic patterns of hydrocarbon chain dynamics in the surfactant interphase, for example. Our ^{13}C NMR data
for several AOT ternary systems, as described below, provide insight
in this area. The use of ^{13}C NMR makes it possible to probe the
dynamic processes and conformational preferences at individual AOT
carbons, without the requirement of adding a probe molecule such as
a spin-labeled species. In addition, ^{13}C chemical shifts make it
possible to draw certain conclusions about water penetration into
the interphase.

The preferred conformations in the succinic acid portion of
aggregated AOT molecules has also been investigated using ^{1}H NMR[5,6];
the three possible staggered conformations about the C_1-C_1, bond
are shown in Scheme I. Rotamers I and II are favored.

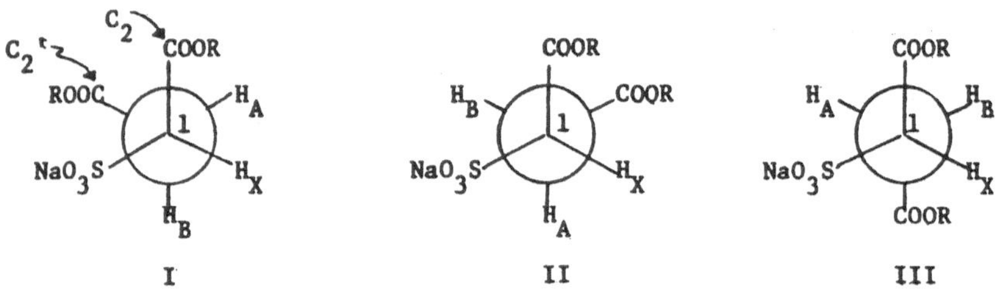

Scheme I

Maitra and Eicke[5] make a persuasive case for rotamer I having
the most pronounced amphiphilic character and thus enhancing hydrogen-bond formation with solubilized water; it is the major conformer in isooctane. On the other hand, rotamer II is substantially
populated in CDCl$_3$; its importance increases as the R value increases. Eicke has proposed that AOT molecules are relatively flexible in the w/o microemulsion domain; the energetic preference for
rotamer I (which arises at low R because of its more pronounced
hydrogen-bonding ability) is lessened, and rotamer II's population
increases.

A comparison of an AOT molecule possessing the S-configura-
tion at C_1 and an sn-3 phosphatidylcholine (Scheme II) reveals certain structural similarities[7]. The conformational possibilities at
C_2-C_3 in V are analogous to those for AOT at C_1-C_1,; the rotameric
preferences have been found by Hauser, et al[8] to be independent of

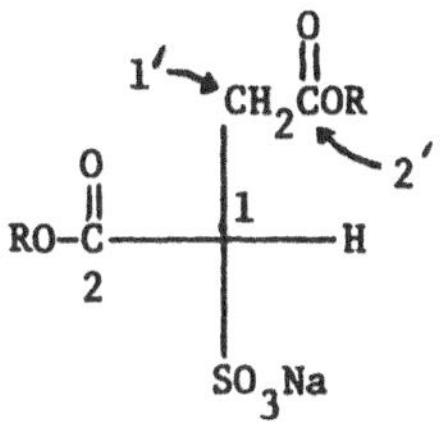

IV
Aerosol OT

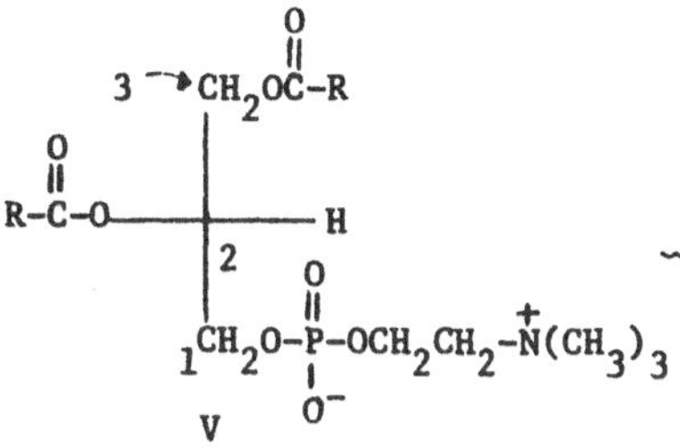

V
an sn-3 L-phosphatidylcholine

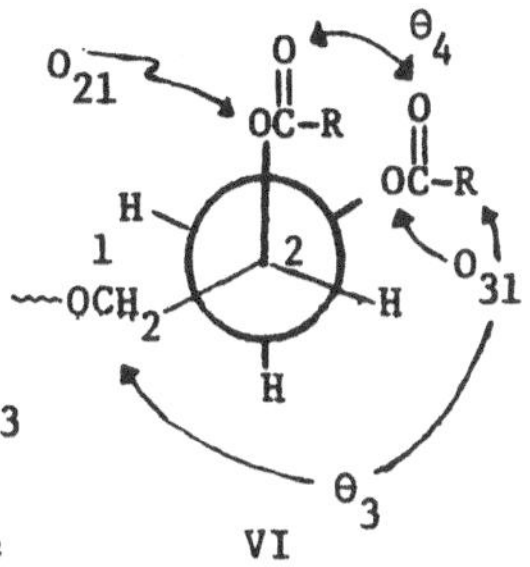

VI

Scheme II

V's state of aggregation (normal or reversed micelle, bilayer, etc.). The two torsion angles[9] θ_3 and θ_4 (θ_3 refers to the C_1–C_2–C_3–O_{21} and θ_4 to the O_{21}–C_2–C_3–O_{31} moiety) are identified in VI: the rotamer having θ_3 anti and θ_4 +gauche has a fractional population of 0.60 (this is analogous to II for AOT), while the rotamer having θ_3 –gauche has a fractional population of 0.33 (this is analogous to I for AOT).

The ^{1}H NMR work of Dennis et al[10] has established that the two protons on the sn-2-α-methylene carbon of phosphatidylcholines are magnetically non-equivalent in micellar form (either pure or mixed micelles), while the sn-1-α-methylene protons are magnetically equivalent and are located in a more hydrophobic environment than the sn-2-α-methylenes. These results may be explained in terms of phosphatidylcholine conformational preference; the sn-2 chain begins parallel to the micelle (or bilayer) surface, while the sn-1 chain is perpendicular. This preference is supported by X-ray diffraction studies[11] of phospholipid bilayer crystals and by ^{2}H NMR studies. As VII indicates, the sn-1 chain extends considerably further into the bilayer in this arrangement. Inspection of the molecular models for the phospholipids' rotamers I and II indicates a θ_3 torsion angle of 180° (i.e. anti) is optimum for the observed arrangement of the sn-2 and sn-1 chains.

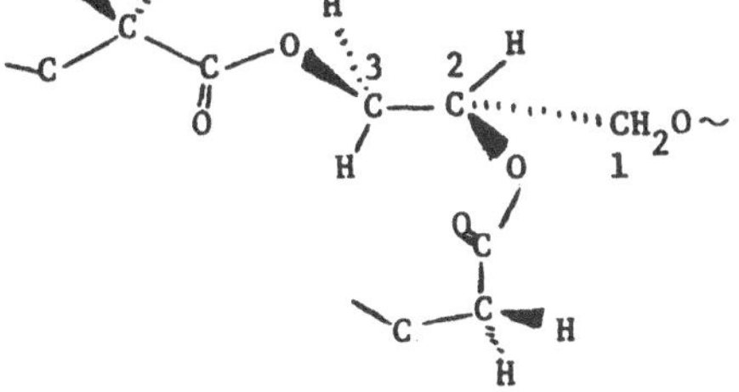

VII

Scheme III

RESULTS AND DISCUSSION

^{13}C NMR spin-lattice relaxation times (T_1's) and chemical
shifts (δ) were measured for AOT reversed micelles and w/o microemulsions in cyclohexane, benzene and carbon tetrachloride. With
cyclohexane it was possible to investigate extensively the w/o
microemulsion portion of the L_2 region (Figure 1) by using special-
ly purified AOT (see the Appendix). Figure 2 shows the proton-de-
coupled spectrum of AOT in cyclohexane, with the resonances assigned
according to Ueno[6]. Since these data have been published elsewhere[13],
only selected topics which bear directly on questions raised in the
Introduction are discussed here.

Chemical Shift Data. The δ's of carbons in the head-group region depend upon R, with the most marked dependence (see Figure 3)
occurring at $C_{2'}$, which is in the ester group analogous to the
sn-1 chain of a phospholipid. Carbonyl chemical shifts are known to
be sensitive to the hydrogen-bonding ability of the medium[14]; $C_{2'}$
is in a more hydrophobic environment than C_2 initially; $C_{2'}$ moves
to a still less aqueous environment as the w/o microemulsion domain
is reached. In phospholipid vesicles, bilayers and micelles the
ester carbonyl of the sn-1 chain has been found also to be the one
less accessible to water[15]. C_2's δ being relatively independent of

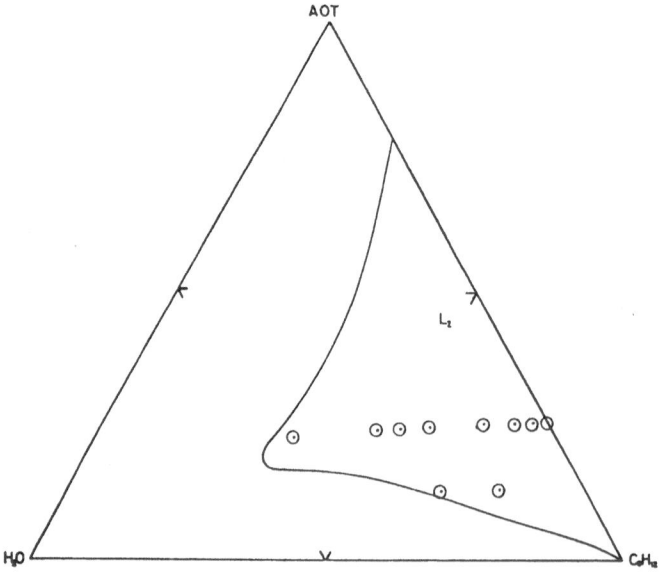

Figure 1. *Ternary phase diagram of AOT/cyclohexane/H$_2$O at 30 ± 1°C
showing the inverted micelle w/o microemulsion region (L$_2$). The
circles indicate the composition of the sample studied by ^{13}C NMR.*

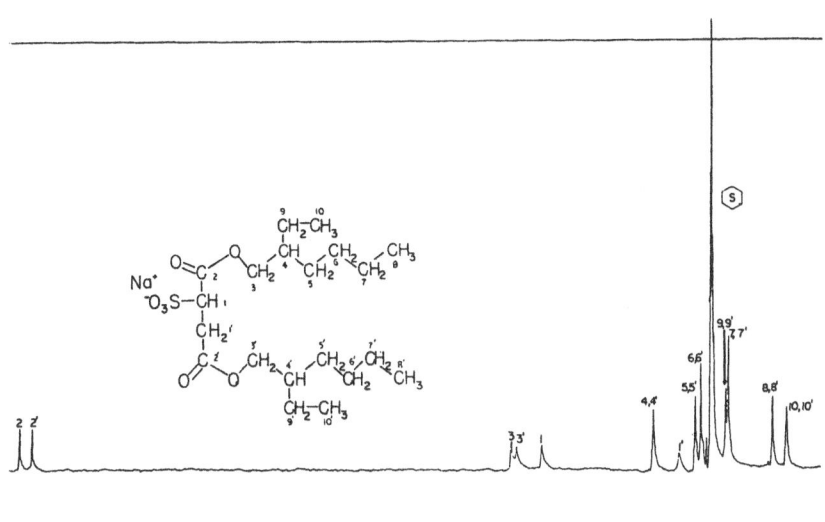

Figure 2.   *15.1-MHz* ¹³*C NMR spectrum of Aerosol OT/cyclohexane/H_2O showing the assignments given by Ueno.*

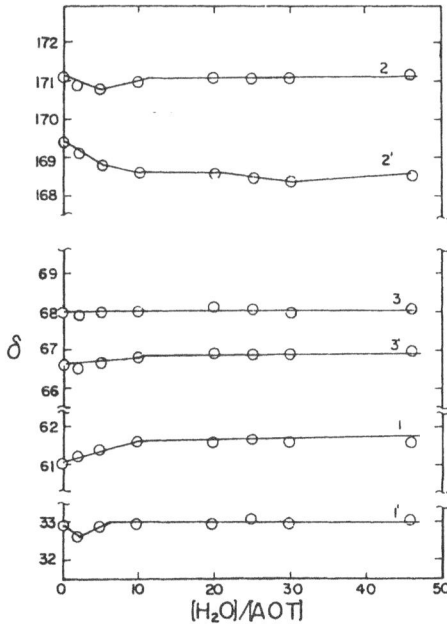

Figure 3.   *Change in* ¹³*C chemical shifts of carbons 1-3 as a function of added water concentration in cyclohexane.*

Table I. NT_1 (SEC) OF AOT/CYCLOHEXANE/WATER SOLUTIONS AS A FUNCTION OF R AT 15.1 MHz

C	R=	0.5 M AOT								0.25 M AOT		0.5 M AOT degassed
		0	2.2	5.3	10.2	19.9	25.2	29.6	46.3	26.1	46.3	26.0
10,10'		1.50	2.10	3.24	3.51	3.57	3.42	3.96	3.75	3.12	3.30	3.33
9,9'		0.34	0.56	0.66	1.06	0.88	0.80	0.76	1.00	0.61	0.59	0.82
8,8'		3.18	4.44	5.94	6.27	5.61	5.82	6.39	6.75	5.13	4.65	5.91
7,7'		0.84	1.28	1.82	1.44	2.04	2.20	2.22	1.92	1.58	1.77	1.54
6,6'		0.50	0.68	1.00	0.96	1.14	1.00	1.46	1.26	1.27	1.32	1.28
5,5'		0.40	0.42	0.44	0.62	0.38	0.54	0.62	0.44	0.54	0.40	0.62
4,4'		0.044	0.058	0.15	0.17	0.16	0.19	0.22	0.18	0.16	0.13	0.19
3		0.044	0.090	0.12	0.16	0.17	0.15	0.22	0.13	0.12	0.14	0.17
3'		0.062	0.088	0.13	0.18	0.16	0.17	0.24	0.13	0.14	0.14	0.16
2[a]		0.54	0.72	1.23	1.14	1.54	1.70	2.03	2.21	1.07	1.74	2.00
2'[a]		0.74	0.97	1.69	1.93	2.65	2.60	2.53	3.25	2.08	2.90	3.18
1		—[b]	0.026	0.045	0.068	0.10	0.061	0.056	0.045	0.067	0.076	0.046
1'		—[b]	—[b]	0.062	0.10	0.10	0.064	0.070	0.030	0.070	0.068	0.042

a. T_1 is reported

b. Significant line broadening resulted in insufficient S/N to measure T_1 for these resonances.

R value while C_2,'s δ moves upfield as R increases is consistent
with rotamer II's fractional population increasing at the expense
of rotamer I. The same chemical shift behavior was observed at C_2
and C_2, in benzene and CCl_4 as R increased, so that an apolar sol-
vent-induced conformational preference about the C_1-C_1, bond in AOT
cannot by itself explain the observed differences in AOT's water-
solubilizing ability. Our ¹³C NMR data agree with Eicke and Maitra's
¹H NMR data[5] in predicting that rotamer II increases in importance
as R increases.

Spin-Lattice Relaxation Times at 15.1 MHz. NT_1's at 15.1 MHz
are shown in Table I for the cyclohexane systems. All NT_1's increase rapidly up to R = 5-10, indicating the increased mobility
of an AOT molecule in the reversed micelle. For carbons 4,4'-10,10',
NT_1 remains constant or increases slightly at higher R. Segmental
motion is observed at all water concentrations. Effective correlation times (τ_{eff}) were computed using Equation 1.

$$\frac{1}{NT_1} = \frac{\gamma_H^2 \gamma_C^2 \hbar^2}{10r^6} \left[\frac{\tau_{eff}}{1 + (\omega_H-\omega_C)^2\tau_{eff}^2} + \frac{3\tau_{eff}}{1 + \omega_C^2\tau_{eff}^2} + \right.$$

$$\left. \frac{\tau_{eff}}{1 + (\omega_H + \omega_C)^2\tau_{eff}^2} \right]$$

Equation 1

The extent of segmental motion can be represented by the ratio,
$\tau_{eff}(4)/\tau_{eff}(8)$. Initial addition of water results in a sharp decrease in this ratio (Figure 4). Segmental mobility then gradually
decreases to R = 30 and increases at the highest R. Changes in segmental mobility may be due to changes in the relative probability
of gauche conformations occurring at individual C-C bonds (an equilibrium property of the system), or to changes in the rate of
gauche-trans interconversion (a dynamic property of the system)
which may occur either at individual C-C bonds or as part of a correlated motion involving several C-C bonds[16]. Correlated motions
are discussed in the next section.

The large values of $\tau_{eff}(4)/\tau_{eff}(8)$ observed at low R indicate
a higher probability of gauche conformations; the lower ratio brought
about by addition of water reflects the increasing probability of
trans conformers along the chain. This is supported by a calculation
of A_0, the surface area at the surfactant/oil interface (dotted line
in Figure 4), using Equation 2,[17] where r_h is the radius of an aggre-

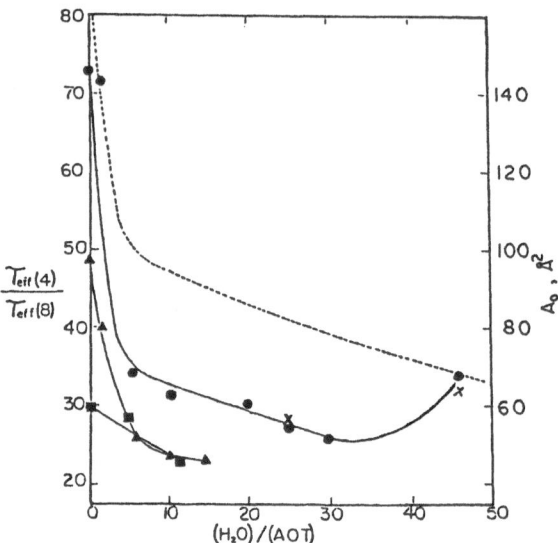

Figure 4. *Extent of segmental motion, expressed as τ(4)/τ(8), as a function of R: (●)0.5M in cyclohexane, (X)0.25M in cyclohexane, (▲) in CCl₄, (■) in benzene. The effective area (Aₒ) at the surfactant/oil interface is given by the dotted line.*

gate[18], 1 is the length of the surfactant molecule (11 Å), and A_w is the surface area at the surfactant/water interface.

$$[r_h/(r_h - 1)]^2 = A_o/A_w \qquad \textit{Equation 2}$$

If the increase in segmental motion at R = 46 were due to aggregate crowding, resulting in more gauche conformers, dilution of the samples should decrease $\tau_{eff}(4)/\tau_{eff}(8)$. This is not observed (cf. 0.25 M AOT, Figure 4). If the time each carbon-carbon bond spends in a trans conformation increases, segmental mobility will increase[16]. Internal crowding of the alkyl chains causes this behavior.

Note that at R = 0, $\tau_{eff}(4)/\tau_{eff}(8)$ decreases in the order $C_6H_{12} > CCl_4 > C_6H_6$, while at R = 5 and 10, the ratio is approximately the same in CCl_4 and C_6H_6. Benzene and CCl_4 are more polarizable than C_6H_{12}, so that some interaction of these former two solvents with the sulfonate and ester groups of AOT is reasonable. At low R values, C_6H_6 and CCl_4 may penetrate the surfactant interphase[19], while when C_6H_{12} is the solvent, the voids at the surfactant/oil interface are filled by the hydrocarbon chains themselves. This constitutes a reasonable explanation for the preference for gauche

conformers in cyclohexane at low R (and hence the large observed
τ_{eff} ratio). It also verifies on a molecular level that the organic
solvent-polar group interaction may be partly responsible for the
observed differences in AOT's water-solubilizing ability in the
three solvents.

Relaxation of carbons near the head group should be dominated
by overall aggregate reorientation. Rotational correlation times
for the aggregates in cyclohexane were calculated using Equation 3
and data on hydrodynamic radii from the work of Day et al [20], Zulauf
and Eicke[4a], and Gulari et al[21]. Since these investigators used

$$\tau_R = 4\pi r_h{}^3 \eta/3kT)$$ Equation 3

several different solvents, quantitative agreement between τ_{eff} and
τ_R is not expected. Figure 5 demonstrates[22] that undoubtedly for C_1
$\tau_{eff} \simeq \tau_R$ up to R = 20-25. At higher R values, rigid body motion of
the whole AOT molecule (rotational and/or lateral diffusion) within
the aggregate provides additional mechanisms for reorientation of
the C-H vectors at or near the head group. For example, lateral diffusion of phosphatidylcholines across the surface of a lipid bilayer
has been observed with τ_d's $(\equiv L^2/6D_L)$ on the order of $10^{-7}s$[23].

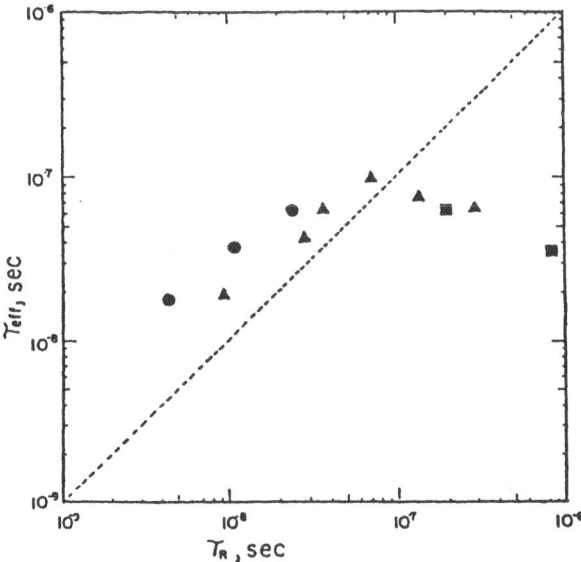

Figure 5. *Comparison of $\tau_{eff}(1)$ calculated by using Equation 1
with τ_R calculated by using Equation 3: (●) r_h in cyclohexane,
R = 2,5,10 left to right (Day et al); (▲) r_h in isooctane,
R = 2,5,10,20,25,30 left to right (Zulauf and Eicke); (■) r_h in
heptane, R = 30,46 (Gulari et el). The dotted line has a slope
equal to one.*

Employing Equation 4 permits us to separate τ_{eff} into contributions for overall reorientation of the aggregates and internal motions. The resulting dependence of τ_{int} on R is shown in Figure 6;

$$1/\tau_{eff} \approx 1/\tau_R + 1/\tau_{int} \qquad\qquad Equation\ 4$$

for each alkyl tail carbon it decreases in the reversed micelle region and then levels off in the w/o microemulsion domain.

Frequency-Dependent T_1's in Cyclohexane. Measurements were performed at 100.6 MHz for samples with R = 0 and 30. At R = 0, all NT_1's are larger at the higher frequency, including values for carbons in the alkyl tails where τ_{eff} formally satisfies extreme narrowing conditions. For R = 30, NT_1's of carbon 6,6'-10,10' are equivalent at 15.1 and 100.6 MHz. The alkyl chains in the aggregates evidently experience correlated (i.e., interdependent) motions[16,24]; with respect to internal motion, these may include kink formation, annihilation and diffusion. Motions occurring towards the chain ends depend on those occurring further in (this can be quantified by using correlation factors[16]), so that the measured T_1's are reflecting the system's order as well as rates of internal motion.

CONCLUSIONS

Chemical shift data for AOT's carbonyl oxygens and for the ABX proton system at C_1 and $C_{1'}$ can be interpreted in terms of a water-dependent rotational isomerism about the C_1-$C_{1'}$ bond. The conforma-

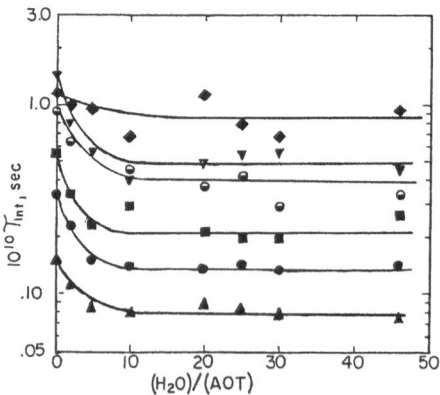

Figure 6. *Internal correlation times (τ_{int}) calculated for C-5,5' to C-10,10' as a function of R: (◆) 5,5'; (●)6,6'; (■)7,7'; (▲)8,8'; (▼)9,9'; (⊙)10,10'.*

tion favored in the microemulsion regime is analogous to the one found at C_2-C_3 for phosphatidylcholines in micellar and bilayer form.

Carbon -13 T_1 data for AOT reversed micelles and w/o microemulsions have been used to obtain information on the solvent dependence of the hydrocarbon chain dynamics in the aggregates. It is observed that solvent penetration into the surfactant interphase decreases in the order $C_6H_6 > CCl_4 > C_6H_{12}$.

ACKNOWLEDGEMENT

Acknowledgement is made to the donors of the Petroleum Research Fund, administered by the American Chemical Society, for the support of this research.

APPENDIX

Since salts profoundly affect the water solubilizing capacity of Aerosol OT in cyclohexane[3b], the surfactant was purified as follows. This is a combination of other procedures[3b,25,26] and was the only method to give consistent results.

AOT (50-60 g, Aldrich) was stirred overnight with 5-10 g of coconut charcoal in 500 ml of benzene. After filtration through a 0.5-μm FH Millipore filter, 25 ml of water was added, the solution was shaken, and the flocculant was allowed to settle. The benzene solution was decanted and the procedure was repeated with 15 and 10 ml of water, followed by removal of the benzene. The solid was then dissolved in 400 ml 1:3 methanol/water and washed 3 times with 100 ml of 60-80 ligroine to remove any remaining organic impurities. The aqueous methanol was then carefully evaporated (to avoid foaming), after which solid AOT was dried overnight at 75°C in vacuo. Dissolution in anhydrous methanol and filtration through a 0.08-μm polycarbonate filter (Nucleopore) removed any additional salts. Methanol was stripped off and the solid dried 48 h in vacuo. Percent recovery was 50-60%; a solution of 10% AOT in cyclohexane incorporated >70 mol of water per mol of AOT. The absence of residual acidic materials was checked by using 1-methyl-4-(cyanoformyl)pyridinium oxide (COP)[3a].

REFERENCES AND NOTES

1. H.F. Eicke, Top. Current Chem., 87 : 86 (1980) and references therin.

2. a. K. Kon-no and A. Kitahara, J. Colloid Interface Sci.,
         35 : 636 (1971).
      b. P. Ekwall, L. Mandell and K. Fontell, ibid., 33 : 215 (1970).
      c. M. Wong, J.K. Thomas and M. Grätzel, J. Am. Chem. Soc.,
         98 : 2391 (1976).
      d. J. Sunamoto, T. Hamade, T. Seto and S. Yamamoto, Bull. Chem.
         Soc. Jpn. 53 : 583 (1980).
 3. a. L.J. Magid, K. Kon-no and C. Martin, J. Colloid Interface
         Sci., 83 : 307 (1981).
      b. H. Kuneida and K. Shinoda, ibid., 70 : 577 (1979).
      c. S.M.F. Tavernier and R. Gijbels, Talanta, 28 : 221 (1981).
 4. a. M. Zulauf and H.F. Eicke, J. Phys. Chem., 83 : 480 (1979).
      b. H.F. Eicke, Pure Appl. Chem., 53 : 1417 (1981).
      c. R. Kubik, H.F. Eicke and B. Jönsson, Helv. Chim. Acta,
         65 : 170 (1982).
 5. A.N. Maitra and H.F. Eicke, J. Phys. Chem., 85 : 2687 (1981).
 6. M. Ueno, H. Kishimoto and Y. Kyogoku, Chem. Lett., 599 (1977);
    J. Colloid Interface Sci., 63 : 113 (1978).
 7. As drawn, these molecules may be described as having the L con-
    figuration at C_1 and C_2 respectively.
 8. H. Hauser, W. Guyer, P. Skrabal and S. Sundell, Biochemistry,
    19 : 366 (1980).
 9. See M. Sundaralingam, Annals N.Y. Acad. Sci., 195 : 324 (1972)
    for more information on this type of nomenclature.
10. a. M.F. Roberts, A.A. Bothner-By and E.A. Dennis, Biochemistry,
         17 : 935 (1978) and references therein.
      b. J. DeBony and E.A. Dennis, ibid., 20 : 5256 (1981).
11. P.B. Hitchcock, R. Mason, K.M. Thomas and G.G. Shipley, Proc.
    Natl. Acad. Sci., 71 : 3036 (1974).
12. A. Seelig and J. Seelig, Biochemistry, 13 : 4839 (1974);
    Biochim. Biophys. Acta, 406 : 1 (1975).
13. C.A. Martin and L.J. Magid, J. Phys. Chem., 85 : 3938 (1981).
14. G.E. Maciel and J.J. Natterstad, J. Chem. Phys., 42 : 2752
    (1965).
15. C.F. Schmidt, Y. Barenholz, C. Huang and T.E. Thompson,
    Biochemistry, 16 : 3948 (1977).
16. R.E. London and J. Avitabile, J. Am. Chem. Soc., 99 : 7765
    (1977).
17. L.M. Prince, In "Emulsions and Emulsion Technology",
    K.J. Lissant, Ed.; Marcel Dekker: New York, 1974; Vol. 6,
    pp. 125-177.
18. r_h is taken from Ref. 4a; A_w from H.F. Eicke, J. Rehak, Helv.
    Chim. Acta, 59 : 2883 (1976).
19. E. Keh and B. Valeur, J. Collid Interface Sci., 79 : 465 (1981).
20. R.A. Day, B.H. Robinson, J.H.R. Clarke and J.V. Doherty,
    J. Chem. Soc., Faraday Trans. 1, 75 : 132 (1979).

21. E. Gulari, B. Bedwell and S. Alkahafaji, J. Colloid Interface Sci., 77 : 202 (1980).

22. Note that the assumption that τ_{eff} increases as T_1 increases means that $(\omega_C + \omega_H)^2 \tau^2 \gg 1$. See D. Doddrell, V. Glushko and A. Allerhand, J. Chem. Phys., 56 : 3683 (1972)

23. P. Devaux and H.M. McConnell, J. Am. Chem. Soc., 94 : 4475 (1972).

24. G. Levy, M.P. Cordes, J.S. Lewis and D.E. Axelson, J. Am. Chem. Soc., 99 : 5492 (1977).

25. H.F. Eicke and H. Christen, J. Colloid Interface Sci., 48 : 281 (1974).

26. F.M. Menger and G. Saito, J. Am. Chem. Soc., 100 : 4374 &1978).

SMALL ANGLE NEUTRON SCATTERING BY MICELLAR NON-IONIC SURFACTANTS IN APOLAR MEDIA

J.C. Ravey and M. Buzier

Laboratoire de Biophysique Moléculaire ERA CNRS No.28
Université de Nancy I,   BP 239
54506 Vandoeuvre les Nancy, France

INTRODUCTION

The amphiphile aggregation phenomenon in non-polar organic solvents is at present a subject largely open to discussion. Although
in many cases[1-5] the existence of aggregates is generally admitted,
the physico-chemical factors responsible for that aggregation in
non-aqueous systems are far from being well identified and understood[1,2]. And as far as the morphological parameters of the aggregates are concerned, no direct evidence of the true structure has
ever been obtained[4,5].

In the present paper we report the results of a structural
investigation by small angle neutron scattering of some non-ionic
amphiphiles in various apolar solvents at different temperatures.
It will be shown that there is an interesting general pattern for
the aggregation phenomena in these systems which should be of great
help in the understanding of the influence of the various parameters
governing that process. We shall also report on the influence of the
addition of relatively small quantities of water on the structure of
the aggregates. In fact, a correlation can be drawn between the
amount of water solubilized ("swollen micelles"), the properties of
the non-polar binary mixtures and the overall composition of the
corresponding ternary system.

EXPERIMENTAL

Chemical: The non-ionic surfactants were polyoxyethylene glycol
dodecyl ethers $C_{12}(EO)_n$, with n = 4,5,6,from Nikko Chemicals (Japan).
They were dried with molecular sieve materials, as was the hydro-
genated decane (Merck, puriss). The deuterated solvents (ethanol,

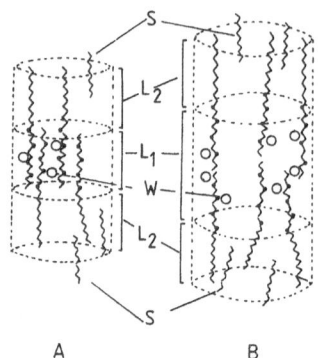

Figure 1. Hank-like (A) and lamella (B) structure of the aggregate in non-polar media.
L_1 : *hydrophilic polyoxyethylene chain of the surfactant molecule*
L_2 : *hydrophobic chain of the surfactant molecule*
S : *solvent molecule*
W : *water molecule*

chloroform, benzene, heptane, cyclohexane, decane, hexadecane) were obtained from CEA (France) or Merck-Sharp-Dohme (Canada). Partially deuterated decane was prepared by Dr. Rouviere (Montpelier, France).

Small angle neutron scattering: The measurements were carried out at the I.L.L. in Grenoble using a small angle scattering instrument D 17. The general Procedure described elsewhere was employed here,[6] which can be summarized as follows. Theoretical spectra were calculated for a given overall composition of the system. After a proper choice of values for all the morphological parameters of a given structure (see Figure 1), a numerical integration was performed on the computer which uses the average of all the orientations of the scatterers. Then the best fit between the theoretical and experimental curves for the whole range of the q values allowed the determination of the most probable set of the parameter values.

RESULTS AND DISCUSSION

I. Binary Systems

The concentration range of all non-ionic surfactants studied in the present investigation was 1-20% by weight. The temperature was varied between 10°C and 45°C.

From the experimental results and depending on the temperature, the amphiphile-solvent systems can be divided into three classes, based on the behavior of the I(q) intensities, scattered for q → 0

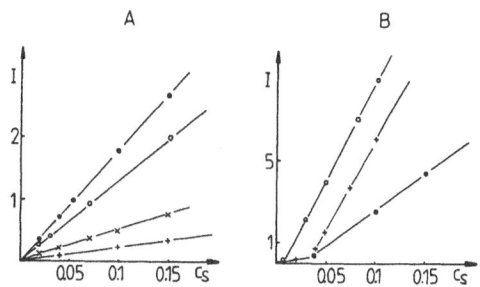

*Figure 2. Plots of I (scattered intensity extrapolated at q = 0)
versus C_S (concentration of surfactant g/g). Figure A: $C_{12}(EO)_4$ at
20°C with + chloroform, x benzene, o cyclohexane, • heptane.
Figure B: $C_{12}(EO)_6$ + decane at • 45°C, + 20°C, o 10°C.*

as a function of the total surfactant concentration (Figure 2).

a) In the first class of solvents (methanol, chloroform and
benzene) (Figure 2 A) there exist only monomers or possibly a few
dimers over the whole range of concentration (at 20°C).

b) For the system $C_{12}(EO)_4$ + heptane or cyclohexane at 20°C the
curves (Figure 2 A) do not exhibit a sharp transition, but still
reveal the presence of very small aggregates whose aggregation number noticeably increases with surfactant concentration.

c) Curves in Figure 2 B exhibit a more or less sharp transition
in the range of 1 to 5% by weight. Then for each of the systems of
this third class there exists a pseudo-critical concentration C_1
which could be compared to a critical micellar concentration: below
it, only monomers or possibly dimers can be detected, and above it,
we measure aggregation numbers very noticeably higher than those
existing in systems of class two. These aggregation numbers and the
type of the aggregates (see Figure 1) are given in Table I for the
last two groups of systems.

The results can be summarized as follows :

 - Effect of the hydrophillicity of the surfactant: When the length
of the polyoxyethylene chain increases, the aggregation numbers also
increase, with a correlated small decrease in the critical concentration, C_1.

 - Effect of the temperature: A decrease in temperature always
results in an increase of the average aggregation numbers (and a
decrease in C_1) which correspondingly may allow for a transition
Hank-lamella.

Table I :   Mean Values of the Morphological Parameters of
            Reverse Micelles

Solvent	Surfactant	T (^{O}C)	C_1	N	Shape: $(EO)_N$ extended
Hexadecane	$C_{12}(EO)_6$	20	0.032	40-50	Lamella
	$C_{12}(EO)_5$		0.035	12	Hank(~ lamella)
	$C_{12}(EO)_4$		0.040	10	Hank(- lamella)
Decane	$C_{12}(EO)_6$	10	0.010	24-30	Lamella
		20	0.034	20-25	Hank(-lamella)
		45	0.036	10	Hank
	$C_{12}(EO)_4$	20	0.045	9	Hank
Cyclohexane	$C_{12}(EO)_6$	10	0.033	8	Hank
	$C_{12}(EO)_4$	20	No	2-4	Hank
Heptane	$C_{12}(EO)_4$	20	No	5-8	Hank

C_1 : pseudo-critical concentration (g/g)

N : aggregation number

- Effect of the solvent: It appears that when the solvents are
linear hydrocarbons, C_7, C_{10}, C_{16}, the aggregation is favored when
the chain length is longer. The effect is small for $C_{12}(EO)_4$ but
quite noticeable for $C_{12}(EO)_6$, where a transition Hank-lamella seems
also to occur.

II. Ternary Systems

Structural investigations of ternary mixtures concern the mixture $C_{12}(EO)_4$-decane-water (the concentration of the surfactant
varies between 7 and 25% and the water content does not exceed 8%).

The most striking feature of the water solubilization process
is the very large increase in the micellar size, till the demixing
line is encountered. Thus the solubilization is not only achieved
by a swelling of the previously existing aggregates, but also water
molecules drastically promote their coalescence. From this point,
we can guess that a spherical shape is very unlikely.

In Figure 3 we depict the phase diagram with the iso-aggregation
number curves. This bundle of lines tends to converge to the "oil
corner", although they are not exactly straight lines. Then an oil-
dilution at constant micellar size does not seem to be possible.

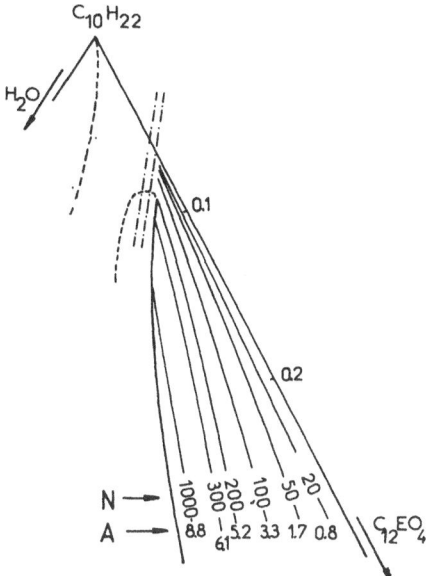

Figure 3. Ternary diagram of $C_{12}(EO)_4$ + $C_{10}H_{22}$ + H_2O) at $20°C$. Iso-aggregation number curves. N = aggregation number; A = number of water molecules per surfactant molecule.

At any rate it is quite apparent from the phase diagram that any attempt to dilute a system containing large aggregates would lead to a demixing phenomenon. For the smaller aggregates this would pro-mote a marked change in structure. Let us note again that these par-ticle sizes cannot be known with high accuracy, since the inter-particle effects can only be approximated.

As far as the morphological parameters are concerned, the results clearly show the existence of layered structures. The polyoxyethylene chains are in the extended conformation, and the water molecules are all bound on the oxyethylene groups. The maximum hydration number per EO group is between 2 and 3, depending on the surfactant con-centration: addition of more water would induce a demixing pheno-menon. At lowest hydration rates the aggregates are Hank-like par-ticles, while in the neighborhood of the demixing line they seem to be changed into lamellas. These results are in agreement with the low conductivities of these systems [7].

REFERENCES

1. E.Ruckenstein and R. Nagarajan, J.Phys.Chem., 84: 1349,(1980)
2. H.F.Eicke and H. Christen, J.Coll.and Interf.Sci.,46:417,(1973)
3. R.C.Little and C.R.Singleterry, J.Phys.Chem.,68: 3453,(1964)
4. A.V.Smirnova, A.F. Koretskii and N.A. Sokolovskaya, Koll.Zhurnal 38: 727, (1976)
5. A.S. Kertes, in "Micellization,Solubilization and Microemulsions" Ed. K.L. Mittal, 445 (1977)
6. J.C. Ravey and M. Buzier, in "International Surfactants in Solution" Lund, (Sweden)(1982).
7. M.H. Boyle, M.P.McDonald, P. Rossi and R.M. Wood, in "Microemulsions", Ed. I.D.Robb, Plenum Press, N.Y., 103 (1982).

ELECTRIC POLARIZATION OF W/O-MICROEMULSIONS:

STUDIES USING ELECTRIC BIREFRINGENCE

Zora Marković

Department of Biophysical Chemistry
Biocenter of the University of Basel
CH-4056 Basel / Switzerland

INTRODUCTION

A water-in-oil microemulsion is a particular kind of stable colloidal system which can be visualized as water droplets suspended in oil with the surfactant molecules accumulated at the oil/water interface. In the case of ionic surfactants, such as Aerosol OT, it is assumed that the mobile counter-ions in the aqueous phase will be concentrated near the amphiphilic ions situated at the interface. These particular liquid structures display strong polarization when subjected to an electric field, as is shown by electric birefringence studies using Aerosol OT/Water/Isooctane solutions. An attempt is made to explain the mechanism by which it is possible to produce such strong electric polarization in water-in-oil microemulsions.

EXPERIMENTAL

i) _System_ The W/O-microemulsion studied was the ternary system Aerosol OT/Water/Isooctane, prepared by dissolving appropriate weights of AOT, sodium-diethylhexyl sulphosuccinate, (Fluka, Switzerland) in isooctane of highest grade quality, 99.5% purity (Fluka) and adding to this solution an exact volume of deionized, twice-distilled water. The surfactant concentration, C_{AOT}, hydrated surfactant concentration, C_{AOT+H_2}, and the molar ratio of water to AOT molecule, $w_o = [H_2O]/[AOT]$, were in the ranges: $2.5 \times 10^{-2} C_{AOT} < 2.0 \times 10^{-1}$ mol dm^{-3}, $4.5 \times 10^{-2} < C_{AOT+H_2O} < 3.6 \times 10^{-1}$ g cm^{-3}, $50 < w_o < 95$.

ii) <u>Electric birefringence (Kerr effect) technique</u> An apparatus for Kerr effect measurements similar to that described by Benoit[1]
was designed and constructed[2], having analyzer and polarizer in the
crossed position. The principle of the measurement is that when a
rectangular electric pulse (a), Figure 1, is applied to the W/O-
microemulsion solution, a time dependent birefringence of the medium, Δn, is produced (b).

The importance of this technique is that the analysis of such
a single birefringence signal offers information about the contribution of permanent and induced moments to particle orientation
(rise part of the curve), about the optical and electrical properties ·of the particle (the steady state part of the curve), and about
the particle dimensions (the decay part of the curve). This technique appears to be particularly suitable for experimental studies
of W/O-microemulsions with high water contents and surfactant concentrations. There is a definite advantage over photon correlation
spectroscopy, with which it is not possible to study droplet sizes
greater than $w_o \overset{\sim}{=} 50$, due to the excessively high intensity of the
scattered light. The method of electric birefringence[2] was used to
determine the hydrodynamic radii of droplets with w_o up to 100.
Whereas photon correlation spectroscopy measures the size of the
droplets without perturbation in the system, the applied electric
field in the birefringence measurements may, under certain conditions, cause deformation of the droplets.

iii) <u>Droplet shape</u> In the presence of an external force, droplets appear to have a slightly prolate shape[2]. From combined measurements of translational and rotational diffusion, the axial ratio was
found to be 2.1 to 2.8 for $50 < w_o < 95$. Further evidence was given by
flow viscosity measurements. It was assumed that the interactions
between the droplets could be neglected.

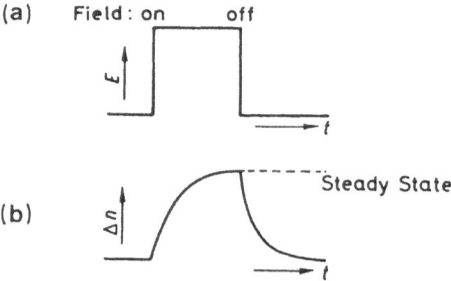

Figure 1. *Time dependence of the electric (a) and birefringence
pulse (b).*

   In the absence of external fields, the droplets are expected
to be spherical, the fields causing the deformations. Provided that
the interfacial tension is $<10^{-3}$ dyne cm^{-1} and particle size is
$\sim 10^{-6}cm$, an electric potential as used (up to 5000 V cm^{-1}) could
be sufficient to cause droplet deformation[4],[5].

RESULTS

   The analysis of the rise and decay parts of the birefringence
curves produced by (i) reversed and (ii) single polarity electric
pulses should give an indication of the relative influence of the
contribution of a permanent or an induced dipole moment on the birefringence of the W/O-microemulsion.

   (i) The reverse-pulse technique is extensively used to identify
the orientation mechanism and to estimate the ratio of the permanent
to the induced dipole moment, provided that the induced polarization
relaxation time is short compared to the disorientation relaxation
time. A typical birefringence signal observed with an $AOT/H_2O/i-C_8H_{18}$
solution, subjected to two consecutive electric pulses of opposite
polarities, is shown in Figure 2. At the change of the electric pulse
polarity, no reorientational decrease/increase of birefringence was

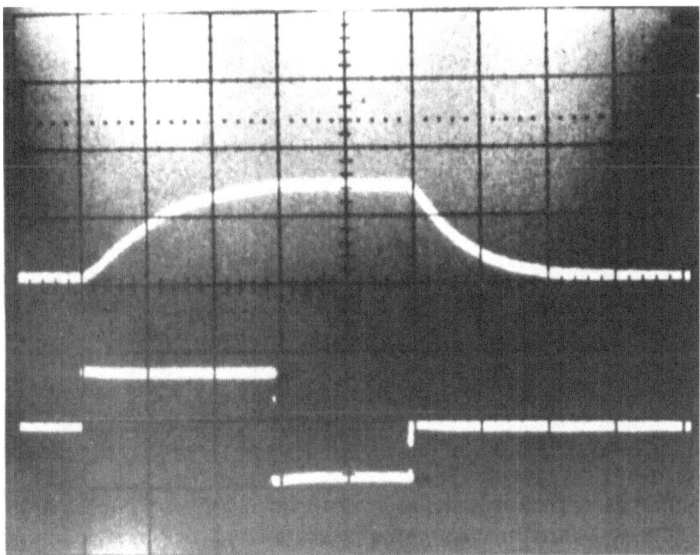

Figure 2. *The photocurrent signal (top curve) produced by an
electric pulse of reversed polarity, E = 2500 V cm^{-1} (lower curve).
Pulse length was 1 ms. Solution: 0.18M $AOT/H_2CO/iC_8H_{18}$, w_o = 80,
T = 22°C.*

observed, as would be expected with a permanent dipole moment. It is concluded that pure induced polarization is responsible for the bi-refringence.

(ii) With a single polarity electric pulse of low strength, for a monodisperse solution of particles, the rise in the birefringence signal will be symmetrical to the decay curve for an orientation that is only due to a pure induced dipole moment[1]. A typical example of this behavior was observed with the birefringence signal of $AOT/H_2O/i-C_8H_{18}$ solution, Figure 3. Symmetrical profiles of a rise and a decay curve, $\tau_r = \tau_d$, and rapid initial rise of birefringence are both strong indications for pure induced polarization.

In order to investigate the origin of the induced polarization, the experimentally observed specific Kerr constant, K_{SP}^{exp}, was compared with the theoretical values due to

a) distortional polarization, K_{SP}^{dis}, or

b) ionic atmosphere polarization, K_{SP}^{ion}.

The specific Kerr constant, K_{SP}, is a measure of the optical, $(g_1 - g_2)$, and electrical, $(g_1^E - g_2^E)$, anisotropy of a particle. In the case of a rigid molecule of volume V with no permanent dipole

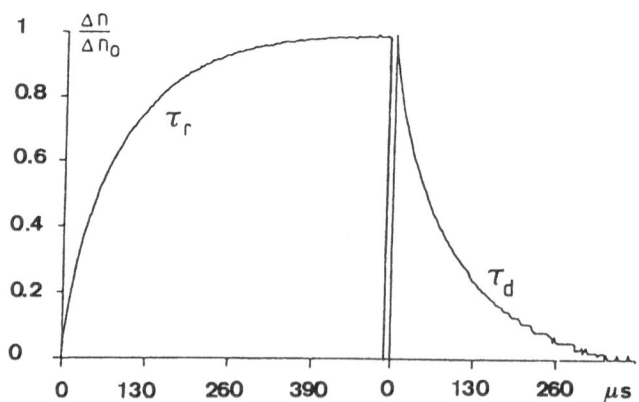

Figure 3. *The normalized birefringence signal as obtained by computer analysis. $\Delta n/\Delta n_o$ against the time (μs) in 0.18M $AOT/H_2O/i-C_8H_{18}$ solution, $w_o = 70$, at T = 30. 0ºC, E = 3100 V cm^{-1}, and pulse length of 519 μs.*

moment, it was shown[6] that:

$$\frac{2 \pi V}{15 n^2 kT} \cdot (g_1 - g_2) (g_1^E - g_2^E) = K_{SP} = \frac{\Delta n_o}{\emptyset n E^2} \qquad Equation\ 1$$

---------------------------------   ------------

theoretical estimation   *experimental*
determination

where Δn_o is the steady state birefringence, E is the electric field
strength, $\emptyset$ is the volume fraction of hydrated surfactant, and n is
the refractive index of the continuous, isotropic medium. The subindices 1 and 2 refer to particle axes parallel and perpendicular to
the applied field, respectively.

Steady state birefringence, Δn_o, of the ternary system
$AOT/H_2O/i\text{-}C_8H_{18}$ was investigated as a function of the applied field
strength, E, and the concentration of hydrated surfactant, Figure 4.

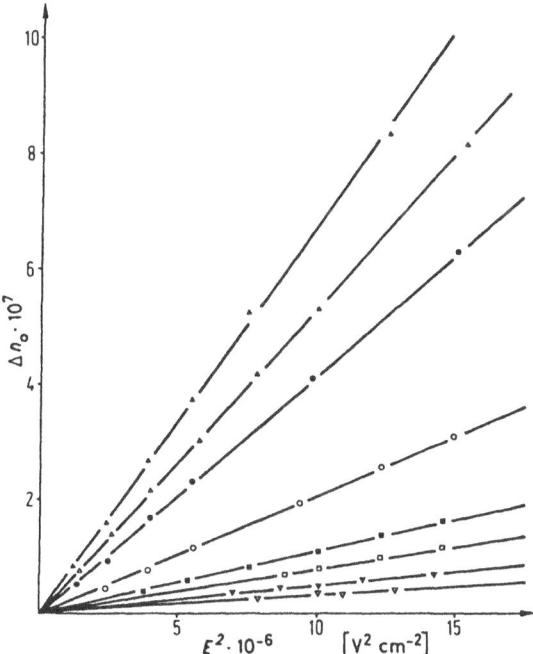

Figure 4. *Steady-state birefringence,* Δn_o, *as a function of the*
square of electric field strength, E^2, *in* $AOT/H_2O/i\text{-}C_8H_{18}$ *solutions,*
$w_o = 74$ *at* 19^oC. *Parameter: concentration of hydrated surfactant,*
C_{AOT+H_2O}, *in g* cm^{-3}: (▲) 0.362; (△) 0.310; (●) 0.272; (o) 0.181;
(■) 0.121; (□) 0.09; (▼) 0.06; (▽) 0.045.

The slopes, $\Delta n_o / E^2$, of the fitted straight lines were used in Equation 1 to calculate K_{SP}^{exp} for various concentrations of hydrated surfactant. The concentration independent intrinsic Kerr constant [K] was then obtained by the extrapolation of K_{SP}^{exp} values to zero concentration, and yielded $[K] = \lim_{c \to o} K_{SP}^{exp} = 4.58 \times 10^{-18} \, m^2 \, V^{-2}$.

a) Estimation of K_{SP}^{dis} by assuming distortional polarization

The optical, $(g_1 - g_2)$, and electrical, $(g_1^E - g_2^E)$, anisotropy factors are related to the properties of the particles and solvent by an expression developed by Peterlin and Stuart[6], which may be simplified as:

$$(g_1 - g_2) = f \, (n_1, \, n_2, \, n, \, p)$$

$$(g_1^E - g_1^E) = f \, (\varepsilon_1, \, \varepsilon_2, \, \varepsilon, \, p)$$

Equation 2

where n_1, n_2, and ε_1, ε_2 are the refractive indices and the dielectric constants, respectively, of the particle.

It can be assumed, in the absence of a permanent dipole moment, that $n_1 = n_2$ and $\varepsilon_1 = \varepsilon_2$, thus neglecting the intrinsic anisotropy of the droplet structure. It follows from Equations 1 and 2* that the main contribution to the birefringence, the so-called form birefringence, is due to the difference between the refractive indices of the droplet, n_1, and solvent, n, and to the axial ratio of the droplet, p, Figure 5. In order to estimate K_{SP}^{dis}, one needs estimates of p, n_1 and ε_1. Hydrodynamic measurements of translational and rotational diffusion constants and viscosity gave a value of $p = 2.7$ for 0.2M $AOT/H_2O/i\text{-}C_8H_{18}$, $w_o = 74$ at 20°C. The refractive index of a microemulsion droplet was obtained using Looyenga's model[7] and n_1 was found to be 1.357[8]. The dielectric constant ε_1, of the droplet was estimated, using Maxwell's relation, $n_1^2 = \varepsilon_1$, and since n_1 was measured at optical frequencies, the ε_1 value yields the distortional polarization. These values can be used to obtain an estimate of $K_{SP}^{dis} = 1.60 \times 10^{-21} \, m^2 \, V^{-2}$. The droplet volume was $1.41 \times 10^{-23} \, m^3$ and the isooctane parameters were $n = 1.389$, $\varepsilon = 1.96$.

However, in order to account for any uncertainty in the estimation of p and n_1, K_{SP}^{dis} was calculated for a range of p and n_1 values, the results being shown in Figure 5. It is seen that K_{SP}^{dis}

---

* The full form of equation 2, together with a typical calculation, is given in Reference 2, pp. 151-158.

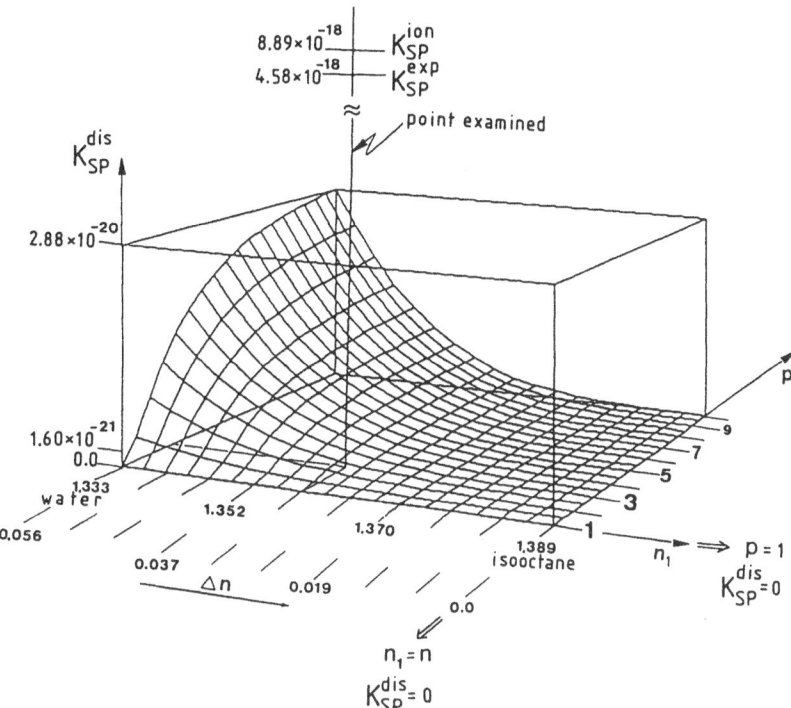

Figure 5. *Variation of form birefringence, expressed as K_{SP}^{dis}, with axial ratio p and $\Delta n = n_1 - n$ ($n_1 = n_{droplet}$, $n = n_{isooctane}$).*

is much smaller than K_{SP}^{exp}, even for extreme values of p and n_1 (10 and 1.333, respectively).

b) Estimation of K_{SP}^{ion} by assuming ionic atmosphere polarization

The model of ionic atmosphere polarization[9] assumes a high surface conductivity of the droplet at the water/oil interface due to the high local concentration of ions. When subjected to an electric field, the ionic atmosphere becomes asymmetrical and consequently the electric anisotropy factor, $(g_1^E - g_2^E)$, increases. An estimate of the order of magnitude of K_{SP}^{ion}, was obtained by assuming that the droplet is a much better conductor than the solvent, $\varepsilon_1 > \varepsilon$. For the same experimental conditions as in a), it was found that

$$K_{SP}^{ion} = 8.89 \times 10^{-18}\ m^2\ V^{-2},$$

and this value is compared with the K_{SP}^{exp} and K_{SP}^{dis} values in Figure 5.

CONCLUSIONS

Distortional polarization alone cannot explain the magnitude of the observed birefringence as is shown by the discrepancy between the theoretically calculated Kerr constant and the experimentally observed one.

It was also observed[10] that the magnitude of the electric birefringence decreases rapidly at low frequencies (<40KHZ) of the applied electric field, a phenomenon which could be expected only at much higher frequencies of the applied electric field if the birefringence were due to distortional polarization alone.

The estimate of the specific Kerr constant obtained by assuming ionic atmosphere polarization is in far better agreement with the experimental findings than is the estimate from distortional polarization.

REFERENCES

1. H. Benoit, Ph D Thesis, University of Strasbourg, France (1950).
2. Z. Markovic, Ph D. Thesis, University of Basel, Switzerland (1980).
3. M. Zulauf and H.F. Eicke, J. Phys. Chem. 83 : 481 (1979).
4. C.G. Garton and Z. Krasucki, Proc. Roy. Soc. A., 280 : 211 (1964).
5. C.T. O'Konski and H.C. Thacher, J. Phys. Chem., 57 : 955 (1953).
6. A. Peterlin and H.A. Stuart, Hand- und Jahrbuch der Chemischen Physik, Vol. 8, Leipzig (1943).
7. H. Looyenga, Physica 31, 410 (1975).
8. R. Kubik, Diplom Thesis, University of Basel (1979).
9. C.T. O'Konski, J. Phys. Chem. 64, 605 (1960); C.T. O'Konski, J. Phys. Chem. 23, 1559 (1955); C.T. O'Konski and A.J. Haltner, J. Am. Chem. Soc. 79, 5634 (1957); M. Eigen and G. Schwarz, J. Colloid Sci. 12, 181 (1957).
10. Z. Markovic, unpublished results.

LIPID POLYMORPHISM, REVERSE MICELLES, AND PHOSPHORUS-31 NUCLEAR MAGNETIC RESONANCE

Joachim Seelig

Biocenter, University of Basel
Klingelbergstrasse 70
CH-4056 Basel, Switzerland

LIPID POLYMORPHISM

Lipid polymorphism is a most unusual phenomenon. Depending on the external conditions such as water content, temperature, presence of proteins or metal ions etc., lipids can form aggregates of quite different <u>long-range</u> orders. Due mainly to the pioneering X-ray diffraction work of V. Luzzati and his co-workers[1] we have obtained a firm understanding of the principles of lipid organization and of the types of possible structures. Two examples are illustrated in Figure 1:

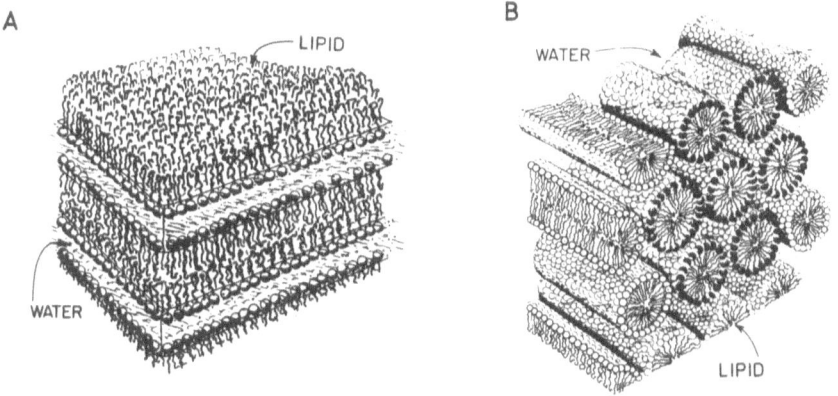

Figure 1. *Lipid polymorphism. (A) Bilayer; (B) cylinders (hexagonal phase). Adapted from reference 32.*

1) The lamellar phase (L$_\alpha$-phase; Figure 1A) is by far the most
dominant lipid structure and is the cornerstone of biological membranes. In this structure the lipids are packed into bilayers which
are separated by layers of water. Such a structural arrangement has
a "linear symmetry" displaying Bragg spacing ratios of 1:1/2:1/3 in
a small-angle X-ray diffraction experiment. 2) As a second example,
Figure 1B provides a picture of lipids arranged in a cylindrical
structure. The rod-like cylinders are surrounded by water shells and
are packed in a hexagonal array. The two-dimensional hexagonal symmetry of this phase (H$_I$-phase) gives rise to Bragg spacing ratios of
$1 : \frac{1}{\sqrt{3}} : \frac{1}{\sqrt{4}} : \frac{1}{\sqrt{7}}$. An inverted hexagonal phase (H$_{II}$) with the same spacing ratios has also been observed. In this type of mesophase a
cylindrical core of water molecules is surrounded by a network of
lipids. This inverted hexagonal phase is closely related to reverse
micellar structures and will be discussed in more detail below.

The transition from a bilayer to a hexagonal phase requires
little energy[2,3] and is indicative of a rather subtle balance of a
variety of different forces (hydrophobic interactions, hydration
forces, electrostatic interactions). Shifting this balance only
slightly as, for example, by the addition of metal ions may induce
a transition from one mesophase to the other.

A necessary prerequisite of lipid polymorphism is the flexibility of the lipid hydrocarbon chains. Only if the hydrocarbon chains
are fluid-like can they adapt easily to the various geometries imposed by the molecular packing. This short-range disorder is also
essential in accommodating the chemical heterogeneity of naturally
occurring lipids.

PHOSPHORUS-31 NUCLEAR MAGNETIC RESONANCE

The present understanding and classification of lipid polymorphism is based almost exclusively on a systematic analysis of X-ray
diffraction patterns of pure lipid model compounds. Indeed, X-ray
diffraction is still the only unambiguous method for the exact determination of the packing arrangement of lipids. However, due to
a number of intrinsic limitations, it is difficult to apply this
method to biological membranes. First, it is necessary to produce
stacks of oriented membranes in order to create a lattice of sufficient size for X-ray scattering experiments. Secondly, even if the
preparation of oriented membranes is experimentally feasible it
would still be difficult to detect the presence of a second phase
as a small 'impurity' in a predominantly bilayer arrangement.

These disadvantages are not encountered with the application

of ^{31}P-nuclear magnetic resonance (NMR) which in the hands of
P. Cullis and B. de Kruijff has led to a new blossoming in the
field of lipid polymorphism[4]. These researchers realized that the
long-range order of lipids is also reflected in the [31]-P-NMR spectra
of non-oriented lipid dispersions. Basically three types of spectra
can be distinguished and are displayed in Figure 2. The simplest
spectrum is encountered with lipid phases in which 'isotropic' motions occur e.g. detergent solutions, micelles, reverse micelles,
small single-walled vesicles, cubic phases. Under these conditions
the intrinsic chemical shielding anisotropy of the phosphate segment is averaged out completely and the NMR spectrum consists of a
single narrow resonance (Figure 2A bottom). In contrast, if the
same lipid molecule is incorporated into a lipid bilayer, the motion of the polar group is more restricted and the chemical shielding anisotropy is only partially averaged out. Coarse dispersions
of lipid bilayers are characterized by a rather broad spectrum as
shown in Figure 2C (top). The separation between the weak low field
shoulder and the intense high field peak is called residual chemical shielding anisotropy $\Delta\sigma$ and amounts to ca.-50 ppm for most bi-

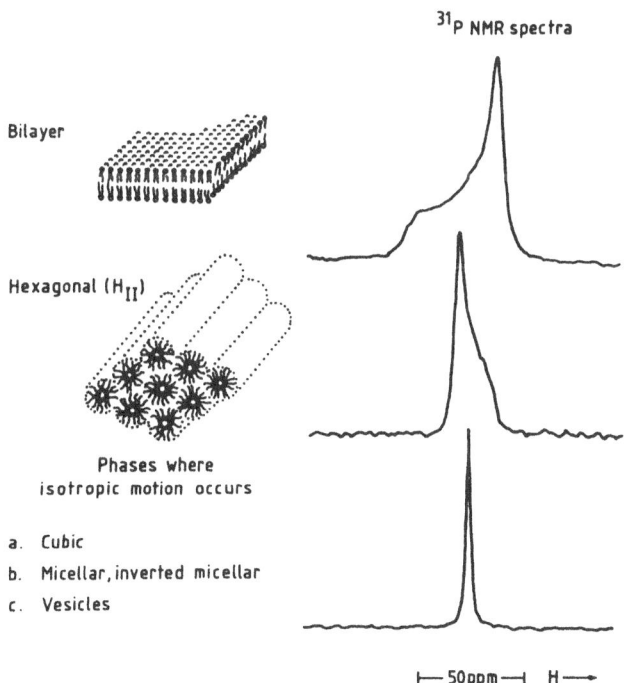

^{31}P NMR spectra

Bilayer

Hexagonal (H$_{II}$)

Phases where
isotropic motion occurs

a. Cubic
b. Micellar, inverted micellar
c. Vesicles

⊢—50ppm—⊣ H—→

Figure 2. ^{31}P-NMR spectra (at 36.4 MHz) characteristic of different lipid phases. Adapted from reference 4. From top to bottom:
C, B, A (see text).

layers and biological membranes investigated to date[5]. Finally, if
there is no change in the headgroup structure, but the lipid is incorporated into a hexagonal phase, the spectrum 2 B (shown in the
middle) is observed. Compared to the lipid bilayer, the chemical
shielding anisotropy of the hexagonal phase has changed its sign
and is reduced in size by exactly a factor of two. The origin of
these ^{31}P-NMR spectra is explained in more detail in references 5
and 6. For practical applications it suffices to note that the three
spectra shown in Figure 2 are simply used as the characteristic
^{31}P-NMR signature of the corresponding phases.

However, it should also be noted that these empirical assignments are not without pitfalls. This is due to the fact that the
appearance of the ^{31}P-NMR spectrum is not only determined by the
packing geometry of the lipids but also by the average orientation
and conformation of the phospholipid head groups[5]. Model calculations have demonstrated that all three types of spectra shown in
Figure 2 can be generated by bilayer dispersions simply by changing
the orientation of the choline dipole with respect to the bilayer
normal[7]. On the other hand, in defense of the approach suggested by
Cullis and de Kruijff it should also be realized that in all cases
where the packing geometry is known from X-ray diffraction the
^{31}P-NMR spectra do conform to the pattern of Figure 2. All experimental evidence obtained with ^{31}P-NMR so far points to a remarkable
constancy of the head group orientation, at least as far as the phosphate segment is concerned.

Compared to X-ray diffraction, the use of ^{31}P-NMR has the advantage that (i) no membrane orientation is required, (ii) the coexistence of a minor isotropic compound in a predominant bilayer-like
structure is easily detected because of the rather narrow resonance,
(iii) the measuring time is short, and (iv) with appropriate precau-
tions the measurements can be made with intact living systems.

TRANSITION FROM A BILAYER TO A HEXAGONAL PHASE

This type of transition is most commonly encountered with unsaturated phosphatidylethanolamines and is illustrated here for two
different examples.

Temperature-induced transition. Dispersions of 1,2-dielaidoyl-
sn-glycero-3-phosphoethanolamine (DEPE) in buffer exhibit two types
of phase transitions. At 35°C they undergo a relatively sharp gel-
to-liquid crystal phase transition which essentially corresponds to
a melting of the hydrocarbon chains. In addition, the system exhibits a more gradual transition from a lamellar to a hexagonal phase

at temperatures between 45 and 60°C[2,8]. This transformation is quite obvious from the ^{2}H- and ^{31}P-NMR spectra shown in Figure 3 (from reference 8). At low temperatures (35–45°C) the ^{31}P-NMR spectra are characterized by a large negative shielding anisotropy and are typical for a lipid bilayer. With increasing temperature this signal gradually disappears and is replaced by a spectrum in which the chemical shielding anisotropy has a positive sign and is reduced in size by a factor of 2, the typical spectrum of a hexagonal phase. The coexistence of a lamellar and a hexagonal phase is revealed even better by the ^{2}H-NMR spectra of Figure 3. The deuterium label is attached at the <u>trans</u> double bond of both elaidic acyl chains of DEPE and at low temperatures only one pair of lines is observed with a separation of about 25 kHz. In the intermediate temperature range

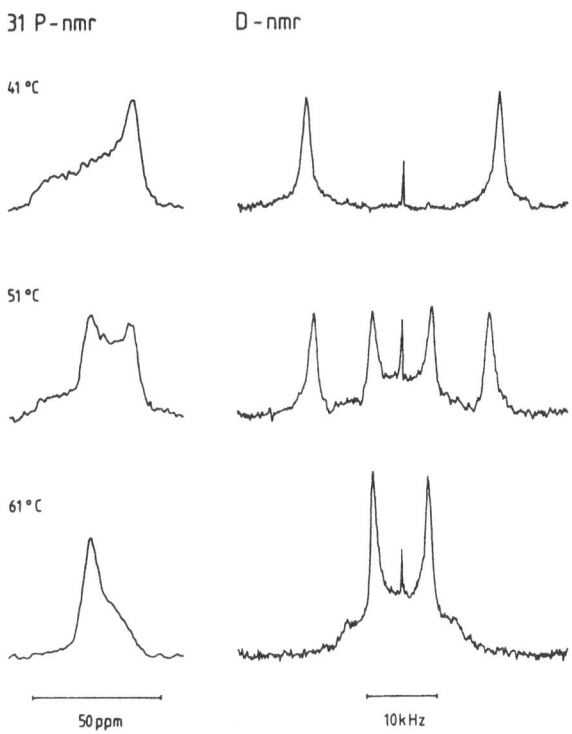

Figure 3. *Lamellar ⟶ hexagonal liquid-crystal phase transition for synthetic 1,2-dielaidoyl-sn-glycero-3-phosphoethanolamine deuterated at carbon atoms 9 and 10 of both fatty acyl chains. The figure shows ^{31}P- and ^{2}H-NMR spectra: 41°C lamellar phase, 51°C coexistence of lamellar and inverted hexagonal phase; 61°C inverted hexagonal phase. From reference 8.*

(45-60°C) the spectrum consists of two quadrupole splittings and the intensity of the smaller splitting of the hexagonal phase grows at the expense of the lamellar signal. Above 60°C only the quadrupole splitting associated with the hexagonal phase is seen. Theoretically, the lamellar → hexagonal transition should reduce the quadrupole splitting by exactly a factor of 2 if the average conformation of the hydrocarbon chains remains unaltered[9]. The experimentally observed reduction is, however, larger and amounts to almost a factor of three. The hydrocarbon interior of the hexagonal phase thus appears to be less ordered than that of the lamellar phase, at least at the level of the trans double bond. In contrast, at the level of the phosphate the reduction in size of the residual chemical shielding anisotropy corresponds to exactly the theoretical factor of 2. This difference can be explained by the different packing constraints encountered in lamellar and hexagonal phases. X-ray investigations suggest that unsaturated phosphatidylethanolamines have a strong tendency to form inverted hexagonal phases[10]. As mentioned above, a cylindrical core is made up of water molecules and the inner aqueous cylinder is surrounded by the lipid polar groups with the fatty acyl chains facing outward, forming a semi-liquid hydrocarbon environment between the aqueous rods. The anchoring of the polar groups in the lipid-water interface can be expected to be similar or identical in both the lamellar and the inverted hexagonal phases, but the hydrocarbon chains of the hexagonal phase must be packed less regularly than those of the bilayer phase. This gain in configurational freedom is reflected in the reduction of the deuterium quadrupole splittings beyond the geometric factor 2.

Protein-induced transition. The modulation of the lipid phase structure by membrane-bound proteins poses an interesting problem. Reconstitution studies with cytochrome c oxidase[11-13] and sarcoplasmic reticulum ATPase[14] using unsaturated phosphatidylcholines have led to the conclusion that the lipids in the reconstituted membranes retain the bilayer organization even at the highest protein concentrations. However, quite different results have been obtained with a small hydrophobic pentadecapeptide, gramicidin, and corresponding ^{31}P-NMR spectra are shown in Figure 4 (taken from reference 15). The lipid used in this study is 1,2-dierucoyl-sn- glycero-3-phosphocholine (22:1$_c$, 22:1$_c$-PC) which in pure form adopts the bilayer structure at all temperatures investigated (Figure 4A). When gramicidin is incorporated in dispersions of this lipid (1:10 molar ratio), it induces a hexagonal phase organization (presumably a H_{II}-hexagonal phase) as is evidenced by the ^{31}P-NMR spectra of Figure 4B. On the other hand, when gramicidin was added to bilayers of dimyristoylphosphatidylcholine (1:10 molar ratio), only ^{31}P-NMR spectra typical of the bilayer organization were observed[16]. These experiments suggest that the balance between the length of the

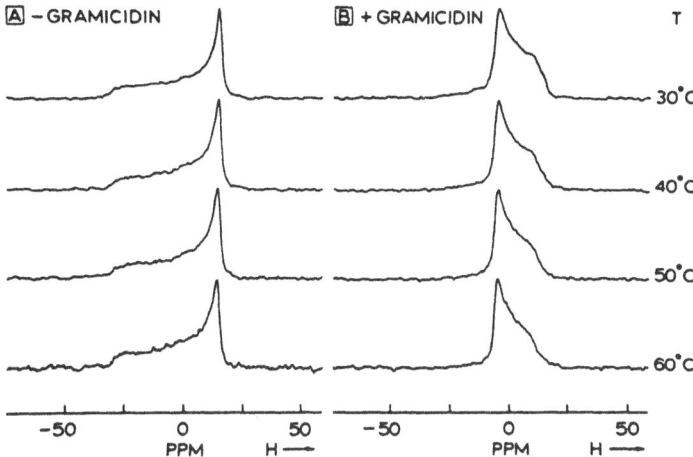

Figure 4. *Protein-induced hexagonal phase. ³¹P-NMR spectra of unsaturated phosphatidylcholine dispersions with and without gramicidin. From reference 15.*

hydrophobic domain and the membrane thickness is of critical importance for the organization of the lipids.

TRANSITION FROM A BILAYER TO AN ISOTROPIC PHASE

This type of transition was first observed around 1978 in ³¹P-NMR studies of microsomal membranes isolated from rat, bovine and rabbit liver[17,18]. The membrane vesicles were derived from the metabolically very active endoplasmic reticulum of the liver tissue and had a very high content of highly unsaturated phospholipids. The ³¹P-NMR spectra suggested that a large part of these phospholipids was undergoing isotropic motion. Based on these findings it was further speculated that the formation of non-bilayer structures, in particular the formation of reverse micelles, would provide a mechanism for the transport of lipids and other substrates across the membrane and could also play an important role in the fusion of two bilayer membranes. This hypothesis was strengthened by the discovery of so-called 'lipidic particles' in pure lipid membranes by electron microscopy[19,20]. Freeze-fracture studies demonstrated that membranes made of lipids, which were capable of forming an inverted hexagonal phase (H_{II}), had pits and particles on their freeze-fracture faces. These lipidic particles with a diameter of 9-11 nm were interpreted as inverted cylindrical micelles (cf. Figure 5c) since it was difficult to reconcile the presence of extended areas of hexagonal phase (H_{II} phase) with the barrier function of the membrane.

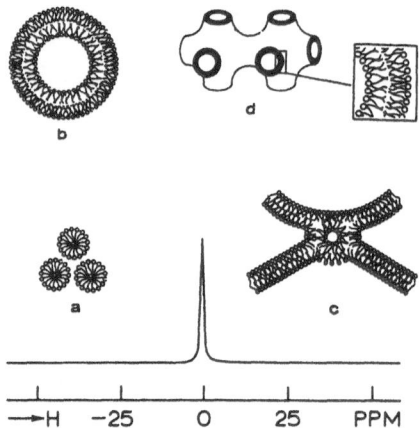

Figure 5. *Phospholipid arrangements which give rise to a narrow*
31P-NMR signal. (a) Micelle; (b) sonicated vesicle; (c) reverse
micelle; (d) cubic phase. From reference 21.

Figure 5 (taken from reference 21) then summarizes a variety
of structures which could give rise to a narrow ('isotropic')
31P-NMR signal. The simplest case are phospholipid micelles
(Figure 5a). Micelles are small and highly dynamic structures and
the chemical shielding anisotropy of the phosphate segment is easily averaged out. Micellar structures are formed from phospholipids
with rather short hydrocarbon chains (< 8 carbon atoms chain length)
or by lysolipids. However, it should be noted that lysolipids mixed
together with equimolar amounts of free fatty acids form stable bilayer structures[22]. Extensive sonication of phospholipid dispersions produces small single-walled bilayer vesicles of about 20-30 nm
diameter. The lateral diffusion around the highly curved bilayer surface as well as the rapid tumbling of the vesicle itself (Figure 5b)
also lead to an isotropic 31P-NMR signal. A detailed study of the
variation of the shape of the 31P-NMR signal with the size of the
vesicles has been published[23]. Further possibilities for isotropic
averaging of the 31P shielding tensor are the inverted micelle
(Figure 5c) and the cubic phase (Figure 5d)[24]. Figure 5 therefore
demonstrates that the observation of an isotropic 31P-NMR signal
does not allow an unambiguous interpretation in terms of one particular lipid structure.

Isotropic 31P-NMR lines have now been observed under a variety
of experimental conditions. Figure 6 shows a comparison between
1,2-dipalmitoyl-sn-glycero-3-phosphocholine (DPPC) and 1-palmitoyl-
lyso-phosphatidylcholine (taken from reference 25). Dispersions of
DPPC in water are always organized in the bilayer structure and the

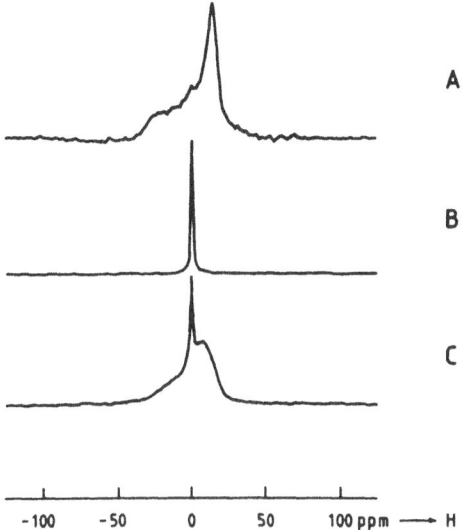

Figure 6. *Isotropic phospholipid motion ³¹P-NMR spectra (at 81 MHz) Aqueous dispersions of (A) 1,2-dipalmitoyl-sn-glycero-3-phospho-choline at 45°C; (B) 1-palmitoyl-lyso-sn-glycero-3-phosphocholine at 25°C; (C) 1-palmitoyl-lyso-sn-glycero-3-phosphocholine at -10°C. From reference 25.*

corresponding spectrum (Figure 6A) has the characteristic signature of a fluid lipid bilayer. In contrast, when lysophosphatidylcholine is dispersed in water, this leads to the formation of micelles resulting in a narrow ³¹P-NMR signal, as shown in Figure 6B. However, if lysophosphatidylcholine is cooled to -10°C, the resulting ³¹P-NMR spectrum can be regarded as a superposition of a bilayer spectrum and an isotropic signal (Figure 6C). The coexistence of an extended bilayer structure with an isotropic phase has also been encountered in many other lipid model systems. Unfortunately, the data are not easily summarized, since the phenomenon of lipid polymorphism is apparently dependent on many different parameters, the significance and interrelationship of which is still not understood. References 26-28 may be used as key literature for earlier work in this field.

BIOLOGICAL RELEVANCE OF NON-BILAYER STRUCTURES

The use of ³¹P-NMR has made it possible to investigate a large variety of phospholipid model membranes, lipid mixtures, and biological membranes, much more than would ever have been possible with X-ray diffraction. These studies may be summarized as follows:
1) The phenomenon of lipid polymorphism has been clearly established

for pure lipid model systems. A large variety of conditions are
known to date which induce non-bilayer structures. Both the hexagonal phase and the isotropic phase(s) have been observed with lipid-
water dispersions. The hexagonal ^{31}P-NMR signals are generally as-
cribed to an inverted hexagonal phase (H_{II}), the isotropic signal
to inverted micelles or cubic structures. Both assignments have a
high degree of probability but are not unambiguous.

2) The existence of non-bilayer structures in biological membranes
is by no means established, in spite of earlier claims to the contrary. A priori one would expect that most of the membrane bound
phospholipids are organized in a bilayer structure since, after all,
the main function of a membrane is that of a barrier. For this
reason, the H_{II} phase appears to be the least attractive for a biological membrane, since it destroys the barrier properties over an
extended area. In fact, no experimental evidence for the occurrence
of an hexagonal phase (H_{II}) in a biological membrane has been reported to date.

3) From a geometric point of view reverse micelles would be ideal
candidates to rationalize membrane metabolic processes such as membrane fusion, exocytosis, endocytosis,etc. Indeed, all non-bilayer
^{31}P-NMR signals observed so far in biological membranes exihibit the
characteristics of isotropically moving phospholipids, suggesting
the involvement of reverse micellar structures. However, for the
same reasons as given above, these defects should occur at relatively low concentrations in order not to destroy the integrity of the
membrane barrier. Some of the early observations that a large percentage of the membrane lipids are participating in non-bilayer
structures are difficult to understand.

4) A major difficulty in obtaining reliable ^{31}P-NMR spectra of metabolically active cells is the relatively rapid decomposition of the
cell membrane. ^{31}P-NMR studies of rat liver mitochondria first suggested that an appreciable amount of the phospholipids was undergoing isotropic motion[29]. When these experiments were repeated under
more carefully controlled conditions, neither isotropic nor hexaag-
onal type of phospholipid ^{31}P-NMR signals were observed[30].

5) Based on freeze-fracture electron microscopy, it has recently
been proposed that the so-called 'tight junctions' between adjacent
cells of epithelial tissue do not consist of proteins but of lipidic
particles. It is postulated that the individual tight junction
strands are pairs of inverted cylindrical micelles sandwiched between linear fusions of the external membrane leaflets of adjacent
cells, as shown diagrammatically in Figure 7[31]. This is an interesting speculation, but clear experimental proof is still missing.

In conclusion, the ^{31}P-NMR studies have led to a much deeper

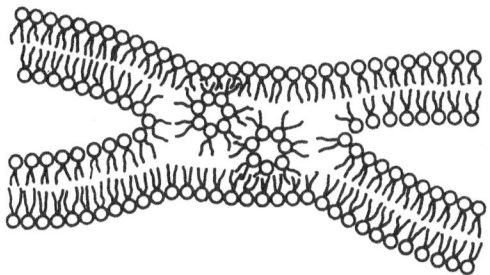

Figure 7. *Proposed organization of the phospholipids at a tight junction strand. From reference 31.*

understanding of lipid polymorphism in lipid model systems. For biological membranes, the picture remains diffuse and more work is needed before the existence of non-bilayer structures is convincingly demonstrated in intact membranes.

REFERENCES

1. V. Luzzati, Biological Membranes Vol. 1, 71 (D. Chapman, ed.) Academic Press, New York (1968).
2. J. Stenius, J.B. Rosenholm and M.J. Hakala, Colloid Interface Sci. (Proc. Int. Conf.) 2 : 397 Academic Press, New York (1976).
3. P.R. Cullis and B. de Kruijff, Biochim. Biophys. Acta 513 : 31 (1978).
4. P.R. Cullis and B. de Kruijff, Biochim. Biophys. Acta 559 : 399 (1979).
5. J. Seelig, Biochim. Biophys. Acta 505 : 105 (1976).
6. W. Niederberger and J. Seelig, J. Amer. Chem. Soc. 98 : 3704 (1976).
7. A.M. Thayer and S.J. Kohler, Biochemistry 20 : 6831 (1981).
8. H.U. Gally, G. Pluschke, P. Overath and J. Seelig, Biochemistry 19 : 1638 (1980).
9. J. Seelig, Quart. Rev. Biophys. 10 : 353 (1977).
10. R.P. Rand, D.O. Tinker and P.G. Fast, Chem. Phys. Lip. 6 : 333 (1971).
11. A. Seelig and J. Seelig, Hoppe Seyler's Z. Physiol. Chem. 359 : 1747 (1978).
12. L.K. Tamm and J. Seelig, Biochemistry, 22 : 1474 (1983).
13. S.Y. Kang, H.S. Gutowsky, J.C. Hsung, R. Jacobs, T.E. King, D. Rice and E. Oldfield, Biochemistry 18 : 3257 (1979).
14. J. Seelig, L. Tamm, L. Hymel and S. Fleischer, Biochemistry 20 : 3922 (1981).

15. C.J.A. van Echteld, N. de Kruijff, A.J. Leunissen-Bijvelt and J. de Gier, Biochim. Biophys. Acta 692 : 126 (1982).

16. D. Rice and E. Oldfield, Biochemistry 18 : 3272 (1979).

17. A. Stier, S.A.E. Finch and B. Bösterling, FEBS Lett. 91 : 109 (1978).

18. N. de Kruijff, A.M.H.P. van den Besselaar, P.R. Cullis, H. van den Bosch and L.L.M. van Deenen, Biochim. Biophys. Acta 514 : 1 (1978).

19. A.J. Verkleij, C. Mombers, J. Leunissen-Bijvelt and P. Ververgaert, Nature 279 : 162 (1979).

20. A.J. Verklij, C. Mombers, W.J. Gerritsen, L. Leunissen-Bijvelt and P.R. Cullis, Biochim. Biophys. Acta 555 : 358 (1979).

21. C.J.A. van Echteld, Ph. D. Thesis, University of Utrecht (1982).

22. M.K. Jain, C.J.A. van Echteld, F. Ramirez, J. de Gier, G.H. de Haas and L.L.M. van Deenen, Nature 284 : 486 (1980).

23. E.E. Burnell, P.R. Cullis and N. de Kruijff, Biochim. Biophys. Acta 603 : 63 (1980).

24. G. Lindblom, K. Larsson, L. Johansson, K. Fontell and S. Forsén, J. Amer. Chem. Soc. 101 : 5465 (1979).

25. C.J.A. van Echteld, B. de Kruijff, J.G. Mandersloot and J. de Gier, Biochim. Biophys. Acta 649 : 211 (1981).

26. B. de Kruijff, A.J. Verkleij, J. Leunissen-Bijvelt, J. Hille, C.J.A. van Echteld and H. Rijnbout, Biochim. Biophys. Acta 693 : 1 (1982).

27. C.P.S. Tilcock, M.B. Bally, S.B. Farren and P.R. Cullis, Biochemistry 21 : 4596 (1982).

28. T.F. Taraschi, A.T.M. van der Steen, B. de Kruijff, C. Tellier and A.J. Verkleij, Biochemistry 21 : 5756 (1982).

29. P.R. Cullis, B. de Kruijff, M.J. Hope, R. Nayar, A. Rietveld and A.J. Verkleij, Biochim. Biophys. Acta 600 : 625 (1980).

30. B. de Kruijff, R. Nayar and P.R. Cullis, Biochim. Biophys. Acta 684 : 47 (1982).

31. B. Kachar and T.S. Reese, Nature 296 : 464 (1982).

32. F.B. Rosevear, J. Soc. Cosmetic Chem. 19 : 581 (1968).

ON THE STRUCTURE, DYNAMICS AND POSSIBLE FUNCTIONAL ROLES OF

INVERTED MICELLES IN BIOLOGICAL MEMBRANES

Mauricio Montal

Departments of Biology and Physics
University of California San Diego
La Jolla, California 92093 USA

In the past few years, we have focused our studies on the structure/function correlates in membrane receptor proteins. At the core of this endeavor reside the fundamental consequences of the interaction between the different membrane constituents namely proteins, phospholipids, counter ions, and water. The distinct modalities of the structure and function of biological membranes arise from the dynamics and stability of the interactions between these distinct components.

Lipids and proteins, the two most important membrane constituents, share in common the property of amphipathy, which in turn arises from the properties of water (cf. 1). The mode of association between them thus depends on the regions of the molecules that are in contact. It is known that electrostatic interactions between the lipid polar head groups and the charged groups of a protein depend on ion-pair formation; likewise, short range hydrophobic associations result from the juxtaposition of the lipid hydrocarbon chains to the nonpolar side chains of proteins (cf. 2). These interactions are known to give rise to the formation of stable complexes recognized as lipid protein complexes. Lipid protein complexes in membranes can be disassociated with detergents or by modifications of the ionic environment, such as changes in ionic strength or by the presence of divalent cations.

In 1916, Clowes[3], established that an alteration of the ratio of univalent to divalent cations in favor of the divalent ones results in a phase inversion from an "oil in water" state to a "water in oil" state. This phase inversion suggests that transi-

tions in the location of a lipid-protein complex could be modulated
by changes in the electrostatic and hydrophobic interactions between
the protein and the lipid, bringing about effectively a partition of
the lipid protein complex (as inverted micelle) into the hydrophobic
core of the membrane lipid bilayer. It is known that ions[4] and ion-
pairs[5] exist in hydrocarbon solvents. Hence, it is likely that ion-
pairs between lipids and proteins exist in the apolar regions of
membranes. Furthermore, considering that the recombination rates of
ion-pairs in liquids of low dielectric constant are in the nanosecond time range[6], such transitions (lipoprotein → proteolipid) in
location of the lipid protein complex could be extremely fast and
thus provide a mechanism for the regulation of membrane functions,
as will be discussed below.

This brief speculative essay is centered around the following
question: Do inverted micelles of lipid and protein exist in biological membranes? We have obtained evidence in favor of this hypothesis by studying a model system: Integral membrane proteins associated with phospholipids are transferable into organic solvents by
the use of suitable counter ions, which serve to neutralize the
complex (for a recent review see Ref. 7). Typically, the positive
ions have been divalent cations, primarily calcium, which reacts
with the negatively charged phospholipids. A general strategy was
devised for the extraction of protein lipid complexes into organic
solvents and has been applied to several membrane proteins among
them cytochrome-c-oxidase[8], the visual pigment rhodopsin from verte-
brate[9] as well as from invertebrate[10] photoreceptor membranes, bac-
teriorhodopsin[11,12], and photosynthetic reaction centers[13-15].
These complexes have been used to form planar lipid bilayers and
lipid vesicles[8], cf.[7] in order to study the mechanism of action of
the protein at the membrane level.

The rhodopsin phospholipid complex and the reaction center
phospholipid complex in organic solvents have been studied in extensive detail concerning their formation and their photochemical
characterization. It was demonstrated that rhodopsin in hexane conserves the characteristic spectral features it displays in native
retinal rod disc membranes[16]: The difference spectra between dark
and irradiated samples showed a wavelength of maximum absorbance at
∿500 nm and an isosbestic point ∿415 nm. Furthermore, bleached
rhodopsin in hexane, when supplemented with 11-cis retinal, regen-
erated isorhodopsin with a yield ≥ 90%. Using low temperature
spectroscopy to study the bleaching sequence, the following intermediates were observed: bathorhodopsin, lumirhodopsin, metarhodopsin I, metarhodopsin II and metarhodopsin III[16]. The results established the spectral integrity of the rhodopsin lipid complex in
hexane. Purified reaction centers from photosynthetic bacteria were

transferred as a complex with phospholipids into hexane and characterized photochemically[13]. In the visible and infrared regions, the
absorbance spectrum of the hexane extract was similar to that of reaction centers in micellar detergent solution. The light-induced absorbance changes were only $\sim$5% of the expected values; however, the
photoactivity was restored by adding exogenous ubiquinone UQ-10, indicating that endogenous quinones were extracted by hexane resulting
in an apparent loss of activity[13-15]. EPR spectroscopy was also used
to evaluate the activity of the reaction center lipid complex in
hexane. It was found that the light-induced EPR absorbance changes
and the low temperature kinetics of the back reaction in the dark
were similar to those measured for reaction centers in the chroma-
tophore membrane or when the reaction centers were purified[13].

It became evident that supplementary phospholipids and overall
charge neutralization were required in order to favor the partition
of functional proteins into organic solvents [8,9,13, cf.7]. This
suggested that the transfer process occurs through a phase inversion where the hydrophilic sections of the lipid and protein, which
interacted before with the bulk aqueous phase, are enclosed in a
polar core that traps water and is surrounded by the phospholipid
acyl chains, which in turn are in contact with the organic solvent[8,9, cf.7].

Recently, we obtained information about the structure of these
protein lipid complexes in hexane which supports the view that the
process of extraction proceeds through a phase inversion and vali-
dates the notion that inverted micelles of membrane proteins and
phospholipids indeed exist in organic solvents[17]. Briefly, a small
angle X-ray scattering study of the rhodopsin lipid complex in
hexane demonstrated the presence of two distinct particle populations with corresponding radii of gyration of approximately 22 Å
and 160 Å. The radius of gyration of a lipid complex in hexane,
which was prepared in a manner identical to the rhodopsin lipid
complex but from which the protein was omitted, was about 22 Å.
This radius of gyration is comparable to that of the first component obtained from the rhodopsin lipid complex in hexane. It is
known that both the rhodopsin lipid complex in hexane as well as
the lipid extract prepared in the absence of rhodopsin, contain
divalent cations and water[16,13]. Therefore, the most plausible
structure for phospholipid complexes in hexane is that of an inverted micelle, that is a core of water and ions surrounded by a shell
of the phospholipid phosphates, with the phospholipid hydrocarbon
chains exposed to the apolar solvent. This structure is schematically illustrated in Figure 1A. In terms of this model, since the
main contribution to contrast arises from the phosphate groups,
the radius of gyration would be essentially the radius of the

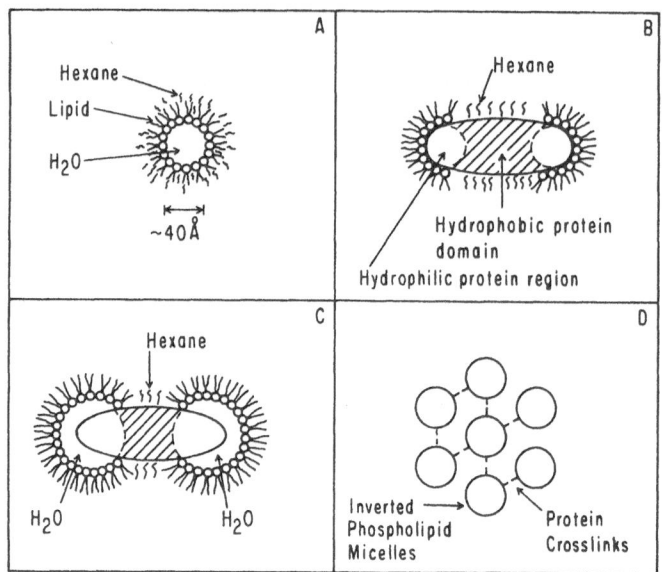

Figure 1. *Schematic representation of the putative structures existent in the hexane extract. A) Inverted phospholipid micelles with a hydrophilic core. B) Anhydrous complexes of a rhodopsin monomer and lipids where polar moieties of both constituents inter-act and where the protein hydrophobic domain is in direct contact with hexane. C) Hydrated rhodopsin-lipid complexes where the pro-tein acts as a crosslink between two inverted phospholipid micelles. D) Aggregates depicted at low resolution where the bars represent the protein crosslinks between the inverted phospholipid micelles (illustrated as circles)[42].*

phosphate shell. This evidence is indicative of the presence of in-verted micelles of a radius of $\sim$20 Å. It corresponds to $\sim$100 lipid molecules per micelle, considering a packing density of $\sim$1 lipid molecule per 60 Å^2 of surface area[18]. This model is consistent with the composition of the sample: the calculated weight ratio of lipid to water for the model is 2.8 assuming a molecular weight for the phospholipid of 750; the measured weight ratio was 2.6.

The presence of the protein results in the observation of a second component in the Guinier plot, which corresponds to parti-cles that have a root mean squared radius of gyration greater than 160 Å. The simplest plausible model that explains both the chemical composition and the known functional properties of the rhodopsin lipid complex in hexane was presented[17]. It was considered that the protein acts as a cross-link between inverted phospholipid micelles as illustrated in Figure 1C. Since a given micelle could contain more than one protein molecule, the formation of large aggregates

was readily accounted for: they could be formed by a large number of
micelles linked by protein cross-links (Figure 1D). A lower limit for
the radius of gyration of the aggregates is about 160 Å. This corresponds to a volume of 1.7×10^7 Å^3. Considering that an individual
micelle has a radius of about 20 Å, and that the hydrocarbon chains
project out another 20 Å, a typical value for biomembranes, it was
estimated that each micelle had a radius of ∿40 Å. The number of such
micelles in an aggregate would be about 200 to 300. In this model the
protein molecule is locally in a bilayer environment, with the hydrophilic domains buried in the micellar aqueous core and the hydrophobic sections in the apolar region between micelles. This would explain why the spectral characteristics of rhodopsin in the rhodopsin
lipid complex are remarkably well preserved. The model accounts for
the large lipid to protein ratio necessary for the extraction of the
protein to the solvent and for the relatively large water content of
the complex in hexane (see also 13). Thus, the model based on the
existence of inverted lipid micelles surrounding the protein polar
moieties while the protein hydrophobic domains act as cross-links
between the inverted micelles[17] accounts for the properties known
for the rhodopsin-[16] and for the reaction center-[13] lipid complexes
in hexane.

Thus, we established the existence of inverted micelles of membrane proteins and phospholipids in organic solvents and that the
complexes exhibit in the organic solvent functional properties which
are associated to the protein in the intact biological membrane.
Therefore, one is tempted to speculate on the dynamics and the possible functional roles of inverted micelles in biological membranes.

It has been recognized for several years that the negatively
charged acidic phospholipids, phosphatidylserine, phosphatidylinositol, and cardiolipin, could undergo phase inversions in the presence of both monovalent and divalent cations[19-21]. This observation
led to the proposal that acidic phospholipids could operate as ion-
ophores or specific transporters of ions across the membrane[19-22].
In particular, calcium ions have been invoked to be involved in such
a process. Gangliosides, which are water soluble sialomucolipids,
are acidic and are known to bind calcium, were shown to form ion-
pairs with sodium and with potassium[23,24]. Ion-pair formation occurs
with the carboxyl-group of the N-acetylneuraminic acid residues[23,24].
Galactolipids, which account for ∿75% of the total lipid of the
chloroplast membrane, have been demonstrated to form inverted micelles[25].

Considerable attention has been focused on the notion that the
particles observed in freeze fracture electron micrographs of lipid
vesicles containing particularly acidic phospholipids may represent
lipidic particles which actually are intrabilayer inverted micelles.

This can be considered as the morphological evidence in favor of the
existence of inverted micelles in membranes the origin of which
might likely be the formation of ion-pairs[26-31]. Recent freeze fracture studies on the nature of the tight junction strands have suggested that in the formation of tight junctions a phase transition
of the planar lipid bilayer might be involved in which inverted
cylindrical micelles are formed[32]. This suggestion is in contrast
to the widely held view that intramembraneous particles present in
freeze fractured replicas of tight junction strands represent integral membrane proteins[33]. The lipidic nature of the tight junction
strands and the occurrence of particles in liposomes formed from
exclusively phospholipids and other lipids, suggest that the existence of inverted micelles in membranes may be more conspicuous
than previously realized. As has been proposed previously by Cullis
et al.[31], inverted micelles may be involved in a variety of cellular
membrane functions such as exo- and endocytosis, membrane fusion, the
movement of lipids across the bilayer, the transport of ions across
the bilayer and perhaps many other processes yet undetected.

The possible regulatory role in cellular functions that inverted micelles may play is suggested by the association of phosphatidylinositol with a variety of calcium dependent cellular responses.
As reviewed by Michel[34,35] and Putney[36] there appears to be a correlation between the breakdown of the phosphorylated forms of phosphatidylinositol and the mobilization of calcium in hormone or neuro-
transmitter stimulated cells. Although this view has been ques-
tioned[37], it is conceivable that the phosphorylated forms of phosphatidylinositol in the presence of calcium form inverted micelles
which, in due turn, could regulate the transport of calcium or other
ions across the membrane, or through ion-pair formation with regulatory proteins would affect the extent of embedding of the membrane
protein in the hydrophobic core of the bilayer thus resulting in a
net change in the activity of the particular protein. Candidates
among the proteins considered to be regulated by phospholipids and
calcium are adenylate cyclases, G-proteins, the phosphorylation of
ion channels and the phosphorylation of hormone receptors, neuro-
transmitter receptors and photoreceptors (e.g. 38).

Another possible functional role for inverted micelles is in
the process of protein insertion into membranes. The insertion of
bacterial toxins into phospoholipid bilayers depends on the occurrence of specific phospholipids in the bilayer. For example, the
insertion of diphtheria toxin into and across planar phospholipid
bilayers depends exquisitely on the location of phosphatidylinositol
or phosphatidylinositol phosphate in the compartment opposite to
that in which the toxin is initially added[39]. In contrast, tetanus
toxin is incorporated into planar bilayers strictly under conditions

in which the putative lipid receptor, gangliosides or a special ganglioside GD1b[40], is a component of the monolayer facing the compartment to which tetanus toxin is initially added[41]. These results indicate that phospholipids can also play the role of receptors and through strong interactions with proteins, such as the toxins, can mediate the specific insertion of proteins into and across membranes. The impact of these observations may become more apparent in the near future.

CONCLUDING REMARKS

In this brief essay, the evidence for the formation and the occurence of inverted micelles of lipids and membrane proteins was reviewed. X-ray scattering studies demonstrated the occurrence of inverted micelles of rhodopsin and lipid in hexane, under conditions where the protein retains its known spectroscopic behavior. A variety of possible functional roles for inverted micelles in biological membranes was considered. It emerges that inverted micelles in biomembranes may be of more significance than previously recognized.

ACKNOWLEDGEMENTS

The model presented in Figure 1 and the experiments that led to it were done in collaborations with V. Ramakrishnan and A. Darszon. This investigation was supported by a research grant from the National Institutes of Health (EY-02084).

REFERENCES AND NOTES

1. C. Tanford (1980) The Hydrophobic Effect: Formation of Micelles and Biological Membranes. New York, Wiley-Interscience, 2nd Ed.
2. M. Montal, Ann. Rev. Biophys. Bioengineer 5 : 119 (1976).
3. G.H.A. Clowes, J. Phys. Chem. 20 : 407 (1916).
4. A. Gemant, Ions in Hydrocarbons. New York, Wiley, pp. 261 (1962).
5. J. Smid, Angew. Chem. Int. Ed. Engl. 11 : 112 (1972).
6. P.K. Ludwig, J. Chem. Phys. 50 : 1787 (1969).
7. M. Montal, A. Darszon and H. Schindler, Quart. Rev. Biophys. 14 : 1 (1981).
8. M. Montal, In: Perspectives in Membrane Biology (ed. S. Estrada-O and C. Gitler). New York, Academic Press, pp. 591-622 (1974).
9. A. Darszon, M. Philipp, J. Zarco and M. Montal, J. Membrane Biol. 43 : 71 (1978).
10. A. Darszon, C.A. Vandenberg, M. Schönfeld, M.H. Ellisman, N.C. Spitzer and M. Montal, Proc. Natl. Acad. Sci. USA 77 : 239 (1980).

11. S.-B. Hwang, J.I. Korenbrot and W. Stoeckenius, J. Membrane
    Biol. 36 : 115 (1977).
12. E. Bamberg, N.A. Dencher, A. Fahr and M.P. Heyn, Proc. Natl.
    Acad. Sci. USA 78 : 7502 (1980).
13. M. Schönfeld, M. Montal and G. Feher, Biochemistry, N.Y.
    19 : 1535 (1980).
14. M.W. Kendall-Tobias, H. Celis, S.A. Celis and A.R. Crofts,
    Biochim. Biophys. Acta 6535 : 585 (1981).
15. N.K. Packham, C. Packham, P. Mueller, D.M. Tiede and P.L. Dutton,
    FEBS Lett. 110 : 101 (1980).
16. A. Darszon, R. Strasser and M. Montal, Biochemistry N.Y.
    18 : 5205 (1979).
17. V.R. Ramakrishnan, A. Darszon and M. Montal, J. Biol. Chem.
    258 : 4857 (1983).
18. D.O. Shah and J.H. Schulman, J. Colloid Interface Sci. 25 : 107
    (1967).
19. H.L. Rosano, J.H. Schulman and J. Weisbuch, Ann. N.Y. Acad. Sci.
    92 : 457 (1961).
20. M.P. Blaustein and D.E. Goldman, J. Gen. Physiol. 49 : 1043
    (1966).
21. J.H. Moore and R.S. Schechter, Nature (London) 222 : 476 (1969).
22. D.E. Green, M. Fry and G.A. Blondin, Proc. Natl. Acad. Sci. USA
    77 : 257 (1980).
23. H. McIlwain, Biochem. J. 78 : 24 (1961).
24. M.W. Spence, Can. J. Biochem. 47 : 735 (1969).
25. A. Sen, W.P. Williams, A.P.R. Brain, M.S. Dickens and P.J. Quinn,
    Nature (London) 293 : 488 (1981).
26. B. de Kruijff, et al., Biochim. Biophys. Acta 555 : 200 (1979).
27. B. de Kruijff, P.R. Cullis and A.J. Verkleij, Trends Biochem.
    Sci. 5 : 79 (1980).
28. A.J. Verkleij, C. Mombers, W.J. Gerritson, J. Leunissen-Bihuelt
    and P.R. Cullis, Biochim. Biophys. Acta 555 : 358 (1979).
29. A.J. Verkleij, C. Mombers, J. Leunissen-Bijuelt and
    P.H.J. Th. Ververgaert, Nature (London) 279 : 162 (1979).
30. P.R. Cullis and M.J. Hope, Nature (London) 271 : 672 (1978).
31. P.R. Cullis, B. de Kruijff, M.J. Hope, R. Nayar and S.L. Schmid,
    Can. J. Biochem. 58 : 1091 (1980).
32. B. Kachar and T.S. Reese, Nature (London) 296 : 464 (1982).
33. L.A. Staehelin, Int. Rev. Cytol. 39 : 191 (1974).
34. R.H. Michel, Trends Biochem. Sci. 4 : 128 (1979).
35. R.H. Michel, Nature (London) 296 : 492 (1982).
36. J.W. Putney, Life Sci. 29 : 1183 (1981).
37. J.N. Hawthorne, Nature (London) 295 : 281 (1982).
38. C.A. Vandenberg and M. Montal, Biophys. J. 37 : 195a (1982).
39. J.J. Donovan, M.I. Simon and M. Montal, Nature (London)
    298 : 669 (1982).
40. W.E. Van Heyningen, J. Gen. Microbiol. 31 : 375 (1963).

41. H. Borochov-Neori, U. Staerz, E. Yavin and M. Montal, Biophys. J. 41 : 381a (1983).
42. From Ramakrishnan, Darszon and Montal (1983), reproduced with permission of the American Society of Biological Chemists, Inc.

THE FORMATION OF REVERSE MIXED MICELLES CONSISTING OF MEMBRANE

PROTEINS AND AOT IN ISOOCTANE

J. Wirz and J.P. Rosenbusch

Biozentrum der Universität Basel

CH-4056 Basel (Switzerland)
and
European Molecular Biology Laboratory
D-6900 Heidelberg (Germany)

INTRODUCTION

Membrane proteins are amphipathic molecules with hydrophobic and hydrophilic surface areas. In vivo, they are integrated into membranes with their hydrophobic surfaces exposed to the core of lipid bilayers (Figure 1A). The domains protruding on either or both sides of the membrane are in contact with the aqueous phase; they are likely to exhibit polar character. Due to their hydrophobicity, integral membrane proteins are insoluble in aqueous media and aggregate to form precipitates. They can be solubilized in water by detergents with which they form complexes with overall polar properties (Figure 1B), whereby solubilization is defined as dissociation into monodisperse, solvated units. In principle, it should be possible to achieve this also in nonpolar solvents (such as isooctane) with detergents known to form reverse micelles. Aerosol OT (AOT) has been used extensively, and its reverse micelles have been characterized[1]. A membrane-spanning protein with 2 polar caps as part of reverse micelles may be visualized as shown in Figure 1C.

Studies of proteins in reverse micelles have been numerous, but they have focused primarily on the activities of the enzymes dissolved in the aqueous inner phase of reverse micelles[2,3]. A characterization of the membrane proteins within reverse micelles could have the following advantages.

231

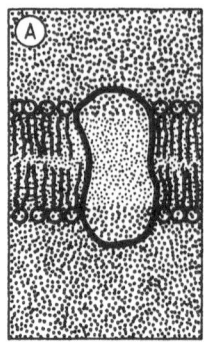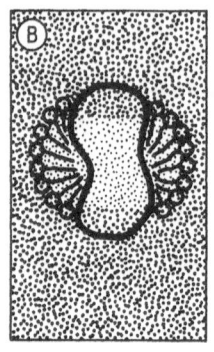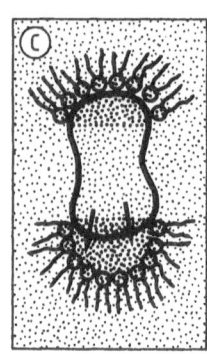

Figure 1. *A transmembrane protein (porin) under various conditions. Lightly shaded areas indicate hydrophobic, darker areas polar regions of protein, amphiphiles and solvents. A, a transmembrane protein integrated into a lipid bilayer. The protein, shown as a dumbbell, is surrounded by phospholipids, with their alkyl chains in the membrane core and their polar heads (indicated by spheres) at the interface. B and C, solubilized protein-detergent complexes. B, in water with octyl-oligooxyethelene (octyl-POE) as detergent. C, in the apolar solvent isooctane, with AOT as surfactant. Evidence for the structure shown in B has been provided by quasielastic light- and neutron-scattering studies[4,5]. Model C is a tentative concept, serving as a working hypothesis in the present study. The amount of water trapped (arrows) is unknown.*

     i) The hydrophobic surfaces of integral membrane proteins are buried in an aqueous phase within membranes or detergent micelles. The chemical modification of the membrane proteins is therefore limited. In reverse micelles, these areas would be exposed to the bulk phase, if solubilization can be achieved in such a system.

     ii) Knowledge of the forces governing the internal structure of membrane proteins is still rather limited. Usually they are conceived to be hydrophobic; the role of electrostatic interactions is unclear.

     iii) Crystallization of membrane proteins could be attempted from organic solvents.

     iv) Several membrane proteins form channels whose states (opened or closed) depend on transmembrane potentials. On a molecular level, this is likely to reflect the effect of an electric field on a dipole moment inherent in the protein. It would be very interesting, therefore, to determine whether dipole moments of such proteins change under various conditions. Dielectric dispersion measurements would allow the determination of dipole moments as well as the size and shape of protein-detergent complexes. We have

chosen porin, a protein spanning outer membranes of E.coli as the
model system to study the properties of mixed protein-detergent complexes in organic solvents. The reasons for this choice are manifold.
The protein is well-characterized. Pertinent in the present context
are the following aspects: in an aqueous phase, the protein forms
monodisperse complexes with detergents (e.g. octylpentaoxyethylene).
The model given in Figure 1B is based on experimental results using
both conventional techniques and quasielastic laser and neutron
scattering on pure micelles and protein-detergent complexes[4],[5].
Such complexes have been crystallized successfully[6], yielding structural resolution to 2.9 Å. Its function, the formation of water-
filled pores across outer membranes, has been studied in detail:
its channels exist in at least two configurations, open and closed,
which are in equilibrium with each other. The equilibrium is dependent on the transmembrane potential[7]. Furthermore, the protein
is very stable; if prepared in homogeneous form by a negative purification scheme[8], it exists as protein aggregates which are functional[7] even after treatment with a large number of organic solvents[9], such as alcohols, ethers, alkanes, benzene, toluene, acetone, DMSO etc. (the only known exception being phenol).

The immediate questions raised here are the following:
i) Can the protein be dissolved in isooctane using AOT or other
detergents?
ii) If so, is the resulting solution monodisperse?
iii) Can the dipole moment of the protein be measured in this
system, and how does it compare to that obtained in aqueous conditions? The following is a progress report of our endeavors to answer
these questions.

EXPERIMENTAL PART

Preparation of porin   In aqueous solutions, porin was purified to
homogeneity in the non-ionic detergent octyl-POE (octyl-oligoxy-
ethelene) as described in detail elsewhere[10]. The resulting solutions contained the protein in monodisperse trimeric state[11]. The
final detergent concentration was 1%. Electrolytes were removed by
extensive dialyses of protein samples (1 ml with 10-20 mg protein)
against water (100-200 ml) containing 1% octyl-POE, and 10-20 g
Bio Rad AG-X 501.After overnight dialysis at room temperature, conductance values of the dialysis buffer were < 50μS. Preparation of
protein in isooctane was performed by adding AOT to a final concentration of 1% to 1 ml protein (1-5 mg/ml) in water containing
1% octyl-POE. Solutions were extensively dialyzed against tri-
distilled water containing 1% AOT. Samples were then lyophilized
to apparent dryness, and an equal volume of isooctane was added.

Solubilization was aided by sonication, resulting in clear solutions.
The protein concentration was determined, and its native state was
assessed by polyacrylamide gel electrophoresis in SDS with or without boiling samples in the SDS sample buffer[8].

Dielectric dispersion measurements   These were performed on two
impedance bridges (Boonton Electronics, Parsipanny, N.Y.). For low
frequencies (5-500 KHz), a direct capacitance bridge, model 75C,
was used. An admittance bridge, model 33 A/1, covered the high frequency range (1-50 MHz). Both bridges were used for measuring the
capacitance as well as the conductance of the solutions. Samples
were measured in a cylindrical dielectric cell with exchangeable
inner electrodes[13]. All measurements were performed at 25°C in a
thermostatted cell. The cell constants for conversion of the
measured capacitance to the dielectric increment were established
by meausring solvents of known dielectric constants and conductivities. In all experiments, protein solutions as well as solvents
alone were measured. Determinations of the dispersion effect in
water with or without 1% octyl-POE did not reveal significant differences in conductance or dispersion effects. Dielectric increments
of matrix porin solutions were measured over a frequency range of
5 KHz to 50 MHz. The concentration range of protein was governed by
the samll dielectric increment (which necessitated that the protein
be present at high concentrations), and the limited conductanc
range (1 mS of the bridge 75C) which imposed an upper limit. The
best results were obtained with concentrations of 10-20 mg/ml. After
dialyses of protein solutions, described above, the conductance of
the solvent was 20-50µS, while that of protein solutions was
300-1000µS, depending on protein concentration. Occasionally, precipitation of protein occurred during dialysis due, presumably, to
the closeness of the pH of deionized water to the isoelectric
point[8] of the protein (pH 5.9). Precipitation seriously affected
measurements; it decreased the apparent dielectric increment, since
large aggregates, unlike solubilized trimers, did not orient in the
electric field. The apparent increase of capacitance in the low
frequency range is due to electrode polarization (cf. Results). It
was pronounced at high conductance and low frequencies. The effect
of electrode polarization may be magnitudes higher than that due
to the presence of the protein[14]. It was corrected for according
to Oncley[14] with an exponent determined experimentally[13]. In
measurements in isooctane/AOT, the electrode polarization effect
could be neglected, due to the very low conductance of isooctane.
Estimations of Stokes radii and dipole moment were performed as
described in detail[13]. The major difficulty in these measurements
was attaining a good solubilization of matrix porin in isooctane/
AOT.

<u>Estimation of dipole moment</u> The theory of Onsager[15] was used, taking only the protein into account, as detergents are unlikely to contribute significantly to the measured increments. For the calculation of the dipole moment, μ, the mass[8] of the protein was taken to be 110,000 d. The dielectric constants used were 78.5 for solutions containing 1 % octyl-POE, and 2.4 for 1% AOT in isooctane. A value of 2.5 was used as the inner refractive index of the protein[16].

RESULTS

   The dielectric dispersion of matrix porin in water (containing 1% octyl/POE) is shown in Figure 2A for the sake of comparison with results in the reverse micelle system. The Stokes radius of the

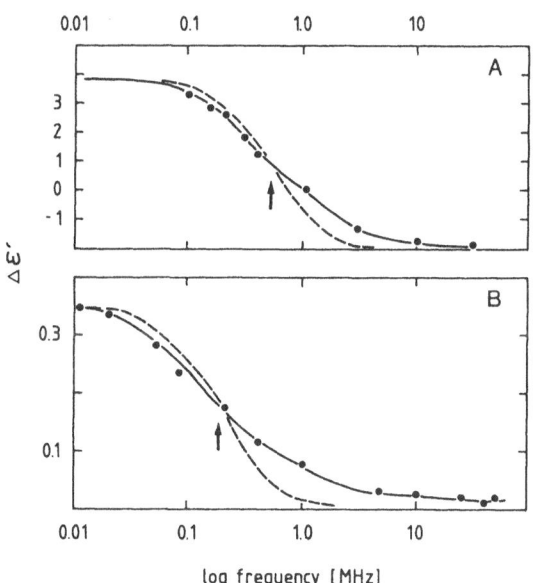

Figure 2. *Dielectric dispersion (Δε') of matrix porin in water (A), and in isooctane (B). The critical frequency, νc (indicated by an arrow), at which Δε' is equal to Δε'$_{1/2}$, yields a value of ∿ 500 KHz in water. The dotted line is a simulation of the dispersion of spherical particles (Debye dispersion) with the same total dielectric increment and the same critical frequency ν_c. In the presence of AOT in isooctane, (B), the protein concentration was 0.1 mg/ml. The experimental curve also deviates from pure Debye dispersion (broken line). Compared with the measurements for water solutions, the critical frequency, ν_c, (arrow) is shifted to a lower frequency (0.13 MHz), and enhanced relaxation times.*

protein-detergent complex was calculated from ν_c to be 50+ 5Å, and
is in good agreement with several independently obtained values[4].
The dipole moment of porin, obtained from $\Delta\varepsilon'$, was estimated to be
375+ 35 Debye (D). This estimate assumes that there are trimers
with 3 subunits, all pointing in the same direction, an assumption
strongly supported by structural studies[17].

Measurements in isooctane/AOT exhibited much more favorable
signal-to-background ratios than in the water/octyl-POE system. Due
to the very low conductance of isooctane, electrode polarization
could be neglected. Figure 2B shows the dispersion curve of matrix
porin in isooctane. Calculating the Stokes radius from the relaxation frequency, a radius of 73 Å was obtained, using the measured[13]
solvent viscosity of 0.49 cP. From the total dielectric increment,
we have estimated a dipole moment of about 1800 D. The protein remained native, as determined by its electrophoretic mobility, which
is known[18] to correspond to its channel-forming ability.

DISCUSSION

The validity of the method used can be judged from the results
obtained with measurements in the aqueous phase. The hydrodynamic
radius calculated compares well with results obtained by several
independent techniques[4]. The dipole moment of 375D/trimer is rather
low, but it is within the range of those measured with soluble
proteins[14]. We estimate[13] the accuracy of the measurement to be
within 20%. The secondary structure of porin appears to be mostly
β-pleated sheets[8]; this could be a reason for the low value of the
dipole moment. The data for isooctane/AOT, taken at face value,
yield good results, due to the low conductivity of the solvent and
the absence of electrode polarization effects. The shape of the
dispersion curve suggests reverse micelles with a Stokes radius of
about 73 Å. This value, though larger than that obtained for water
solutions, indicates that the protein is not inserted into microdroplets, but rather that local reverse micelles form, as suggested
in Figure 1C. The protein retains its native state. The dipole
moment is substantially larger than the one determined for aqueous
solution, but we know too little about the state of the protein-AOT
complex in isooctane to draw clear conclusions. The difficulties
with evaluating the results for isooctane/AOT are mainly twofold.
First, the extent of solubilization in this system was variable.
This may have trivial origins, but it requires more careful examination. The second problem consists of the fact that AOT is an ionic
detergent. Proteins such as porin are asymmetrically arranged within
the membrane, regardless of their overall shape. Results from chemical modification experiments[19] have shown, for instance, that a

single amino group per polypeptide is available at the outer face of the protein, but none at the inner. Such differences may well be accompanied by different association numbers of AOT with the polar surfaces at the two sides of the membrane. A difference of a single unit charge across a membrane of 50 $\overset{\circ}{A}$ thickness has been calculated[13] to yield a dipole moment of about 240 D. It would therefore be advantageous to use non-ionic amphiphiles which can form reverse micelles. Preliminary experiments have been unsuccessful.

Nonetheless, and with all the necessary reservations, we may conclude that porin can be dissolved in a reverse micelle system without denaturation. The micelles appear to be monodisperse, with a Stokes radius supporting the highly schematic representation of reverse mixed micelles in Figure 1C. Due to the stability of porin in organic solvents, it may be possible to study its structural properties under these conditions, in order to probe the forces which hold subunits together, and to further study the dielectric properties of porin.

ACKNOWLEDGEMENTS

We are indebted to Professor G. Schwarz for his advice and for making available unpublished procedures. We acknowledge expert secretarial help by Marianne Remy. This work was supported by Grant 3.656.80 from the Swiss National Science Foundation.

REFERENCES

1. M. Zulauf and H.F. Eicke, J. Phys. Chem. 83, 841 (1979).
2. P. Douzou, Adv. Enzymol. 51, 1-74 (1980).
3. C. Grandi, R.E. Smith and P.L. Luisi, J. Biol. Chem. 256, 837 (1981).
4. M. Grabo, J.P. Rosenbusch and M. Zulauf, submitted (1983).
5. M. Zulauf and J.P. Rosenbusch, J. Phas. Chem., in press (1982).
6. R.M. Garavito, J. Jenkins, J.N. Jansonius, R. Karlsson and J.P. Rosenbusch, J. Mol. Biol., 164 : 313 (1983).
7. H. Schindler and J.P. Rosenbusch, Proc. Natl. Acad. Sci. USA, 75, 3751 (1978).
8. J.P. Rosenbusch, J. Biol. Chem. 249, 8019 (1974).
9. J.P. Rosenbusch and R. Müller, in "Solubilization of Lipoprotein Complexes", H. Peters and J.P. Massué (eds), European Press, Gent, pp. 59 (1977).
10. J.P. Rosenbusch, R.M. Garavito, D.L. Dorset and A. Engel, in "Protides of the Biological Fluids", 29th Coll., Pergamon Press, Oxford, pp. 171 (1982).

11. A. Lustig and J.P. Rosenbusch, submitted (1982).

12. G. Schwarz, unpublished procedure.

13. J. Wirz, Diploma Thesis, University of Basel (1981).

14. J.L. Oncley, J. Am. Chem. Soc., 1115 (1938).

15. G. Schwarz, "Dielectric and Related Molecular Processes",
    Vol. 1, The Chem. Soc. London (1972).

16. R. Pethig, "Dielectric and Electronic Properties of Biological
    Materials", J. Wiley, New York (1979).

17. D.L. Dorset, A. Engel, A. Massalski and J.P. Rosenbusch,
    J. Mol. Biol., in press (1983).

18. J.P. Rosenbusch, A.C. Steven, M. Alkan and M. Regenass, in
    "Electron Microscopy and Molecular Dimensions", W. Baumeister
    and W. Vogeel (eds), Springer Verlag, pp. 1-10 (1980).

19. M. Schindler and J.P. Rosenbusch, J. Cell Biol. 92, 742 (1982).

BILE SALT AGGREGATION IN AQUEOUS AND NON-AQUEOUS MEDIA

Siegfried Lindenbaum and Madhu Vadnere

Pharmaceutical Chemistry Department
The University of Kansas
Lawrence, Kansas 66045 USA

The distribution of four bile salts: sodium cholate
(I), sodium deoxycholate (II), sodium chenodeoxycholate
(III) and sodium ursodeoxycholate (IV) between aqueous
buffer and 1-octanol has been measured as a function of
temperature between 25° C and 55° C and as a function of
bile salt concentration at concentrations below 0.1
mole/liter in the aqueous phase. The distribution isotherms obtained have been explained on the basis of reversible association in the aqueous phase. The treatment
assumes that the bile acid exists as a monomer in the
organic phase, which is verified by vapor pressure osmometry. A graphical method has been employed to estimate
the association constants in the aqueous phase for the
various equilibria encountered. An aggregation number
of four for IV and twelve for I, II and III has been
estimated. From the results, thermodynamic functions
associated with the transfer of each of the bile salts
from water to octanol and those related to association
processes in the aqueous phase were calculated. These
results are consistent with our previous findings that
the premicellar association of bile salts occurs by
hydrophobic interaction. The value of the thermodynamic
parameters for the transfer of bile salts from aqueous
buffer to octanol revealed that there is an unfavorable
enthalpic and favorable entropic contribution for all
four bile salts. However, for IV, which is an epimer of
III, both enthalpic and entropic contributions are reduced, compared to III, suggesting that there is a pro-

nounced effect of stereochemical orientation on the hydrophobic interaction.

   The distribution of deoxycholate (II) between aqueous buffer and more lipophilic organic phases consisting
of 70:30::isooctane:1-octanol (v/v) (system A) or 80:20::
isooctane:chloroform (v/v) (system B) was also studied.
The distribution isotherms suggested that II associates
strongly in the organic systems A and B unlike in pure
1-octanol. The model was modified to include association
of II in the organic phase to describe distribution behavior. The treatment suggests that II exists as a monomer and a dimer in system A with a dimerization constant
of $820\underline{M}^{-1}$. A model consisting of monomer-tetramer-hexamer
in the organic phase best describes the data for system B.
The data support the view that association in the organic
phase is due to hydrogen bonding between bile acid molecules. These studies suggest that partition of bile salts
from an aqueous to a lipid membrane phase would involve
an inversion from hydrophobic "back to back" association
in the aqueous phase to hydrogen-bonded association in
the lipid phase. In a similar fashion the enterohepatic
cycling of bile acids must be accompanied by the inversion of structure, since this process also requires that
bile acid partitions from an aqueous environment into a
membrane lipid bilayer.

INTRODUCTION

   Bile salts (Figure 1) are biological detergents which play an
important role in the dissolution or dispersion of cholesterol and
other lipids[1,2]. The solubility of cholesterol in aqueous solutions
of bile salts and in bile salt-lecithin-water systems has been
studied extensively[3,4]. It has been shown[5] that bile salts are
capable of solubilizing a large number of otherwise poorly water
soluble organic and inorganic compounds. The water solubility of
several drug substances has been shown to increase in the presence
of bile salts[6-11]. There is no doubt that the solubilizing property of bile salts is due to their amphiphilic nature and their ability to aggregate in aqueous solution to form micelles[2,12,13]. One
objective of our studies was to characterize the aggregation and
micelle formation in aqueous solution.

   Bile salts, as a consequence of their micellar properties,
are of great importance in assisting the digestive process by solubilizing and dispersing lipid substances. Bile acids (salts) are

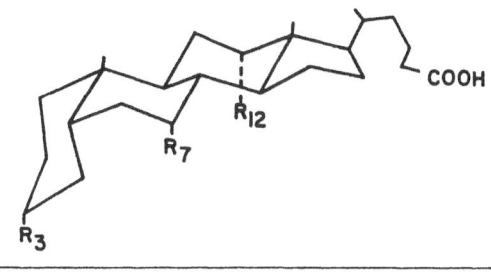

R$_3$	R$_7$	R$_{12}$	COMPOUND
OH	OH	OH	CHOLIC ACID
OH	H	OH	DEOXYCHOLIC ACID
OH	OH	H	CHENODEOXYCHOLIC ACID (3α, 7α)
OH	OH	H	URSODEOXYCHOLIC ACID (3α, 7β)

Figure 1.   *Structure of bile acids.*

also readily reabsorbed through the intestinal membrane, ultimately
to be returned to the liver to replenish the bile-acid pool. This
entero-hepatic cycling[14] conserves bile acids for repeated use in
the digestive process and in preventing the precipitation of cholesterol in the gall bladder. This process requires bile acids to
cross lipid membranes. Previous studies[12,15] have suggested that
in aqueous solution bile acids are associated by hydrophobic forces.
Micellar aggreates are thus formed in aqueous solution by association along the hydrophobic or lipid sides of the molecules leaving the hydroxyl and carboxylate moieties free to form hydrogen
bonds with the solvent. In traversing lipid bilayer membranes, as
part of the entereohepatic cycling process, bile acids partition
from an aqueous phase to a relatively less polar phase in the
membrane lipid bilayer. A further objective of this report is to
examine the association pattern of bile salts in an aqueous environment, where hydrophobic forces dominate the interaction between
these amphiphiles, and to determine what changes in association
occur when these substances partition into a lower dielectric, non-
hydrogen bonding medium. Organic liquids of varying polarities and
hydrogen-bonding capabilities are used as models for the change in
environment experienced by bile acids in crossing biological membranes.

The aggregation pattern of bile salts in aqueous and non-aqueous solvents was studied by measuring the partition equilibria of bile salts between aqueous buffer solutions and water immiscible organic solvents.

Experimental Methods

Details concerning the experimental methods for the measurement of partition coefficients have been published[16],[17]. Briefly, partition coefficients for the distribution of bile acids (salts) between aqueous buffer (pH 8) and organic solvents were measured using ^{3}H labelled bile salts. Vials containing 4 ml of organic solvent and 4 ml of aqueous phase with varying amounts of bile salt were shaken for 12 hours in a thermostated water bath. Samples from each phase were withdrawn and count rates were measured in a liquid scintillation counter after adding scintillation cocktail.

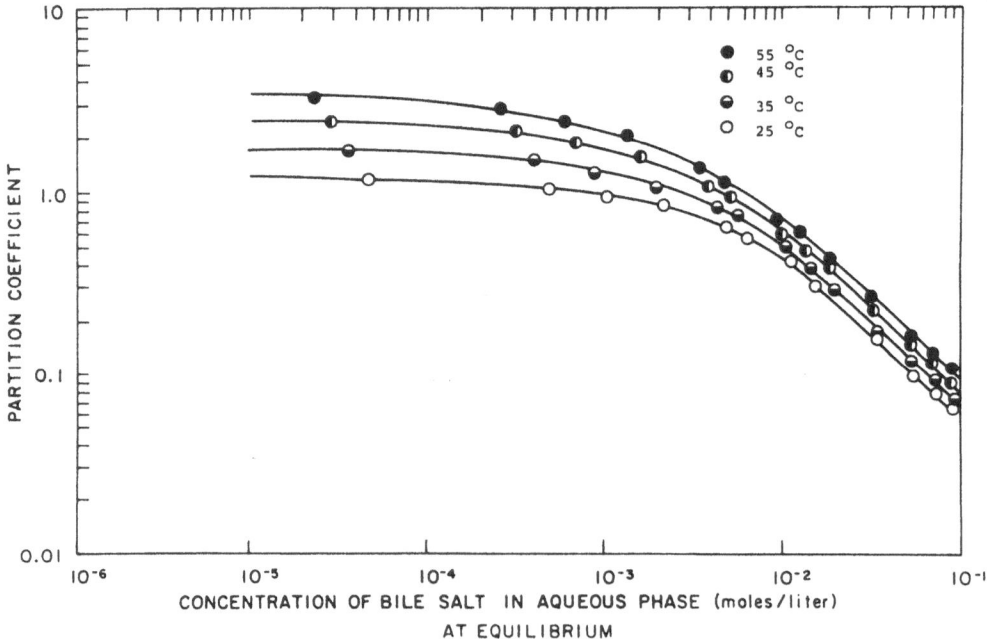

Figure 2. *Distribution isotherms for sodium cholate between 1-octanol and 0.02 M Tris hydroxy-methylaminomethane (Tris) buffer (pH 8). The solid lines are calculated from association constants in Table I (from Ref. 16).*

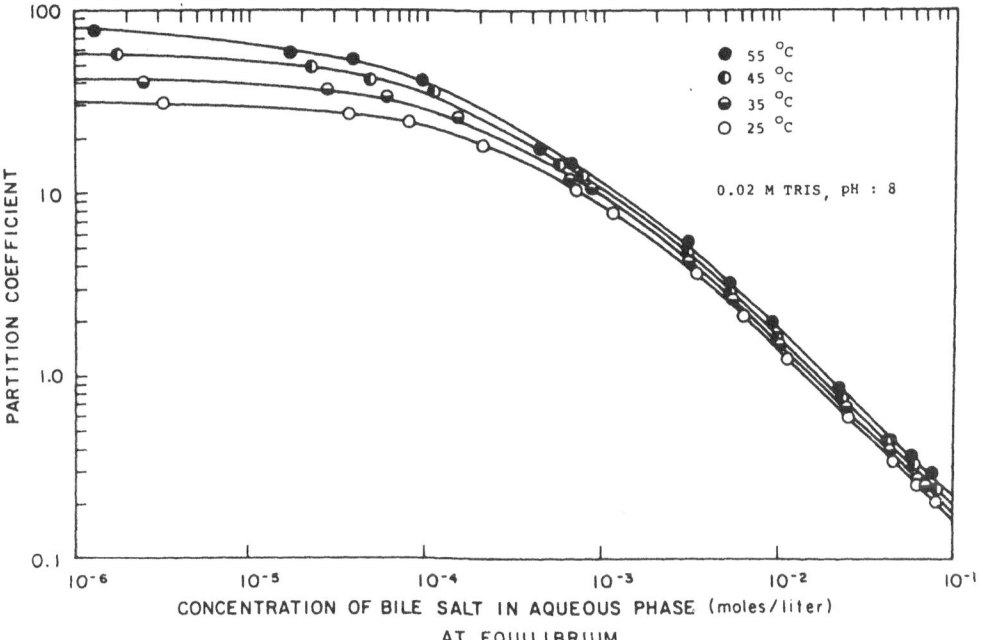

Figure 3. *Distribution isotherms for sodium deoxycholate between 1-octanol and Tris buffer (pH 8) (from Ref. 16).*

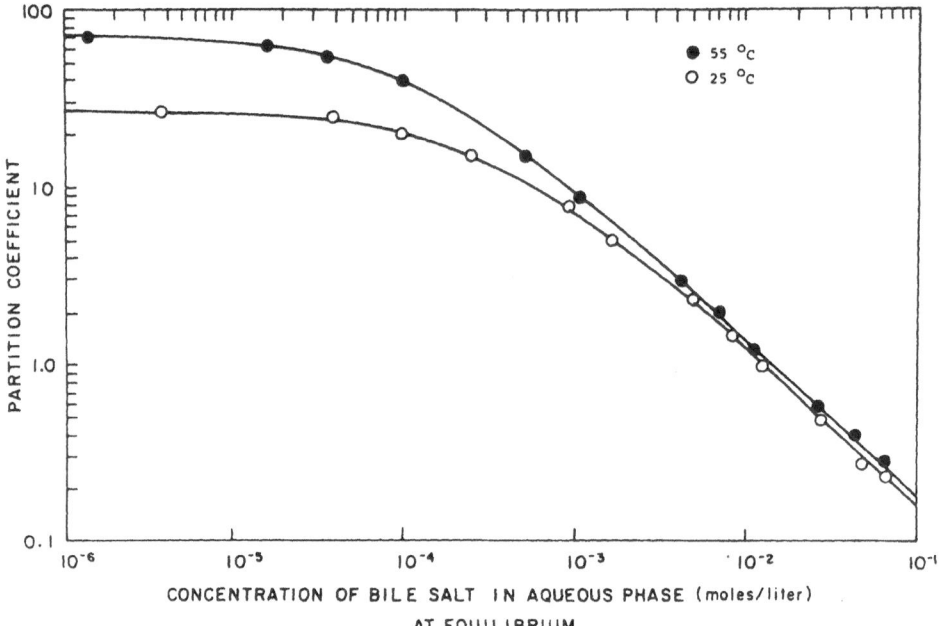

Figure 4.   *Distribution isotherms for sodium chenodeoxycholate between 1-octanol and Tris buffer (pH 8)from Ref. 16).*

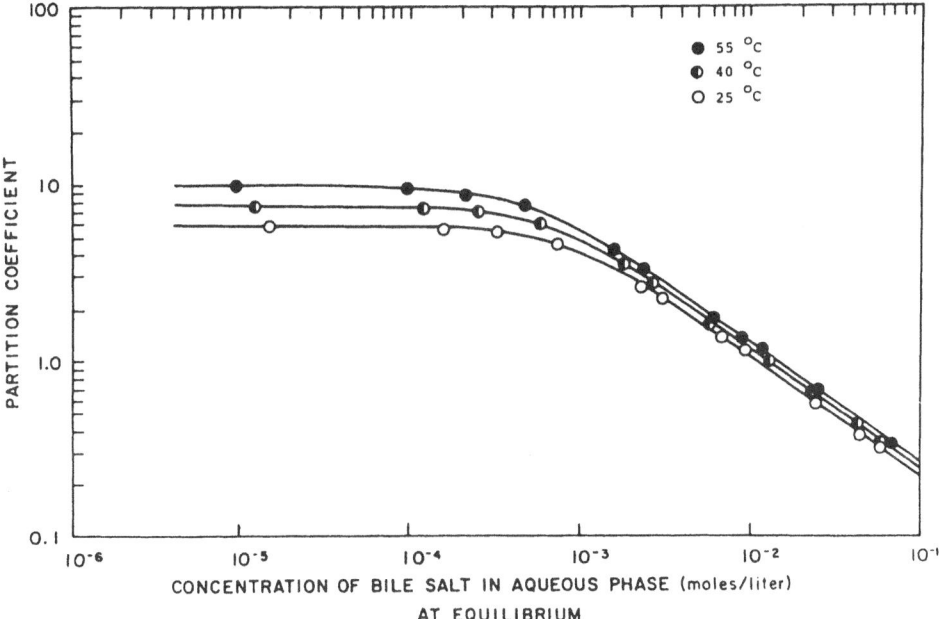

Figure 5. *Distribution isotherms for sodium ursodeoxycholate between 1-octanol and Tris buffer (pH 8) (from Ref. 16).*

Theoretical Considerations

Initial partition experiments were performed with four bile salts, i.e., sodium cholate, sodium deoxycholate, sodium chenodeoxycholate and sodium ursodeoxycholate, between aqueous buffer and 1-octanol. These data are given in Figures 2-5. Whereas previous reports[12],[13] indicated that in the aqueous buffer solution the bile salts would be associated to a considerable extent, it was assumed for this treatment that in the octanol phase, the bile acids were present only as monomers. Vapor pressure osmometry measurements on bile acids dissolved in octanol indicated that in this solvent the bile acids were present almost entirely in the monomeric fully dissociated form[16]. If association occurs only in the aqueous phase, then the equilibria indicated in Scheme I pertain. It is further assumed that at infinite dilution in the aqueous phase, the bile salt is also present as the monomer.

PARTITION EQUILIBRIA

SCHEME I

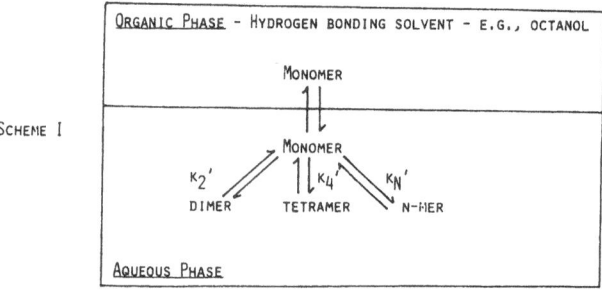

SCHEME II

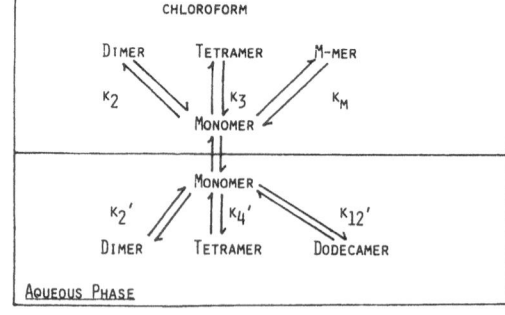

Under these conditions:

$$\frac{K^o_{app}}{K_{app}} = 1 + 2\ K'_2\ [M]_{aq} + 3\ K'_3\ [M]^2_{aq} + \ldots \qquad Equation\ 1$$

where

$$K^o_{app} = \frac{C_{org}}{C_{aq}}_{c \to o} = \frac{Concentration\ of\ bile\ salt\ in\ the\ organic\ phase}{Concentration\ of\ bile\ salt\ in\ the\ aqueous\ phase}$$

$\qquad$ = *partition coefficient at a given pH in the*
$\qquad$ *infinitely dilute solution*

K_{app} = *partition coefficient at the same pH but at a finite*
$\qquad$ *concentration of bile salt (acid).*

$[M]_{aq}$ = *concentration of bile salt (acid) monomer in the*
$\qquad$ *aqueous phase*

$[M]_{aq}$ *is determined by K^o_{app} and the concentration of bile*
$\qquad$ *salt in the octanol phase where $C_{org} = [M]_{org}$; thus:*

$$[M]_{aq} = C_{org}/K^o_{app}$$

The partition coefficient for the partition of bile salts between aqueous buffer and octanol may be fitted to Equation 1 to obtain the values of K'_2, K'_3, etc., the equilibrium constants for the formation of dimer, trimer, tetramer, etc. in the aqueous phase. These equilibrium constant values may then be used to generate species distribution profiles in the aqueous phase.

For non-hydrogen bonding solvents, there is evidence[18,19,20] that bile acids and their esters form dimers and higher aggregates.

For organic solvents in which the bile acid is not present exclusively as the monomer, the equilibria diagrammed in Scheme II apply. In this case it may be assumed that the association equilibria in the aqueous phase will be the same as those determined in the previous study. Therefore, the values of K'_1, K'_2, etc. will be the same.

The concentration of monomeric bile salt anion in the organic phase may be obtained from

$$K^O_{app} = \frac{[M]_{org}}{[M]_{aq}} = \lim_{c \to o} K_{app} \qquad\qquad \text{Equation 2}$$

and the species distribution in the organic phase is then obtained from

$$\frac{[C]_{org}}{[M]_{org}} = 1 + 2K_2\,[M]_{org} + 3K_3\,[M]^2_{org} + \ldots \qquad \text{Equation 3}$$

where K_2, K_3, etc. are the equilibrium constants for the association of bile salt anions in the organic phase. The equilibrium partition data were used to obtain $[C]_{org}/[M]_{org}$ which when fitted to
Eq. 3 yielded K_2, K_3, etc. These were then used to calculate the
species distributions in the organic phases.

RESULTS

1-Octanol-Water System

     Aqueous Phase. The association constants giving the best fit
to Equation 1 are given in Table I. Interestingly, only dimers,
tetramers and dodecamers yielded the best fit to the data. The
aggregation number 12 is in good agreement with the value 13 previously published by Makino et al.[21]. Attempts to include trimers
in the fit failed; this is expected since there is no way for a
trimer structure to form without leaving the back, lipophilic side
of the bile acid anion exposed to the solvent. Association increases
with increasing temperature in all cases, providing further evidence
for hydrophobic interaction between bile salt anions. The temperature variation of K' values is shown for sodium cholate in Figure 6.
Linear Van't Hoff relationships are obeyed for all four bile
acids studied, allowing the calculation of ΔH^O values for the association processes. The calculated ΔH^O values are given in Table II.
In all cases, positive values of ΔH^O are obtained, showing that the
association process is entropy dominated[22]. From the K' values given
in Table I, it is possible to calculate a species distribution for
each of the bile salts in aqueous solution. An example for sodium
cholate is given in Figure 7. A schematic drawing showing plausible
structures for hydrophobically associated dimers and tetramers is
shown in Figure 8. Similar structures have previously been suggested
by Ekwall, Fontell and Sten[12] and Small[13,15].

Table I. Association Constants in the Aqueous Phase.

Bile Salt (Aggregate Number)	Temp.	$\log K_2'$ M/liter^{-1}	$\log K_4'$ M/liter^{-3}	$\log K_{12}'$ M/liter^{-11}
Cholate	25°	2.20	4.00	25.70
(12)	35°	2.39	4.46	26.61
	45°	2.57	4.80	27.48
	55°	2.74	5.30	28.26
Deoxycholate	25°	3.48	10.30	36.78
(12)	35°	3.60	10.54	37.92
	45°	3.72	10.78	39.04
	55°	3.88	11.04	40.08
Ursodeoxycholate	25°	2.35	8.15	--
(4)	40°	2.51	8.45	--
	55°	2.65	8.78	--
Chenodeoxycholate	25°	3.43	10.30	36.78
(12)	55°	3.85	11.60	40.78

Table II. ΔH^O for Association Process[a]

Bile Salt	ΔH^O_{dimer}	$\Delta H^O_{tetramer}$	$\Delta H^O_{dodecamer}$
Cholate	4.0	4.8	3.2
Deoxycholate	3.0	2.8	4.0
Ursodeoxycholate	2.2	2.3	-
Chenodeoxycholate	3.1	4.8	4.9

[a]Kilocalories per mole of bile salt in the aggregate.

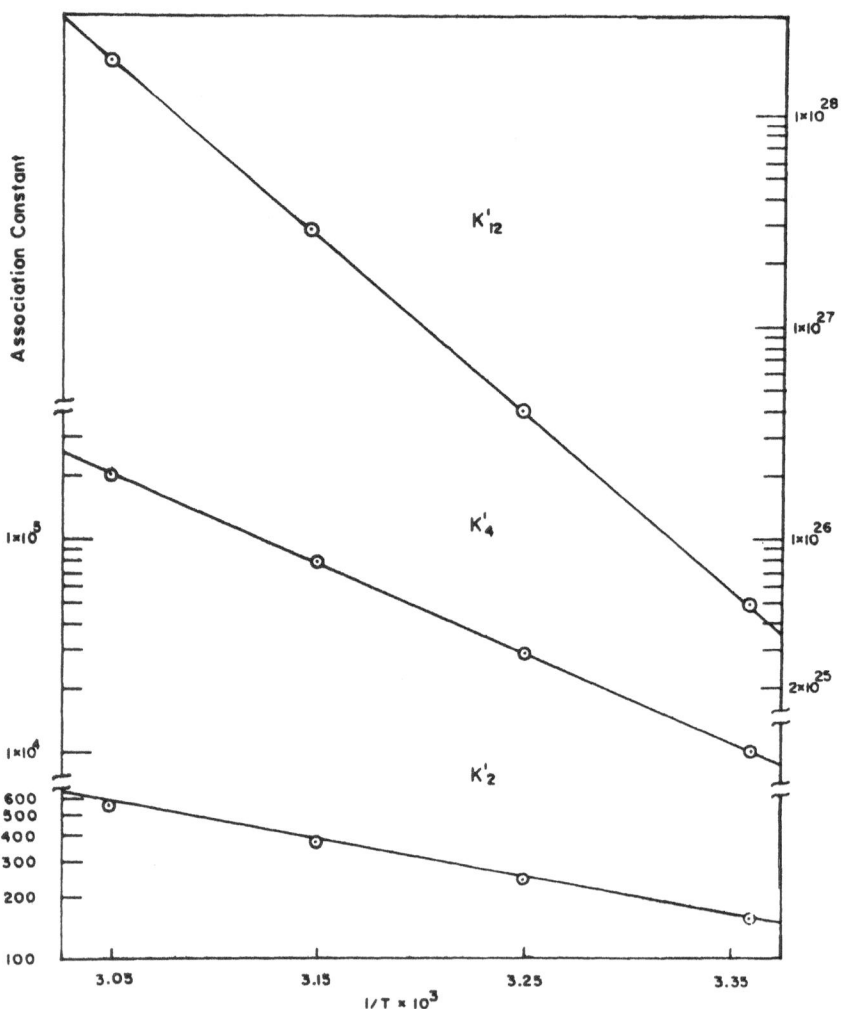

Figure 6. *Temperature variation of aqueous association constants for sodium cholate (from Ref. 16).*

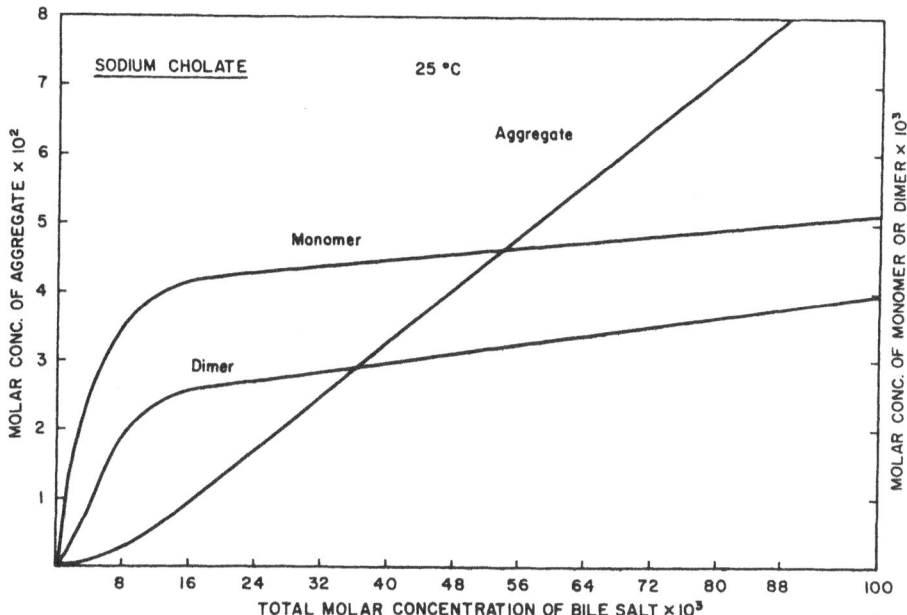

Figure 7. *Species distribution of sodium cholate in aqueous solution. The aggregate refers to the dodecamer species.*

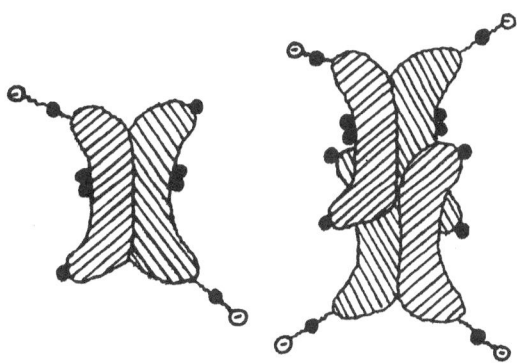

Figure 8. *Proposed structures of dimers and tetramers in aqueous solution. See also Small[13].*

Organic Phase. For this system it was assumed and verified by
vapor pressure osmometry measurements that bile salts are monomeric
in 1-octanol. Partition coefficients were also measured as a function of varying sodium ion concentration in the aqueous phase[16].
The partition was found to be independent of sodium ion concentration suggesting that the species partitioning into the organic phase
at pH 8 is the free acid. Accordingly, a strong variation of K_{app}
with pH was observed; the partition coefficient decreases with increasing pH[16]. From the variation of the partition constants K_{app}
with temperature (Figure 9), ΔH^O for the partition process may be
estimated, and the thermodynamic functions for the transfer from an
aqueous medium to 1-octanol can be evaluated.

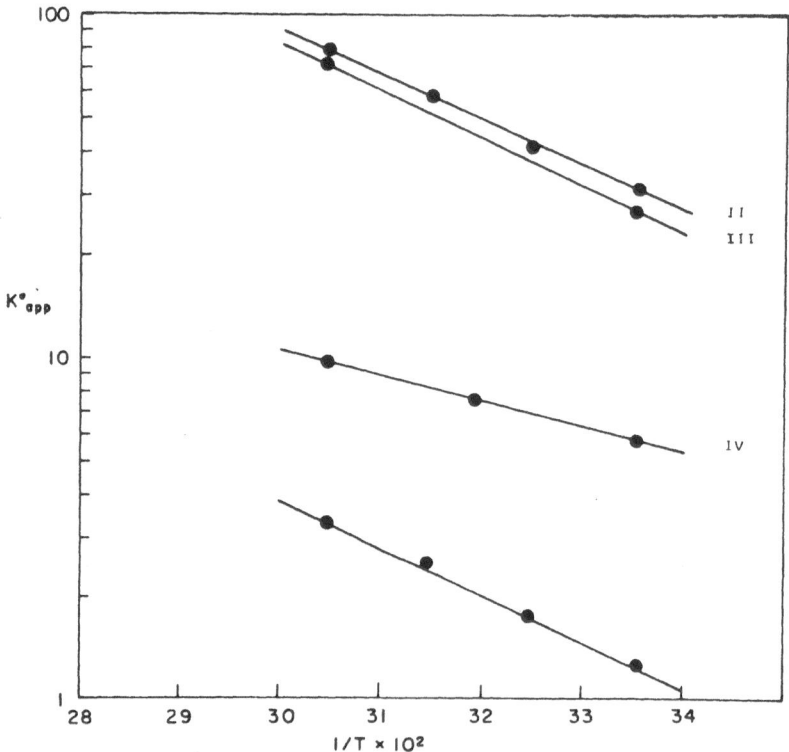

Figure 9.   *Temperature variation of partition coefficients for the
transfer of bile salts from aqueous buffer solution (pH 8) to
1-octanol. Sodium cholate I, sodium deoxycholate II, sodium chenodeoxycholate III and sodium ursodeoxycholate IV (from Ref. 16).*

Table III.   *Thermodynamic Transfer Functions   Water → 1-Octanol*

Bile Salt	K^O_{app}	ΔG^O	ΔH^O	$T\Delta S^O$
	25°	25°	Kcal/mole	
Deoxycholate	32.0	-2.05	+6.0	8.05
Chenodeoxycholate	27.0	-1.95	+6.1	8.05
Ursodeoxycholate	5.8	-1.04	+3.4	4.44
Cholate	1.2	-0.11	+6.3	6.41

The values for the enthalpy of transfer of bile salt from the aqueous to the organic phase are all positive (unfavorable) (Table III). Positive values for ΔH^O of transfer of hydrocarbons and aliphatic alcohols are also obtained[22]. Positive values of ΔH^O for this process are expected on the basis of the hydrophobic properties of these solutes. In aqueous solution the water molecules in the vicinity of these nonpolar molecules or moieties are more strongly hydrogen bonded, so that a region of higher local order exists than in pure water. In partitioning out of the aqueous phase, the normal hydrogen bonded structure is restored, accompanied by a decrease in the amount of hydrogen bonding and a less ordered state. These processes are characterized by an increase in enthalpy (the energy required to break hydrogen bonds) and an increase in entropy, as expected for a loss in order. The data show that the entropy gain is greater for the more hydrophobic molecules. The dihydroxy bile salts, deoxycholate and chenodeoxycholate, show a larger entropy change than cholate, the more polar trihydroxy bile salt. The data for ursodeoxycholate demonstrates that it is not only the number of hydroxyl groups on the molecule that determine its hydrophobic character, the orientation is also important. In ursodeoxycholate the -OH groups are not all in the α configuration; the -OH group in the 7 position is oriented toward the 'back' of the molecule so that there is no longer a clearly hydrophobic side to the structure. As a consequence, even though it is a dihydroxy bile salt, ursodeoxycholate shows a smaller positive change in entropy for the aqueous-organic transfer process than either of the other dihydroxy salts (II and III), or even the trihydroxy salt (I). Even for IV, however, hydrophobic forces appear to predominate, since positive values of ΔH^O and ΔS^O are still obtained for the transfer process.

Less Polar Organic Phases; Isooctane-1-octanol (70:30) System A;
Isooctane---chloroform (80:20) System B.

Aqueous Phase. The association of bile salts in the aqueous
phases in equilibrium with these organic solvent systems is expected to be governed by the association constants given in Table I.

Organic Phase. Preliminary analysis of the partition data indicate that in these less polar solvents, the bile salts can no
longer be assumed to be monomeric as in 1-octanol. Accordingly, the
data were analyzed using Scheme II. The partition coefficients for
the distribution of sodium deoxycholate between aqueous buffer
(pH 8) and solvent systems A and B are given in Figures 10 and 11,
respectively. Analysis of the partition data according to Equation 3
yields values of the equilibrium constants for association in the
organic phases. These association constants are given in Table IV.

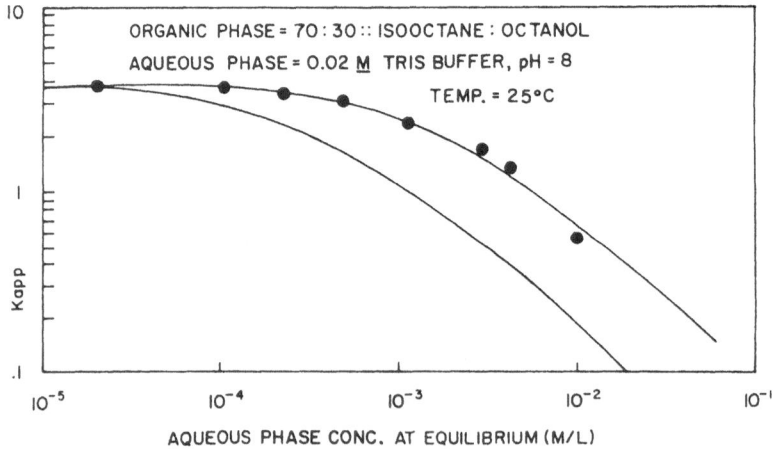

Figure 10. *Distribution isotherm for sodium deoxycholate at 25° C*
for the partition between aqueous buffer pH 8, and solvent system A
(isooctane-1-octanol 70:30 v/v). The lower curve is calculated using
Scheme I showing that association in the organic phase cannot be
ignored. The upper curve is calculated using Scheme II and the con-
stants given in Table IV (from Ref. 17).

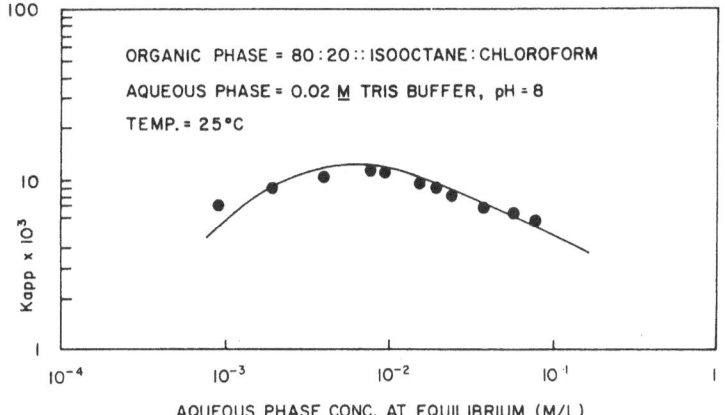

Figure 11. *Distribution isotherm for sodium deoxycholate at 25° C for the partition between aqueous buffer pH 8 and solvent system B (isooctane-chloroform 80:20 v/v). The solid line represents the isotherm calculated on the basis of the 1-4-6 association model using the constants given in Table IV (from Ref. 17).*

Table IV. *Association Constants in the Organic Phases*

Deoxycholate	K_2 (M^{-1})	K_4 (M^{-3})	K_6 (M^{-5})
System A Isooctane Octanol 70:30	8.20×10^2	--	--
System B Isooctane Chloroform 80:20	--	9.96×10^2	4.79×10^{28}

The species distribution profile for deoxycholic acid calculated according to Equation 3 is given in Figure 12. The finding that the association constants are larger and the aggregation number is higher in these less polar solvents suggests that, in these solvents, the bile acids are associated by hydrogen bonding interactions. Such hydrogen bonded structures have been proposed previously. Mazer, Benedek and Carey[18] have suggested that within the interior of a mixed lecithin-bile salt micelle the bile salt is associated by hydrogen bonding between the hydroxyl groups. Oakenfull and Fisher[23] have shown, using molecular models, that the hydroxyl groups on cholate and deoxycholate are favorably situated to form intermolecular hydrogen bonds. They suggested that this model might apply even in aqueous soltuion; however, it now appears evident

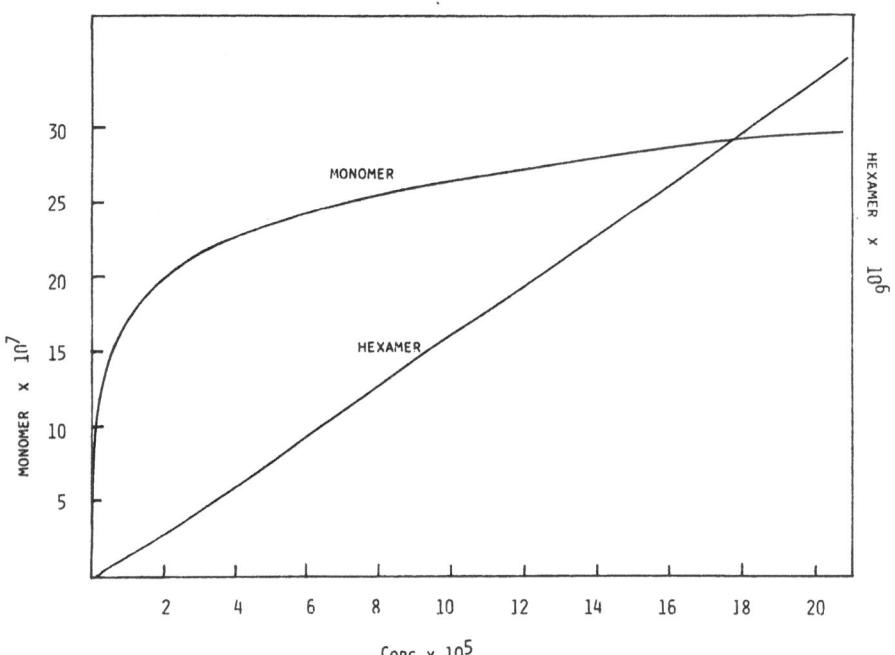

Figure 12. *Species distribution for deoxycholic acid in solvent system B. The concentration of the tetramer is negligible compared to the hexamer (from Ref. 17).*

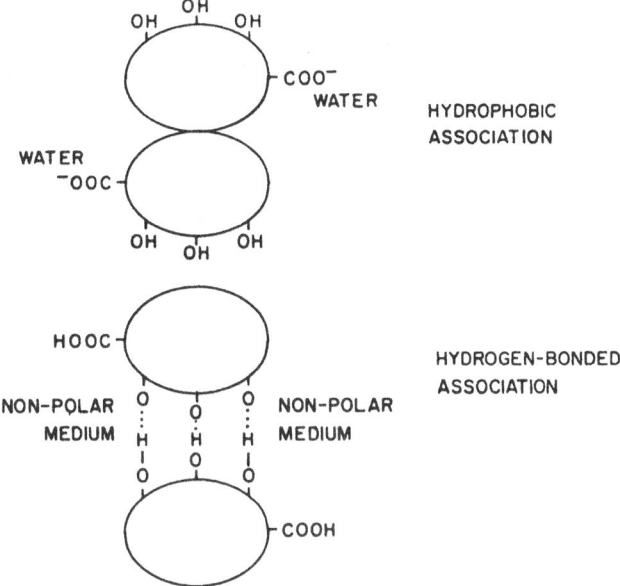

Figure 13. *Schematic diagram showing the change in the association mode for bile acids in going from an aqueous phase to non-polar low dielectric phase.*

that in aqueous solution[16,24] bile salts are associated by hydrophobic forces. The hydrogen-bonded association model, however, seems the most reasonable description of association in non-polar low dielectric media. This would imply that an inversion of structure must occur at the interface when the bile acid transfers across the boundary separating an aqueous medium from a non-polar low dielectric phase. This change in structure of the associated species is illustrated in Figure 13.

CONCLUSIONS

     1. In aqueous solution bile salt anions are associated as dimers, tetramers and dodecamers; the extent of aggregation and the aggregation numbers are a strong function of the number of OH groups on the steroid nucleus. Thus, the deoxycholates are more strongly associated than cholates. The association of ursodeoxycholate is much weaker than chenodeoxycholate; this may be due to the fact that in ursodeoxycholate, which is less distinctly amphiphilic because the −OH group in the 7 position, is on the β side of the ring structure; hence the hydrophilic and hydrophobic parts of the molecule are not as clearly separated as in chenodeoxycholate.

2. The thermodynamic parameters indicate that association in aqueous media is due to hydrophobic forces. The partition into organic media is also favored by a positive entropy change for the aqueous to organic transfer process, providing further evidence for the 'hydrophobic" nature of these bile acids in aqueous solution.

3. In 1-octanol, bile acids exist as monomers; in less polar solvents, bile acids associate, presumably by hydrogen-bonding interactions.

4. The partition of bile salt anions from aqueous media to low dielectric solvents or into lipid bilayers or membranes must be accompanied by an inversion of the structure of associated species. There must be a change at the interface from hydrophobic "back to back" association in the aqueous phase to hydrogen bonded association between hydroxyl groups in less polar membrane phases.

ACKNOWLEDGEMENT

This work was supported by a grant from the National Institutes of Health (AM 18084) and the General Research Fund of the University of Kansas.

REFERENCES

1. P.H. Elworthy, A.T. Florence and C.M. Macfarlane, "Solubilization by Surface Active Agents", Chapman and Hall, London (1968).
2. A.F. Hofmann and D.M. Small, Ann. Rev. Med. 18 : 333 (1967).
3. M.C. Bourges, D.M. Small and D.G. Dervichian, Biochim. Biophys. Acta 144 : 189 (1967.
4. M.C. Carey, J.C. Montet, M.C. Phillips, M.J. Armstrong and N.A. Mazer, Biochemistry 20 : 3637 (1981).
5. F. Verzer, Nutr. Abstr. Rev. 2 : 441 (1933).
6. P. Ekwall and L. Sjoblom, Acta Chem. Scand. 3 : 1179 (1949).
7. P. Ekwall and L. Sjoblom, Acta Endocrinol. 4 : 179 (1950).
8. P. Ekwall, T. Lundsten and L. Sjoblom, Acta Chem. Scand. 5 : 1383 (1951).
9. T.R. Bates, M. Gibaldi and J.L. Kanig, Nature (London) 210 : 1331 (1966).
10. T.R. Bates, M. Gibaldi and J.L. Kanig, J. Pharm. Sci. 55 : 191 (1966).
11. M. Rosoff and A.T.M. Serajuddin, Int. J. Pharm. 6 : 137 (1980).
12. P. Ekwall, K. Fontell and A. Sten, Proceedings of the International Congress of Surface Activity, London, p. 357 (1957).

13. D.M. Small, The Bile Acids, Chemistry, Physiology and Metabolism;
    Vol. 1 Chemistry, P.P. Nair and D. Kritchevsky, Eds., Plenum,
    New York-London (1971).
14. M.C. Carey, The Liver: Biology and Pathophysiology, ed. by
    I. Arias, H. Popper, D. Schacter and D.A. Shafritz, Raven Press,
    New York (1982).
15. D.M. Small, Adv. Chem. Ser. 84 : 31 (1968).
16. M. Vadnere and S. Lindenbaum, J. Pharm. Sci. 71 : 875 (1982).
17. M. Vadnere and S. Lindenbaum, J. Pharm. Sci. 71 : 881 (1982).
18. N.A. Mazer, G.B. Benedek and M.C. Carey, Biochemistry 19 : 601
    (1980).
19. R.O. Zimmerer and S.Lindenbaum, J. Pharm. Sci. 68 : 581 (1979).
20. J. Robeson, B.W. Foster, S.N. Rosenthal, E.T. Adams, Jr. and
    E.J. Fendler, J. Phys. Chem. 85 : 1254 (1981).
21. S. Makino, J.A. Reynolds and C. Tanford, J. Biol. Chem.
    248 : 4926 (1973).
22. C. Tanford, "The Hydrophobic Effect", 2nd ed., Wiley, New York
    (1980).
23. D.G. Oakenfull and L.R. Fisher, J. Phys. Chem. 81 : 1838 (1977).
24. M. Vadnere, R. Natarajan and S. Lindenbaum, J. Phys. Chem.
    84 : 1900 (1980).

NORMAL GANGLIOSIDE MICELLES IN AQUEOUS SOLUTION: INTERACTION WITH PHOSPHOLIPID PLANAR BILAYERS

Franco Gambale[1], Carla Marchetti[2], Cesare Usai[1] and Mauro Robello[3].

1 Istituto di Cibernetica e Biofisica
  Consiglio Nazionale delle Ricerche
  Camogli (Genova) Italy

2 FIDIA Research Laboratories
  Abano Terme (Padova) Italy

3 Istituto di Scienze Fisiche
  Universita degli Studi di Genova
  Genova, Italy

Gangliosides are membrane-bound glycosphingolipids with oligosaccharide chains which contain one or several residues of sialic acid, a negatively charged sugar (Figure 1). In animal tissues the most abundant gangliosides are monosialoganglioside GM1, disialogangliosides, GD1a and GD1b, and trisialoganglioside, GT. Most of them are localized on the outer surface of the plasma membrane and are ubiquitous but minor components of the cell surface, with the major exception of the mammalian central nervous system where they comprise up to 5-10% of the total lipid, and where lipid-bound sialic acid often exceeds that bound to glycoproteins[1,2]. Gangliosides have been shown to interact with several bacterial toxins[3], peptide hormones and other external ligands[4,5] probably through association with membrane proteins[6]. Most likely they are involved in functional activities of the cell surface, such as regulation of growth and structural plasticity. All these functions are likely to be mediated by ganglioside head group dynamics with a strong tendency for cooperative associations with each other and with glycoproteins [6,8,9,10]. This behavior appears to be strongly enhanced by divalent cations and affected by crosslinking agents[6,8], as observed with both native and articial membrances.

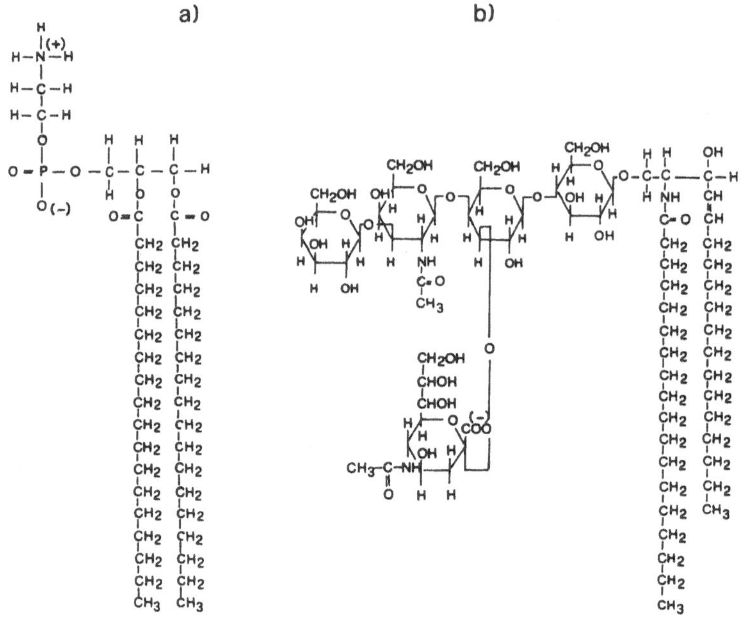

Figure 1. The structures of a phospholipid and a ganglioside: (a) distearoyl phosphatidyl ethanolamine;(b) monosialoganglioside GM1 (N-stearoyl)

The striking feature of gangliosides as lipid components of membranes is their strong amphipathic nature, due to the presence in the molecule of hydrophobic tails and a hydrophilic saccharide chain of comparable dimensions. Phospholipids and gangliosides actually belong to separate lipid classes: the former are "non-soluble swelling amphipaths" which, in an aqueous medium, form liposomal dispersions; the latter are "soluble amphipaths" which disperse in water as micelles[12]. As evidenced by laser light-scattering measurements,micelles appear to have a disc-like shape, a hydrodynamic radius of 60 Å and a molecular weight ranging from 4 to 5 x 10^5 D[2,14]. The critical micellar concentration has recently been shown to be lower than 10^{-6} M,[13,14],with very slow dissociation rates[13]. This last finding suggests that, once formed on the basis of hydrophobic interactions, the ganglioside micellar structure is stabilized by weak bonds between adjacent saccharide chains.

The peculiar behavior of gangliosides may influence membrane organization and interest has been focused on the relationship between gangliosides and the main lipid components of membranes,namely phospholipids. Artificial systems have been widely used to study ganglioside organization in mixed bilayers[8,9,10,11]. Micellar gangliosides have recently been incorporated in the outer surface of phospholipid vesicles[15,16], thus providing a system which better simulates the asymmetrical structure of plasma membranes.

We have investigated the interaction of GM1 ganglioside micelles with planar phospholipid bilayers, formed from a lipid n-decane
solution, following the Mueller-Rudin technique[17]. Gramicidin, a
hydrophobic pentadecapeptide, which forms dimeric channels in lipid
bilayers[18], was used as a conductance probe of the interaction[19].
Gramicidin has a very low solubility in water, with 10^{-10} M as an
upper limit[18]. Therefore, when dissolved in water, it should be
easily incorporated into the hydrophobic core of the ganglioside
micelles. We prepared KCl ionic solutions containing different
gramicidin/micelles ratios and eliminated free gramicidin from the
sample by means of an immersible millipore filter with a cut-off of
10,000 D. Control experiments showed that with our experimental conditions, almost all gramicidin molecules are included in micelles.
Appropriate volumes of freshly filtered solution were added to both
sides of the chamber (containing solutions of the same ionic composition) where the phospholipid membrane was pre-formed. The results
show that the interaction is strongly dependent on the phospholipid
used (see Figure 2). We verified that the phosphatidylserine (PhS)
bilayers need an approxiamtely 100 times higher gramicidin/GM1 ratio
than dioleylphosphatidylcholine (DOPC) membranes to yield comparable
conductance increases. This difference is not seen when free gramicidin from a methanol solution is added to the membrane and confirms
the hypothesis that the incorporation of the antibiotic is mediated by
GM1 micelles. The interaction between GM1 micelles and PhS membranes

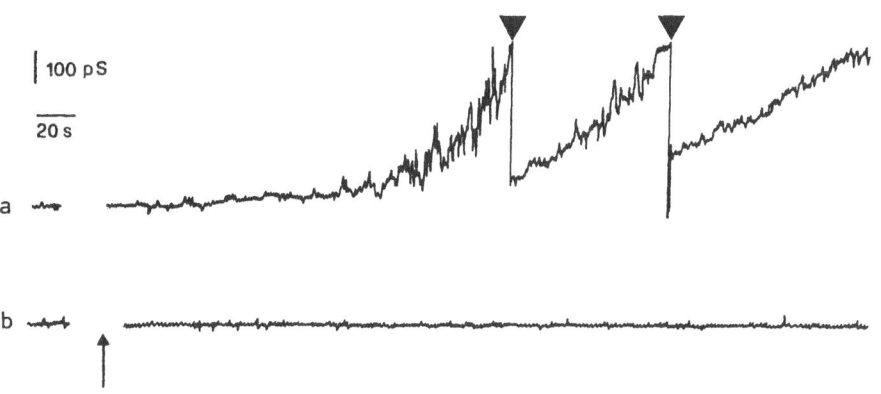

Figure 2.   *Time course of the conductance of a (a) DOPS and (b) a
PhS planar bilayer. Arrow shows where gramicidin: GM1 micelles were
added to a final concentration of $1.5x10^{-5}$ M GM1 (=$5x10^{-8}$ M micelles)
and $5x10^{-9}$ M gramicidin, i.e. a ratio of 1 gramicidin molecule per 10
micelles; KCl=1 M. The ▼ symbols indicate time scale reductions to
0.4 and 1 nanosiemens.*

decreases with increasing ionic strength (10-100 mM KCl) and is enhanced by the addition of divalent cations, either Ca^{2+} or Mg^{2+}
(3-30 mM). These last findings suggest that the interaction is
mainly prevented by electrostatic repulsion between the negatively
charged carboxylic groups of PhS and GM1. This repulsion can be
reduced by increasing the ionic strength or, in a more efficient
manner, by adding divalent cations.

When investigated in more detail, this method is likely to provide not only a technique for asymmetrically inserting gangliosides
in BLM, but also some insights into the question of how gangliosides
interact with natural bilayers and how they are incorporated into
them.

REFERENCES

1. R.W.Ledeen, J. Supram. Struc. 8 : 1 (1978).
2. G.Tettamanti, A. Preti, B.Cestaro, M. Masserini and S. Sonnino
    "Cell Surface Glycolipids" ACS Symp.Ser.No.128,C.C.Sweeley, ed.
    Am.Chem.Soc., (1980).
3. P.H.Fishman, J.Membrane Biol. 69: 85 (1982).
4. P.H.Fishman and R.O.Brady, Science, 194: 906 (1976).
5. S. Hakomori, Annu.Rev.Biochem. 50: 733 (1981).
6. S.W.Craig and P. Cuatrecasas, Proc.Natl.Acad.Sci.USA 72: 3844
    (1975).
7. A. Gorio and G. Carmignoto, "Post-traumatic Peripheral Nerve
    Regeneration", A.Gorio,H.Milesi and S.Mingrino, eds.Raven Press,
    New York (1981).
8. P.M.Lee, N.V. Ketis, K.R.Barber and C.W.M.Grant, Biochim.Biophys.
    Acta 601: 302 (1980).
9. F.J.Sharom and C.W.M. Grant, Biochim.Biophys.Acta 507:280 (1978).
10. E. Bertoli, M. Masserini, S. Sonnino, R.Ghidoni, B.Cestaro and
    G.Tettamanti, Biochim.Biophys.Acta 467:196 (1981).
11. B. Maggio, F.A·Cumar and R.Caputto, Biochem.J. 189:435 (1981)
12. Y.Barenholz, B.Cestaro, D.Lichtenberg, E.Freire,T.E.Thompson and
    S. Gatt, Adv.Exp.Med.Biol. 125: 105 (1980).
13. S. Formisano, M.L.Johnson, G. Lee, S.M. Aloj and H. Edelhoch,
    Biochem. 18: 1119 (1979).
14. M. Corti, V.Degiorgio, R. Ghidoni, S.Sonnino and G.Tettamanti,
    Chem.Phys.Lipids 26: 225 (1980).
15. B.Cestaro, G.Ippolito, R.Ghidoni, P.Orlando and G. Tettamanti,
    Bull.Molec.Biol.Med., 4: 240 (1979)
16. P.L.Felgner,E.Freire, Y.Barenholz and T.E.Thompson, Biochem.20:
    2168 (1981).
17. P. Mueller, D.O.Rudin, H.Ti Tien and W.C.Wescott, Nature,Lond.
    194: 979 (1962).
18. S.P.Hladky and D.A.Haydon, Biochim.Biophys.Acta 274: 294 (1972).
19. N.Duzgunes and S.Ohki, Biochim.Biophys.Acta 640: 734 (1981).

EXPLORING PEPTIDE INTERACTIONS WITH INTERFACIAL WATER USING REVERSED MICELLES

Lila M. Gierasch, Karyn F. Thompson
Jeffrey E. Lacy and Arlene L. Rockwell

Department of Chemistry
University of Delaware
Newark, Delaware 19711 U.S.A.

Water adjacent to biological membranes or macromolecules has properties distinct from bulk water. Many biological processes such as energy transduction or recognition events occur in this interfacial water region. It is critical to understanding these processes that the nature of interactions between polypeptide chains and the interfacial water be elucidated. Water within reversed micelles shows behavior very similar to this biochemically important interfacial water. The small water pools offer an excellent system for studying interfacial water, since there is no large pool of bulk water to obscure the parameters due to the interfacial water. We have been exploring the interactions of polypeptides with interfacial water by solubilizing synthetic model peptides in Aerosol OT (AOT, sodium bis-2-ethylhexyl-sulfosuccinate) reversed micelles and using nuclear magnetic resonance (NMR) and circular dichroism (CD) to analyze the conformations of the peptides. Our results indicate that the model peptides undergo a conformational change in the interfacial water that is induced by the counterions of the AOT headgroups. CD, ^1H and ^{13}C NMR parameters enable conformational monitoring of the peptide upon variation of the amount of water in the internal pools. Furthermore, the influence of the presence of the model peptide in the water pool on the state of the water has been explored using NMR and infrared (IR) spectroscopies. We have found evidence of the perturbation of an amount of water that can be qualitatively related to the nature and size of the surface of the solubilized peptide. These results show promise of using reversed micelles to study polypeptide hydration.

INTRODUCTION

Water in the vicinity of biological macromolecular and membrane surfaces possesses physical properties distinct from those of bulk water[1,2]. This _interfacial_ water has restricted mobility, a depressed freezing point, and unusual nuclear magnetic resonance (NMR) and infrared (IR) spectroscopic parameters, due both to its strong interactions with structural components at the surface (charged polar head groups of membrane lipids or hydrophilic side chains of proteins) and to the disruption of three-dimensional cooperative hydrogen-bonding networks. Many pivotal biochemical processes such as energy transduction and recognition events occur in the interfacial water region, and it is clearly important to elucidate the interactions of biomolecules with such water.

Reversed micelles offer a model system for studies of interfacial water. The water enclosed in these aggregates of amphiphiles in nonpolar media shows behavior very similar to biochemically important interfacial water: restricted mobility, depressed freezing point, and unusual NMR and IR parameters[3-6]. In addition, use of the reversed micellar water pools enables studies of interfacial water _without_ a large adjacent pool of bulk water that hinders the observation of phenomena at the interface. Several reports have appeared in which biomolecules were solubilized in reversed micelle water pools, and conformational and enzymatic activity data were obtained[7-10]. To date, however, no detailed examinations of the nature of the water-solubilizate interaction have been carried out with these systems.

In the present work, a model peptide with a well-defined and limited set of conformational states has been used to explore the influence of interfacial water on polypeptide conformation. The peptide chosen for study is a synthetic cyclic pentapeptide, cyclo (Gly-Pro-Gly-D-Ala-Pro)[11], which is water-soluble and insoluble in n-heptane, n-octane, or benzene. Figure 1 shows the cyclic peptide in its major conformation in a range of solvents[11] and in crystals[12]. Preliminary studies[13,14] revealed that the model peptide could be solubilized by sodium bis-2-ethylhexylsulfosuccinate (AOT) in nonpolar solvents, and that spectral data (circular dichroism (CD) or NMR) could be obtained for these systems. A distinct conformational sensitivity to the amount of water in the reversed

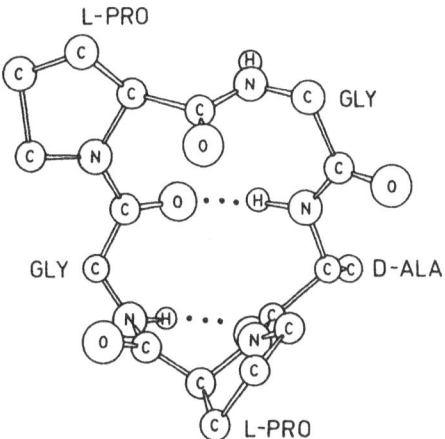

L-PRO

GLY

GLY

D-ALA

L-PRO

Figure 1. *Diagram showing the model cyclic peptide used in these studies. The conformation illustrated, which contains two intra-molecular hydrogen bonds (indicated by dots), is the preferred conformer of the free peptide in bulk solvents and in crystals. Upon formation of an ion complex, the hydrogen bonding is disrupted.*

micelle water pools was noted, but the origin was not clear.

We describe here further investigations of the model peptide in reversed micelles, the results of which implicate a high effective cation concentration as the predominant conformational determinant in the water pools. Additionally, we have been able to characterize changes in the nature of the micellar water pools consequent to the addition of a peptide solubilizate. Our analyses have been based on ^{1}H and ^{13}C NMR, CD, and IR studies of the peptide/AOT system.

EXPERIMENTAL

The synthesis of the cyclic peptide has been reported[11]. AOT was purchased from Fisher Scientific and was purified by dissolving in methanol, filtering, and evaporating the solvent. The resulting white foamy precipitate of AOT was dried <u>in vacuo</u> at 40° for 24 hrs immediately prior to use. Reversed micelle samples were prepared by dissolving dry AOT in the organic solvent (<u>n</u>-heptane, Fisher

Spectrograde for CD and IR; n-octane-d$_{18}$, Cambridge Isotopes, for ^{1}H NMR; or benzene-d$_6$, Aldrich, for ^{13}C NMR). Desired quantities of water (expressed as % v/v) were then added, followed by dry peptide (dried in vacuo at 50° for 12 hrs). Careful measurement of the chemical shift and integrated intensity of the water resonance in ^{1}H NMR yielded a check on the actual water content. ^{1}H NMR experiments with and without peptide were found to be most reproducible when sealed NMR tubes were used. Samples for IR were made up in the same manner, but with 5.5 M H$_2$O in D$_2$O (99.8% D, Bio-Rad) as the added water.

CD spectra were obtained on samples in 1 mm pathlength cylindrical cells with quartz windows, using a Jasco-Durrum J10 spectropolarimeter. NMR spectra were obtained in the Fourier transform mode on a Bruker WM250 spectrometer. Chemical shifts are referenced to internal TMS or dioxane (for ^{13}C NMR spectra in water). Near IR spectra were measured on solutions in 1 cm IR-quartz cells, with a reference sample prepared identically to the observed sample, but lacking the H$_2$O (D$_2$O added instead), using a Cary 17 spectrophotometer purged continuously with dry nitrogen.

RESULTS

The variation of CD spectra of cyclo(Gly-Pro-Gly-D-Ala-Pro) as a function of water content in AOT reversed micelles is shown in Figure 2A. Note particularly the decreased ellipticity at ca. 200 nm at low water content and the concurrent reduction in the magnitude of the negative extremum at longer wavelength. These changes are most pronounced in samples with the smallest amount of water, and the spectra approach that of the peptide in bulk water as the water content increases. An isosbestic point is observed at ca. 225 nm in the water titration. Figure 2B illustrates spectral changes for the cyclic peptide in bulk water as a function of added NaCl. The family of curves generated by varying [NaCl] from 0 M to 5 M bears a striking resemblance to those curves seen for peptide in AOT reversed micelles with varying amounts of water present. A clear isosbestic point is observed in this case as well. Note that spectra at low water content correspond to those seen at high [NaCl]; the CD of the peptide in reversed micelles with 0.5% (v/v) H$_2$O correlates most closely with that measured for 5 M NaCl.

^{13}C NMR spectra of the cyclic peptide within AOT reversed micelles reveal the same conformational trends seen by CD. Again, the observed parameters differ from those in bulk water, and can be induced by addition of NaCl to a solution of peptide in water (Figure 3). Previous work[11] had shown that the cyclic peptide under-

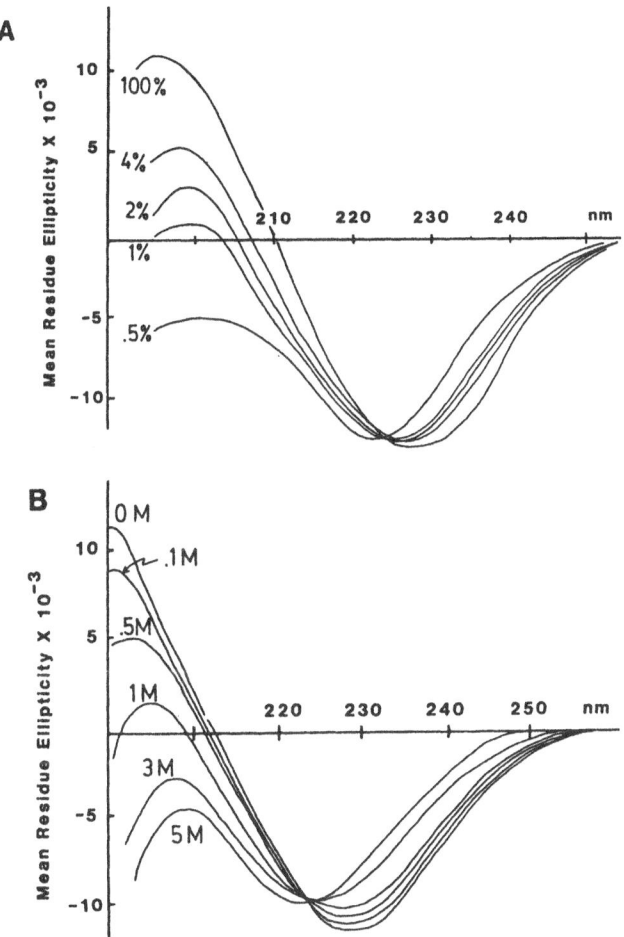

Figure 2. *Circular dichroism spectra of cyclo(Gly-Pro-Gly-D-Ala-Pro), A. as a function of water content in AOT reversed micelles (peptide 1mM in the water phase, AOT 3%, w/v, bulk solvent n-heptane, water concentrations given as % v/v), and B. as a function of NaCl concentration in aqueous solution (peptide 1mM).*

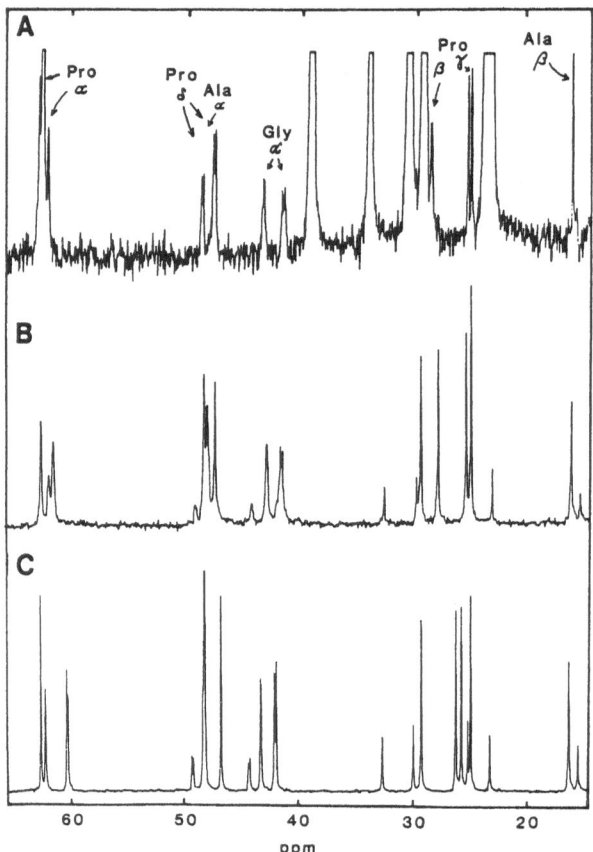

Figure 3. ^{13}C NMR spectra (62.9 MHz) of cyclo(Gly-Pro-Gly-D-Ala-Pro), A. in AOT reversed micelles (peptide 0.74 M in the water phase, AOT 3% w/v, bulk solvent benzene-d_6, 1% water v/v), B. in D_2O with 3 M NaCl, peptide 0.37 M, and C. in D_2O, no salt, peptide 0.37 M. (Note: The doubling of resonances seen in these spectra arises from the ^{15}N-enrichment of the Ala and of one Gly to 95% in this sample of the cyclic peptide. Spin-spin coupling between ^{15}N and ^{13}C leads to the observed splitting, for example, of the Gly C^α resonances). The Pro C^β resonance occurs at 26.22 ppm in C, and at 28.03 ppm in B.

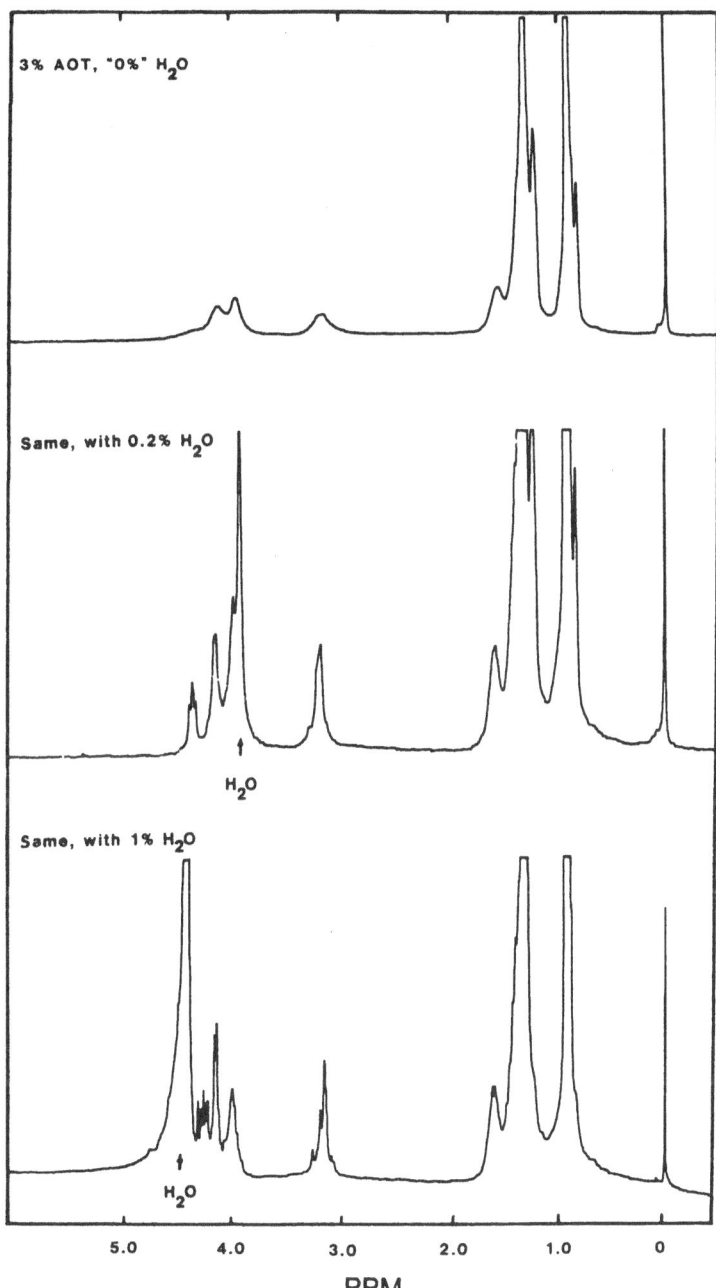

Figure 4. ^1H NMR spectra (250 MHz) of water in AOT reversed micelles as a function of water content (AOT 3%, w/v, bulk solvent octane-d_{18}, water concentrations given as % v/v).

goes a conformational change upon interaction with cations and had
established the spectral alterations associated with this change.
Most diagnostic in the ^{13}C NMR spectrum is the position of one of
the Pro C^β resonances. In the major uncomplexed conformer of the
cyclic peptide in water, this signal occurs at unusually high field
(26.22 ppm from TMS) due to the presence of a γ-turn hydrogen bond
(Figure 3, C). As complexation takes place with concomitant conformational adjustments forming a binding site, the resonance was
seen to shift downfield to a more typical position (28.03 ppm, for
3 M NaCl, Figure 3, B). The extent of this shift can be correlated
with the extent of binding of the peptide and, therefore, can be
used to probe the influence of the micellar microenvironment. Here,
a sample with 1% (v/v) H_2O showed a more strongly shifted C^β resonance position than that of peptide in 3 M NaCl solution in bulk
water (C^β chemical shift in the 1% H_2O in AOT sample 28.62 ppm,
Figure 3, A).

The influence of the presence of the peptide solubilizate on
the water within AOT reversed micelles was monitored using ^1H NMR
and IR spectroscopies of the water. Confirming a previous report[4],
it was found that the ^1H NMR chemical shift of the water resonance
in AOT reversed micelles varied markedly with water content
(Figure 4). At very low water percentages, the water protons re-
sonated at much higher field than in bulk water. Addition of peptide at any particular value of added water caused an upfield
shift of the water signal. Although small, the solubilizate-induced
shift was reproducible in direction and approximate magnitude. A
plot of water resonance chemical shift versus water content, with
and without added peptide, is shown in Figure 5.

IR spectra were obtained in the region of the water OH over-
tone frequency ($2\nu_{OH}$), where a clear distinction between hydrogen-
bonded and "free" OH stretching bands has been reported[15]. The
analysis of this region is simplified by using solutions of H_2O in
D_2O, so that any OH stretches observed are for HOD species[16]. In
this way, coupling of OH stretches within one molecule is obviated.
Figure 6 shows the IR spectral changes in the $2\nu_{OH}$ region as a
function of water content in AOT reversed micelles in n-heptane.
Note the band at 1390 nm at the lowest water content, which undergoes an apparent shift to longer wavelength as the water content
increases. The band at ca. 1420 nm has been assigned to "free"
(not hydrogen-bonded) OH's, while those at 1570 and 1650 nm are
due to hydrogen-bonded species[17,18]. Note also the relative growth
of these latter two bands with respect to the "free" band as water
content increases. Experiments assessing the effect of added peptide are in progress and will be reported elsewhere.

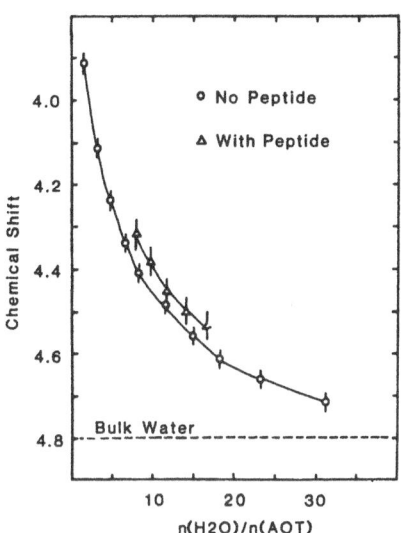

Figure 5. *Chemical shift of H_2O resonance as a function of water content in AOT reversed micelles. Water content is expressed as the molar ratio of water to surfactant, $n(H_2O)/n(AOT)$. Data with added peptide were obtained with 0.74 M peptide in the aqueous phase.*

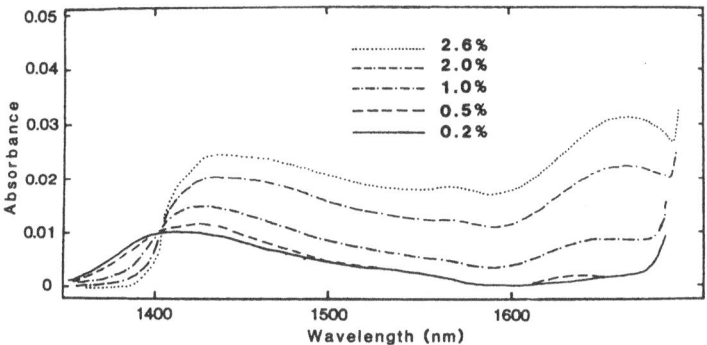

Figure 6. *Near-IR spectra of water in AOT reversed micelles as a function of water content (3% AOT, w/v, bulk solvent n-heptane, water phase 11 M HOD in D_2O).*

DISCUSSION

The model cyclic peptide serves as a sensitive probe of the conformational influence of the water pools inside AOT reversed micelles. It is clear from both CD and ^{13}C NMR data that the peptide undergoes a conformational transition at low water content in AOT samples to an ion-bound form, as is observed in concentrated aqueous salt solutions. A correlation between the effective Na$^+$ concentration in an AOT sample (defined phenomenologically from its impact on the peptide) and a salt concentration in an isotropic solution causing an analogous conformational change (for the titrations monitored by CD) suggests that 0.5% H_2O (v/v), with [H_2O]/[AOT] = 4.1 has the conformational impact of 5 M NaCl; 1% H_2O with [H_2O]/[AOT] = 8.3, that of 1 M NaCl; and 4% H_2O, with [H_2O]/[AOT] = 33.2, that of 0.5 M NaCl. Somewhat different values are obtained in the ^{13}C NMR experiments, where 1% H_2O in AOT correlates with >3 M NaCl. These measurements were performed at significantly higher peptide concentrations in the water pool (0.74 M) than in the CD study (1 mM), and the complexation equilibrium would be expected to be affected. In addition, this large difference in concentration of the solubilizate reflects a correspondingly large difference in the average occupancy of reversed micelles, and potentially significant alterations in the state of the water. The fact that the nonpolar solvent was benzene in the ^{13}C NMR studies may also have influenced the character of the water pool.

The implications of these results for biochemical systems are striking. It should be anticipated that biomolecules interacting in the interfacial water layers around membranes or macromolecules will experience locally high counterion concentrations and undergo consequent conformational changes. Although AOT has a different polar head group from a membrane lipid, there are structural similarities that justify its use as a model, notably the alkyl esters adjacent to an anionic group. Many biological lipids are zwitterionic, but often biomembranes have an excess of anionic lipids (most commonly, phosphatidylserine or phosphatidic acid).

The presence of the peptide solubilizate in the reversed micelle water pool was found by ^{1}H NMR to perturb the physical state of the water. Shifts in the distribution of water from hydrogen-bonded, bulk-like state (lowfield resonance position) to nonhydrogen-bonded (or more weakly hydrogen-bonded) states (highfield resonance position) were induced by addition of the peptide solubilizate. Assuming a two-state model, wherein water molecules are either bulk-like or are in hydration shells allows a qualitative assessment of the perturbation of the water pool by the peptide solubilizate. The shift to higher field, for a given n(H_2O)/n(AOT), seen in the chemical shift versus $n(H_2O)/n(AOT)$ plot upon addition of peptide can be expressed as a $\Delta n(H_2O)/n(AOT)$, or in other words, the incremental increase in water molecules in the nonhydrogen-bonded state (still assuming a two-state model). The amount of peptide added is known, leading to a value of the number of water molecules removed from the bulk (hydrogen-bonded) pool per peptide molecule. The result was ca. 10 H_2O/peptide. The magnitude of this number can be interpreted in terms of specific hydration of polar groups on the peptide and/or in terms of the creation of a surface within the water pool that disrupts the three-dimensional network of water-water hydrogen bonds that characterizes bulk water. Previous measurements of water of hydration of proteins[1,2] have described both effects as contributing to the observed perturbations in water properties. Work is in progress to explore more fully the specific interactions. Present data do not permit a distinction between the two effects. Introducing peptide solubilizates of varying surface area and charge may elucidate the origins of the hydration interactions. Furthermore, it is essential to test the validity of the two-state model for the enclosed water. NMR data reflect a weighted average of all populations of water present, and hence cannot readily be used to distinguish a two-state system from one with more than two states.

IR data, which arise from a measurement with a significantly faster time scale, can be observed for the various states separately. Our preliminary results reveal three states of water that exist in varying proportions as a function of water content in AOT reversed micelles. The first is apparent only below 0.5% H_2O ([H_2O]/ [AOT] = 4.1) and most likely corresponds to waters that solvate the AOT ion pair, but have not yet hydrated the Na^+. Other physical approaches have suggested the existence of such water[19,20]. Once 2-4 [H_2O]/[AOT] are present, the predominant state of the water interacting with the AOT head group corresponds to a hydration shell around the now separated anion and cation, and a shift in the IR band for the nonhydrogen-bonded water is seen. This specific hydration of AOT has been reported to require 8-10 [H_2O]/[AOT][4,21]. Changes in the IR spectrum reflect the build-up of a bulk water pool as the specific hydration interaction is completed. It will be very interesting to study the changes in these water populations upon introduction of a peptide solubilizate.

The potential of the reversed micelle system for studies of hydration of biological molecules is clear. A multifaceted physical approach to characterizing the nature of the enclosed water can be invoked. The interfacial water is seen without a large pool of bulk water that obscures the particular parameters of the populations of water of interest. The present case illustrates the remarkable sensitivity of this approach, inasmuch as solvation of a small solubilizate by a few water molecules could be qualitatively addressed.

ACKNOWLEDGEMENTS

    Financial support for these studies by Merck, Sharp & Dohme, the University of Delaware Research Foundation, the American Cancer Society, and the National Institutes of Health is gratefully acknowledged.

REFERENCES

1. I.D. Kuntz and W. Kauzmann, Adv. Protein Chem. 28 : 239 (1974).
2. R. Cooke and I.D. Kuntz, Ann. Rev. Biophys. Bioeng. 3 : 95 (1974).
3. M. Wong, J.K. Thomas and M. Grätzel, J. Amer. Chem. Soc. 98 : 2391 (1976).
4. M. Wong, J.K. Thomas and T. Nowak, J. Amer. Chem. Soc. 99 : 4730 (1977).
5. M.A. Wells, Biochemistry 13 : 4937 (1974).
6. E. Keh and B. Valeur, J. Coll. Interface Sci. 79 : 465 (1981).
7. P. Douzou, E. Keh and C. Balny, Proc. Natl. Acad. Sci. USA 76 : 681 (1979).
8. J.S. Thompson, H. Gehring and B.L. Vallee, Proc. Natl. Acad. Sci. USA 77 : 132 (1980).
9. C. Grandi, R.E. Smith and P.L. Luisi, J. Biol. Chem. 256 : 837 (1981).
10. S. Barbaric and P.L. Luisi, J. Amer. Chem. Soc. 103 : 4239 (1981).
11. L.G. Pease and C. Watson, J. Amer. Chem. Soc. 100 : 1279 (1978).
12. I.L. Karle, J. Amer. Chem. Soc. 100 : 1286 (1978).
13. J.E. Lacy, K.F. Thompson, P.I. Watnick and L.M. Gierasch, "Peptides: Synthesis-Structure-Function", D.H. Rich and E. Gross, eds., Pierce Chemical Co., Rockford, IL (1981), p. 339.
14. L.M. Gierasch, J.E. Lacy, K.F. Thompson, A.L. Rockwell P.I. Watnick, Biophys. J. 37 : 275 (1982).
15. L.M. Kleiss, H.A. Strobel and M.C.R. Symons, Spectrochim. Acta 29A : 829 (1973).
16. C.A. Swenson, Spectrochim. Acta 21 : 987 (1965).
17. J.D. Worley and I.M. Klotz, J. Chem. Phys. 45 : 2868 (1966).
18. H. Yamatera, B. Fitzpatrick and G. Gordon, J. Mol. Spectry 14 : 268 (1964).
19. R. Zana, "Solution Chemistry of Surfactants, Vol.1", K.L. Mittal, ed., Plenum, New York (1979), p. 473.

20. J. Sunamoto, T. Hamada, T. Seto and S. Yamamoto, Bull. Chem. Soc. Jpn. 53 : 583 (1980).
21. J.H. Fendler, Accts. Chem. Res. 9 : 153 (1976).

THERMODYNAMICS OF NUCLEIC ACIDS ENCLOSED IN REVERSE PHASE VESICLES

Bernhard Brosius, Gerhard Steger,
Wolfgang Hillen*, and Detlev Riesner

Institut für Physikalische Biologie
Universität Düsseldorf
D-4000 Düsseldorf 1, FRG

*Institut für Organische Chemie und Biochemie
Technische Hochschule Darmstadt
D-6100 Darmstadt, FRG

INTRODUCTION

Reverse micelles and vesicles mimic properties of biological membranes in the sense that they include a small water volume in a charged or in an at least polar surface formed by the hydrophilic headgroups of amphiphilic compounds. Membranes with this type of a surface form the cellular compartments which are the reaction vessels for most chemical processes of living organisms. Therefore, it is of basic interest to study the interaction of the membrane surface with the native macromolecular components of the cell, i.e. mainly with proteins and nucleic acids. Whereas the interactions of proteins with lipid membranes and proteins with nucleic acids have been studied in great detail, very little is known about the interaction of nucleic acids and lipid membranes.

In the present work we investigate the influence of a lipid membrane on the thermal stability and cooperativity of the helix-coil transitions of double stranded nucleic acids. The nucleic acids are entrapped in reverse phase evaporation unilamellar vesicles made from natural phospholipids. The system is shown schematically in Figure 1. It is easy to handle because it is stable up to 90°C. It is used in applied research for the infusion of genetic material into acceptor cells.

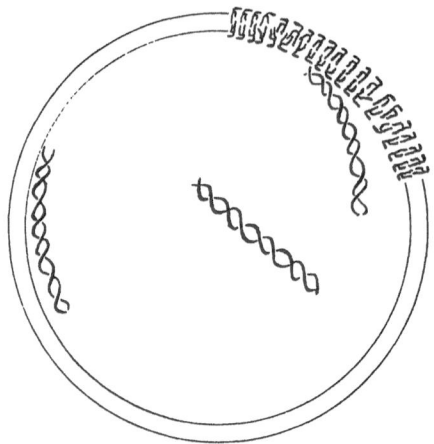

Figure 1. *Schematic diagram of the double stranded nucleic acid*
entrapped in a vesicle. The diameter of the vesicle is about
100-200 nm, the contour length of the nucleic acids is 60 nm for
the 180 base pairs DNA fragment and between 100 and 200 nm for
poly A:poly U. The double layer structure of the membrane and the
double helical structure of the nucleic acid are in an enlarged
scale.

MATERIALS AND METHODS

Chemicals

     Egg-yolk phosphatidylcholine dissolved in chloroform was pur-
chased from Sigma, bovine-brain phosphatidylserine dissolved in
chloroform from Calbiochem, RNase A(EC 3.1.27.5), Pronase and poly
adenylic acid (poly A) from Boehringer Mannheim, poly uridylic
acid (poly U) from PL Biochemicals. All other chemicals were of
reagent grade obtained from E. Merck, Darmstadt.

Preparation of DNA fragment

     The 180 base pairs DNA fragment was purified from a Hae III
digest of the plasmid pVH51 using chromatography on RPC-5 as the
final separation step, following a published procedure[7].

Preparation of vesicles

     The preparation procedure was similar to the method of Szoka
and Papahadjopoulos[1]. A total amount of 1.5 µmole lipid was placed
in a 5 ml round bottom flask and the chloroform was removed under
reduced pressure. The lipid film was redissolved in 1.4 ml diethyl-

ether, and 250 µl nucleic acid containing buffer (1 mg/ml nucleic
acid in 1 mM sodium-cacodylate, 0.1 mM EDTA, pH 7.4, and sodium-
chloride, as specified in the text) was added. Prior to use, poly A
and poly U were mixed in 50 mM ionic strength and diluted to the
final ionic strength in order to guarantee optimal double strand
formation. The solution was sonicated for 90 sec in an ultra sound
bath (Bransonic 220, 50 kHz) and then the ether was removed from the
emulsion under reduced pressure. The lipids were always protected
from oxidation by a nitrogen atmosphere. The suspension of vesicles
containing DNA was centrifuged for 2 min at 16000 g and the supernatant was chromatographed over a Sepharose 2B column (1 cm x 9 cm).
Unilamellar vesicles were eluted in the exclusion volume, well separated from a second peak containing not included DNA. Poly A:poly U
containing vesicles were treated with 10 µl RNase A solution
(2 mg/ml) for 30 min at room temperature, and the digestion was
stopped by addition of 10 µl Pronase solution (66 mg/ml)[2]. After
15 min the vesicles were separated from the fragments of the enzymes and of the not included nucleic acid by chromatography over
Sepharose 2B (0.5 cm x 8 cm).

The yield of lipids was estimated from phosphor analysis after
separation from the nucleic acid. The amount of nucleic acids included in the vesicles was estimated from the hypochromicity measured in the thermal denaturation curve. As a rough average 10 molecules of nucleic acid were included per vesicle.

Thermal denaturation curves

Thermal denaturation curves were recorded in a dual-wavelengths-
spectrometer Sigma ZWS 11 (Biochem/München, Figure 2)[3]. Routinely
the measurements were carried out simultaneously at 260 nm and
320 nm. At 260 nm the sum of the hypochromicity of the nucleic acid
and the change in turbidity of the vesicle was measured, at 320 nm
only the change in turbidity was followed. Because the ratio of the
changes in turbidity at 260 nm and 320 nm was determined in a reference experiment on vesicles without nucleic acids included, hypochromicity curves at 260 nm could be corrected for the contribution
of the turbidity. The corrected curves are presented in the figures.

RESULTS

Resolution of transitions of free and entrapped nucleic acids

The differentiated melting curve obtained for a preparation of
the 180 base pairs DNA fragment in phosphatidylcholine vesicles
shows two clearly separated transitions (cf. Figure 3A). The transition at lower temperature is identical in midpoint temperature T_m

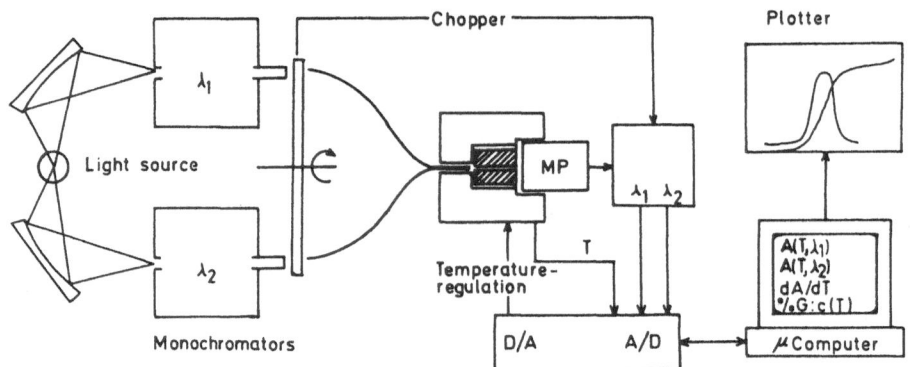

Figure 2. *Instrument for recording thermal denaturation curves at two wavelengths. Light from both monochromators is guided by a fiber optic to the cuvette holder, recorded by one photomultiplier, and processed electronically into two separate channels. The temperature was followed with a NTC-resistor in the thermostated cuvette holder; semimicro cuvettes had 0.6 ml sample volume. The melting curves and the temperature were recorded on a microcomputer (Apple II, interface from Technosystem, Edingen-Neckarhausen) which also generated the temperature program for the thermostating bath.*

and halfwidth $\Delta T_{1/2}$ to that measured for the DNA fragment in the absence of vesicles. Therefore it has to correspond to those DNA molecules not entrapped in the vesicles. The second transition represents the melting of DNA if entrapped in the vesicles. An increase in T_m by 4.0°C and in $\Delta T_{1/2}$ from 0.7°C to 2.7°C was measured. If DNA was added to a preparation of empty vesicles, only the transition of free DNA was detected, showing that the second transition cannot be from an outside binding of DNA to vesicles.

In measurements with poly A:poly U, stabilization and broadening of the transition was found, similar to the results with DNA (cf. Figure 3B). The transition of free poly A:poly U could not be observed in the same experiment, because nucleic acid not entrapped was digested by RNase. Therefore, the transition of free RNA was determined in a separate experiment. In 90 mM ionic strength, under which condition the transition of free nucleic acid is reversible, full reversibility was also found with nucleic acids entrapped in vesicles.

Influence of the lipid composition on the thermal denaturation of the entrapped nucleic acid

Vesicles were prepared from mixtures of phosphatidylcholine and phosphatidylserine. The composition was varied from pure phos-

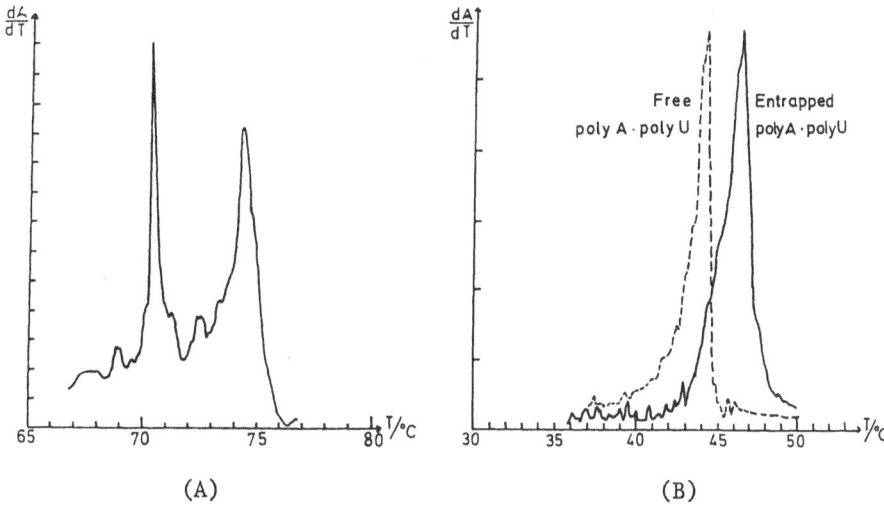

Figure 3. *Differentiated thermal denaturation curves.*
A) 180 base pairs DNA fragment enclosed in phosphatidylcholine
vesicles. $\Delta A/\Delta T$ is in relative units. Smoothing procedure was
carried out as described in (3). Buffer conditions were 1 mM
sodium cacodylate, 0.1 mM EDTA, pH 7.4, 10 mM NaCl.
B) poly A:poly U enclosed in phosphatidylcholine vesicles. 20 mM
NaCl, other conditions as in A). $\Delta A/\Delta T$ scales for both curves
are not identical.

phatidylcholine membranes with an amphoteric surface to pure phosphatidylserine membranes with a negatively charged surface. From
Figure 4 it may be seen that amounts of more than 20% phosphatidylserine lead to a decrease in stabilization. All mixtures of lipids
induce a further increase in $\Delta T_{1/2}$ from 2.7°C (pure phosphatidylcholine) to ∿7°C (20-100% phosphatidylserine).

Influence of ionic strength on the thermal denaturation of entrapped nucleic acid

In order to estimate the electrostatic contribution to the
interaction of nucleic acids with lipid membranes, the ionic
strength of the buffer solution, which is identical inside and outside of the vesicle, was varied between 10 mM Na^+ and 90 mM Na^+.
The helix-coil transitions of the nucleic acids were studied with
poly A:poly U because they occur within a handy temperature range;
with DNA-fragments, the measurements would have to be extended up
to temperatures higher than 90°C. In Figure 5 the ionic strength

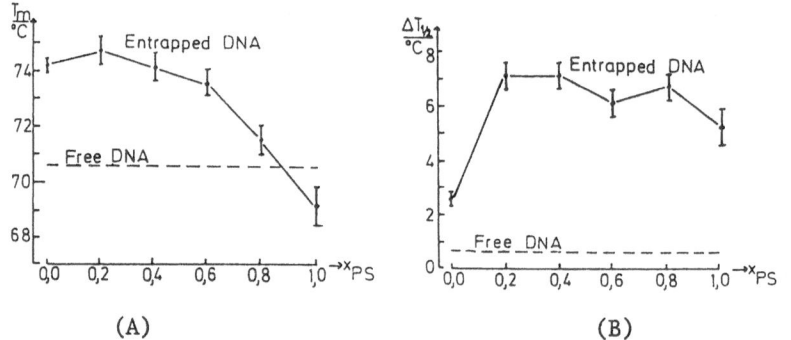

(A)   (B)

Figure 4. *Influence of the lipid composition on T_m-value (A) and $\Delta T_{1/2}$-value (B) of the helix-coil transition of the 180 base pairs DNA fragment. Lipid composition was varied from pure phosphatidylcholine corresponding to $X_{PS} = 0.0$ (molar fraction of phosphatidylserine) to pure phosphatidylserine corresponding to $X_{PS} = 1.0$. Other conditions as in Figure 3A.*

dependence of free and entrapped RNA is depicted. The stabilization due to the enclosure is largest for low ionic strength, decreases up to around 30 mM and shows a slight increase at higher ionic strength.

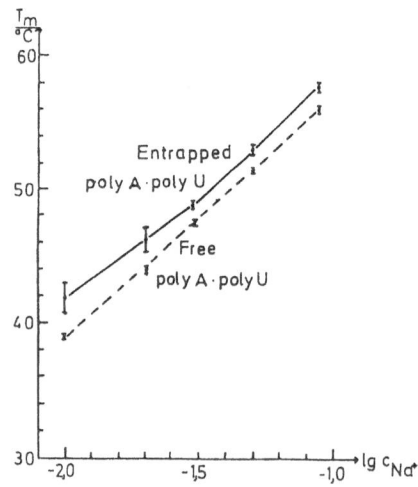

Figure 5. *Ionic strength dependence of the T_m-value of the helix-coil transition of poly A:poly U. Other buffer conditions were as in Figure 3A.*

DISCUSSION

It is shown that the thermodynamic properties of nucleic acids enclosed in vesicles can be followed with sufficient accuracy by optical methods. The enclosure in vesicles affects the stability and the cooperativity of the helix-coil transition of the nucleic acids.

The interaction between the nucleic acid and the membrane surface may make several contributions. Because double strand formation is a second order process, it is concentration dependent, and in the denatured state the enclosure in vesicles guarantees a high effective concentration of single strands. After standard procedures were applied in the case of the DNA fragments[4], this effect was estimated to contribute less than $1^{\circ}C$ to T_m and may account for only a small part of the observed stabilization. It must be expected that there is a stronger influence from electrostatic interactions between the membrane and the nucleic acid. The electrostatic contribution is obvious from two experimental findings. First, as seen from Figure 5, below 30 mM Na^+ the ionic strength dependence of the helix-coil transition is lowered, due to the influence of the membrane. It means that the membrane surface may partially replace the stabilizing counter ions, an effect which is attenuated at higher ionic strength. Because the headgroup of phosphatidylcholine is amphoteric, one has to assume that the stabilizing influence of the ammonium group exceeds the de-stabilizing effect of the phosphate group. Second, increasing the net negative charge on the surface by adding phosphatidylserine decreases the stabilizing effect and results in a slight destabilization. There is, however, also evidence for a direct contact between the nucleic acid and the membrane surface. At ionic strengths above 30 mM Na^+, the stabilization increases slightly, although electric forces are more and more shielded[5]. Furthermore, the drastic loss of cooperativity of the helix-coil transition indicates that the influence of the membrane surface cannot be explained by a mere change of the effective counterion concentration; it is known that the counterion concentration affects the stability more than the cooperativity[6]. At present, we do not understand the molecular basis for the drastic decrease in cooperativity. It shows, however, that due to the contact with the membrane, smaller regions in the nucleic acid are able to open up. From nearly 180 base pairs opening up cooperatively in free nucleic acid, this number is lowered to less than 50, a size in the range of promoter regions. Such a decoupling of different regions on the nucleic acid would facilitate opening particular regions. More experimental data are needed to support these speculations, but at the present it can be said that the biological relevance of the results presented above may lie more in the drastic decrease in cooperativity than in the slight modulation of th stability.

ACKNOWLEDGEMENTS

We thank Profs. C. Nicolau, J. Seelig, W. Vogell and Dr. Schindler for stimulating discussions. The work was supported by grants from the Minister für Wissenschaft und Forschung des Landes Nordrhein-Westfalen and the Fonds der Chemischen Industrie.

REFERENCES

1. F. Szoka and D. Papahadjopoulos, Proc. Natl. Acad. Sci. USA, 75 : 4194 (1978).
2. G.J. Dimitriadis, FEBS Lett., 86 : 289 (1978).
3. J.W. Randles, G. Steger and D. Riesner, Nucleic Acids Res., 10 : 5569 (1982).
4. D. Riesner and R. Römer, in: "Physico-chemical properties of nucleic acids" (J. Duchesne, ed.) Vol. 2, pp 237-318, Academic Press, London and New York (1973).
5. M.T. Record, Jr., C.F. Anderson and T.M. Lohmann, Q. Rev. Biophys. 11 : 103 (1978).
6. O. Gotoh, A. Wada and S. Yabuki, Biopolymers, 18 : 805 (1979).
7. W. Hillen, R.D. Klein and R.D. Wells, Biochemistry, 20 : 3748 (1981).

TECHNOLOGICAL RELEVANCE OF MICROEMULSIONS AND REVERSE MICELLES IN APOLAR MEDIA

D. Langevin

Laboratoire de Spectroscopie Hertzienne de l'E.N.S.
24 Rue Lhomond
75231 Paris Cedex 05

It would be very difficult to make a complete catalog of all the industrial and technological applications of microemulsions and reverse micelles. Rather than giving a prohibitively long list of these applications, we have selected a few important examples, emphasizing the particular role played by the disperse medium in each case. Indeed, if most of the applications of reverse micelles are based upon their ability to solubilize substances, they can also be used as a "reservoir" of monomers or to play special roles in absorption processes. Examples of the relevance of these properties are given for the fields of oil recovery, lubrication, detergents and catalysis. Some speculations will be made which are pertinent to several other fields where the role of these systems is less well understood, such as the preparation of small solid particles and liquid-liquid extraction.

INTRODUCTION

Although the structure of reverse micelles was not described until around 1920 by McBain and coworkers and microemulsions were described by Shulman and coworkers around 1940[1], their practical use, as is often the case, preceded the theory. For instance, a microemulsion formulation was already widely used by Australian housewives more than a century ago. It was a transparent dispersion of eucalyptus oil in water made by adding soap flakes and white spirit. This formulation was very efficient for the washing of wool;

it did not require rinsing and the tensioactive agents present in
eucalyptus oil ensured smoothness and protection for the wool. Several commercial products, such as liquid waxes, cutting oils and
detergent formulations, were patented around 1930[2]. More recently
the importance of microemulsions for tertiary oil recovery was the
origin of the development of a great number of theoretical and experimental studies of these systems. Because of the energy crisis,
photochemistry in micellar systems is actively being investigated
in relation to the storage of solar energy. The potentials for
applications in the field of drug delivery are being considered,
because some specific advantages are offered over vesicles and liposomes[3].

Our purpose in this paper will be to discuss the relevance of
amphiphilic structures in apolar media. We will then exclude the
surfactant systems in which the water content is very large, leading
e.g. to water external micellar solutions. It must be kept in mind
however that these systems are of considerable technological importance in the fields of detergents, food industry and cosmetics. We
will mainly be concerned with reverse micellar systems and microemulsions, although other amphiphilic structures (lamellar, hexagonal, cubic) can be encountered in the water-oil-surfactant phase
diagrams. The viscosity of these phases is generally large (gel-
like) and for this reason they are not very suitable for many practical applications. Indeed, a prerequisite for the elaboration of a
product is to be stable at compositions far removed from those corresponding to gel-like phases.

We now come to the point of the distinction between water external or oil external microemulsions, which is by no means obvious.
Indeed, when the water content is either very low or very high,
the structure of the microemulsions is generally admitted to be
that of spherical droplets of either water or oil dispersed in
either oil (w/o) or water (o/w). When the water content is between
about 10% and 90% the structure is much less clearly understood.
We recently argued[4] that when the droplets have a very fluid and
disordered interfacial surfactant layer, the system may become
bicontinuous above a droplet volume fraction of about 10%. This
type of structure was proposed in 1976 by Scriven[5] and several
theoretical models have since been elaborated[6,7]. When the surfactant layer is more rigid, i.e. when droplets are not allowed to
interpenetrate, the inversion between w/o and o/w is sharper and
the structure may be that of droplets existing in a large range of
water content. In the following, we will include in our discussion
the possible bicontinuous structures.

Another important point is the distinction between reverse micelles and microemulsions[9]. It is generally accepted that the so-called reverse micelles contain amounts of water equal or less than the hydration water of the polar part of the surfactant molecules, whereas in microemulsions the water properties are close to those of bulk water. The solubilization properties of these two systems should clearly be different. However, when the surfactant system plays the role of a reservoir of monomers, the distinction between reverse micelles and microemulsions is less relevant for practical purposes. We will then emphasize this difference only when needed, for example in particular applications where the nature of the microenvironment of micellar cavities plays an important role.

Finally, let us mention that the fields of applications of micelles and microemulsions often correspond to those of vesicles and emulsions. The advantage of the first systems is their thermodynamic stability (for example their physiochemical properties are expected to be unchanged over a temperature cycle) and the very small scale of the microstructure: 10 - 200 Å. The radii of the emulsions' droplets are typically 1μ, which make the system very turbid and unsuitable for some particular applications, for instance photochemistry. The vesicle dispersions are more transparent, but the vesicle radii are never as small as those of micelles (typically 200 Å). This size can be a problem in the field of drug delivery, e.g. when the vesicle is large enough to interact with some small elements of the living cells[3].

In the following we will present some selected examples of the most important applications of reverse micelles and microemulsions. They will not necessarily correspond to the latest developments in the field, which are often found in the patent literature. However, they will allow us to illustrate the different possible roles for the disperse medium.

PETROLEUM INDUSTRY - OIL RECOVERY

Tertiary oil recovery. Due to its enormous economic potential tertiary oil recovery was of interest well before the recent energy crisis[10]. Indeed primary tapping and ordinary water flushing of oil wells allow one to recover only about one half of the total oil content. Several processes have been considered to improve the recovery and have been used in the field since they became profit-earning in recent years. We show in Table I a list of these different processes for oil recovery and the corresponding production in USA in 1980 (for a total production of 8.5 millions barrels/day).

Table I. *Tertiary Oil Recovery U.S.A. 1980*

	Number of Active Projects		Production (Barrels/Day)
Thermal Methods	*133*	*Steam Injection*	*295,000*
	17	*In Situ combustions*	*12,000*
Chemical Methods	*14*	*Microemulsions/Polymers*	*930*
	22	*Polymers only*	*920*
	6	*Caustics*	*550*
Gas Injection	*17*	*CO_2*	*21,000*
	17	*Others*	*53,000*

Source: Oil and Gas Journal, March 31, 1980.

As it can be seen from the table, microemulsion projects are still of trial character, but their economic potentials are extremely interesting for the future, since they allow in theory the recovery of 100% of the oil of the reservoir.

The mechanisms by which microemulsions improve oil recovery have been studied extensively in the past 20 years[10]. At the end of water flooding, the residual oil is believed to be in the form of discontinuous oil ganglia trapped in the smallest pores of the rocks in the reservoir. In order to recover this oil by a flooding process, the interfacial tension between the oil and the injected fluid has to be lower than about 10^{-2} dyne/cm. In this way, the oil ganglia previously trapped by capillary forces will be displaced, with reasonable values of the injection pressure, and all the residual oil will be recovered.

The crude oil-water interfacial tensions are of the order of a few tenths of a dyne/cm. When an appropriate surfactant system is used, the interfacial tension can be reduced to 10^{-3}-10^{-4} dyne/cm. Although the first studies were made with low concentrations of surfactants, several workers suggested that still lower interfacial tensions (IFT) could be obtained with high surfactant concentration (microemulsion) systems. The optimum conditions were shown to be satisfied by microemulsions coexisting with both oil and water: middle phase microemulsions. The role of polymers in the flooding

process is simply to control. Indeed one has to prevent viscous fin-
gering of the drive water into the oil bank, and this is done by
adding water-soluble polymers that increase the viscosity of the
water.

Low oil-water interfacial tension-possible sources. The most
widely studied aspect of the role of microemulsions in tertiary oil
recovery is the lowering of oil-water interfacial tensions. Several
reasons were postulated for the low tensions:

- thin surfactant layer of high surface pressure[11]

- thick surfactant layer of liquid crystalline material[12] or
  associated with the vicinity of a critical consolute point of
  the oil-water-surfactant mixture[13,14].

The first concept seemed very reasonable for microemulsion systems that, according to Shulman, form when suitable surfactant molecules are mixed with oil and water leading to zero interfacial tension:

$$\gamma = \gamma_{ow} - \pi = 0 \qquad\qquad \textit{Equation 1}$$

γ and γ_{ow} being the interfacial tension between oil and water with
or without surfactant respectively. In this simple view, it is assumed that the surfactant is insoluble in bulk oil and water and that
the interactions between different portions of the interface and the
curvature energies are negligible. The fact that the surface pressure can be equal to γ_{ow} at an optimum surfactant concentration
leading to an effective zero tension is a particular feature of oil-
water interfaces. It can be shown for instance, by using the same
assumptions, that when the free surface of water is saturated by
surfactant molecules, the surface tension cannot be lowered below
γ_0, which is the free energy per unit area for the interface sur-
factant-air, which is typically on the order of a few tenths of
dyne/cm[15].

The zero interfacial tension state is never obtained in practice, because the surfactants are always slightly soluble in water
and/or in oil. When the saturation state is reached at the interface, micelles (or other aggregates) begin to form in the bulk phases.
The interfacial tension γ at this point is related to the chemical
potential difference for surfactant molecules in the micelles and in
the interfacial layer. It can be shown that γ can be written as[7,16,17]

$$\gamma = \frac{k}{2RR_0} \qquad\qquad \textit{Equation 2}$$

where R is the micelle radius, k the "rigidity" modulus of the interface and R its spontaneous curvature. Numerical estimations for o/w microemulsions containing ionic surfactants give very small γ values ($\sim 10^{-2}$ dyne/cm)[18].

However, although it was early recognized that oil-water interfacial tensions could be very small around the surfactant "cmc"[19], it seemed that γ could still be much lower at larger surfactant concentrations, even with pure surfactants systems[20]. In principle this can also be accounted for by theory. In fact at large surfactant concentrations,γ contains supplementary terms related to entropic contributions (oil can be solubilized in water and water can be solubilized in oil) and to interactions between droplets of different shapes.   .

Although many crude oil-commercial surfactants-water as well as model systems with pure chemicals have extensively been studied, very few attempts have been made to relate γ to the actual structure of the bulk phases. We will briefly describe in the following the work that we have performed in our laboratory where we have been able to distinguish clearly between the roles of the surfactant film and the microemulsion bulk phases[21].

We have studied a model system: 47% toluene, 47% water + NaCl, 2% sodium dodecyl sulfate (SDS) and 4% butanol. Depending of the salinity of the water S(Wt%) the mixture separates into 2 or 3 phases:

   S < 3.5 a o/w microemulsion in equilibrium with excess oil

   S < 7.5 a w/o microemulsion in equilibrium with excess water

   3.5 < S < 7.5 a middle phase microemulsion in equilibrium
                  with both excess water and oil.

In the two phase region we have been able to dilute the microemulsion phases. By using bulk light scattering techniques we have measured the radii of the micelles and the strength of their interactions (virial coefficient). Some of these results are reported in Table II. For the three phase system we were not able to achieve such a dilution, probably because, as strongly suggested by electrical conductivity measurements[4], the middle phase may be bicontinuous.

We then tried to investigate the influence of the microemulsion phases on γ. The tensions of the initial system are represented in Figure 1 as a function of water salinity. They have been measured with the surface light scattering apparatus built in our laboratory

Table II.

S	R	R_H	B	α
3,5 wt%	85 Å	95 Å	5	-1
5	130	190	3	-5
5,2	155	240	0	-6
8	180	200	2	-7
10	125	150	6	-2

Examples of microemulsion characteristics for different salinities. The radius R and virial coefficient B are related to osmotic compressibility χ through $\chi = (3\,k_BT/4\pi R^3)(1 + B\phi)$, ϕ being the volume fraction of droplets. The hydrodynamic radius and virial coefficient α are related to the diffusion coefficient through $D = (k_BT/6\pi\eta_c R_H)(1 + \alpha\phi)$, η_c being the viscosity of the continuous phase. For hard spheres, $B = 8$, $\alpha = 1.5$. The difference between R and R_H, close to the phase separation boundary $S_1 = 5.4\%$ can be attributed to polydispersity and non-spherical droplet shape.

a few years ago (Surface light scattering and spinning drop apparatus were shown to give surface tension values in close agreement[22]). Figure 1-b represents the tension values obtained when the micelles are removed from the microemulsion phase in the two phase equilibria, i.e. by replacing the microemulsion by the continuous phase of its droplets. The tension values in the three phase system were measured between the two excess phases that contain low surfactant concentration ($\sim$cmc) but no micelles. The striking conclusion from the comparison between the two figures is that γ in the two phase region is the same when micelles are or are not present. Consequently, as in usual micellar solutions, γ is only related to the properties of the surfactant layer at the interface. This can be confirmed theoretically by numerical calculations of curvature entropy and interaction terms by using the data of Table II[18]. The conclusion is less clear for the three phase region but it is evident that the largest of the two tensions γ_{mo} and γ_{mw} (Figure 1) and in particular the tension at the optimal salinity S^* ($\gamma_{mw} = \gamma_{mo}$) are not related to the middle phase microemulsion structure.

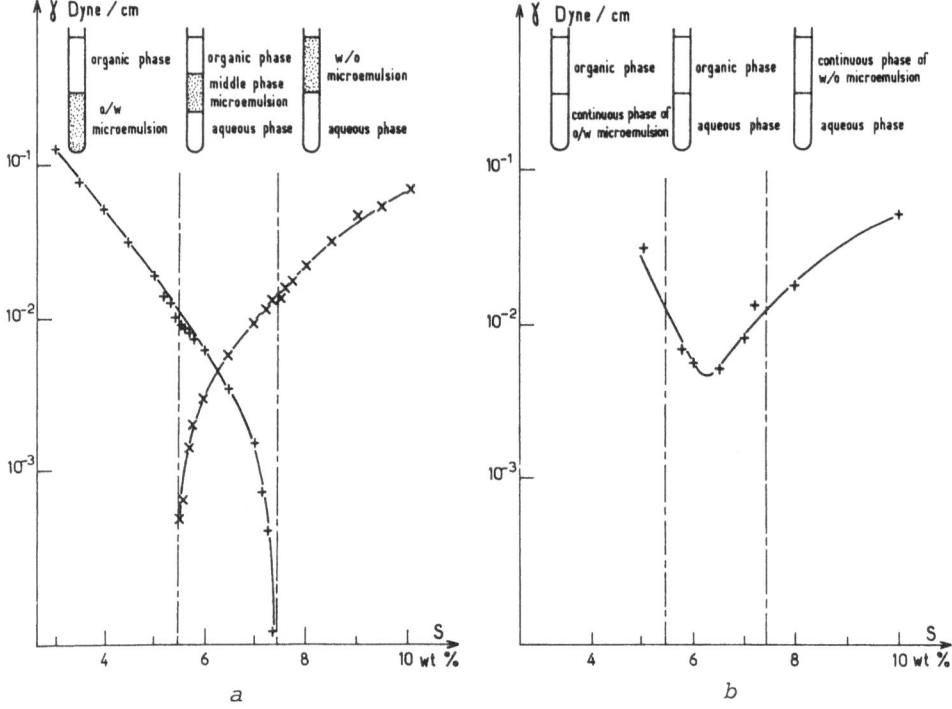

Figure 1. *a) Interfacial tension between the different phases versus salinity, + = γ_{mo}; x = γ_{mw}. b) Interfacial tension between organic and aqueous phases containing no micelles.*

To the contrary, the lowest tensions in the three phase region are probably strongly affected by interactions and (or) entropy effects since evidences of the vicinity of critical consolute points were found close to the salinity values where these tensions are the smallest ones (phases boundaries). As expected these low tensions were associated with thick interfaces[23].

CONCLUSIONS

As far as oil recovery is concerned, the optimal recovery is obtained when all the tensions are below 10^{-2} dyne/cm, i.e. in practice at the optimal salinity. Thus in principle the microemulsion can be removed and low surfactant concentrations could be used without lowering the efficiency. But in practice the surfactants adsorb on the porous rocks and the interfacial tensions become rapidly large. It is then very interesting to use microemulsions in between the oil and water, (the microemulsions playing the role of surfactant reservoirs).

It must also be recalled that microemulsions might play a role in other phenomena involved in oil recovery, namely:

- wetting of the porous rock (solid-liquid surface tensions)

- oil bank formation (coalescence of oil ganglia)

- influence of concentration, pressure and temperature changes during the flooding process

- emulsification processes

- interactions with polymers.

All of these problems are extremely complex and still poorly understood[10].

Let us finally mention that the important research in the field of micellar systems and microemulsions made by the oil companies have led to a parallel development of other technological applications, such as improvement in octane, pollution abatement and petrochemistry[24].

LUBRICATION

Reverse micelles play a very important role in corrosion inhibition. When certain types of surfactants (petroleum sulfonates for instance), are dissolved in motor oils, they can solubilize corrosive oxidation products and prevent them from reacting with engine parts. The simplest mechanism for achieving this aim is micellar solubilization[25], and as expected, certain inhibitors become effective above their cmc. It must be pointed out that some surfactants are not effective at all, indicating that solubilization can be selective, and more surprisingly others are effective only below the cmc, where other mechanisms must be found.

The lubrication properties of mineral oils can be improved by adding water that acts like a coolant. This idea is used in cutting oils where the oil lubricates the cutting surface and the water, because of its much greater thermal conductivity, removes the produced heat; hence the surfactant plays the double role of emulsion stabilizer and corrosion inhibitor. The first cutting oils were emulsions, but as the cutting fluid must be recovered and re-circulated, after a few cycles the emulsion composition varies and its efficiency decreases. Since 1930 a great improvment has been

achieved through using microemulsions because of their stability.
The typical formulation consists of a petroleum sulfonate as corrosion inhibitor and emulsifier, a second emulsifier of the soap
class, a coupling agent like diethylene glycol, an antifoam agent,
water and mineral oil[2].

DETERGENTS

The field of detergents deals with the removal of unwanted material from fabrics or metal surfaces. The possible mechanisms contributing to this process are mainly[26]:

- wetting: adsorption of surfactants on fabrics changes the contact
  angle between the soil particle and the fabric surface (because
  surface tensions are changed). This can lead, if a suitable surfactant is used, to a great reduction in adhesion of the soil to
  the fabric;

- solubilization: ability of the surfactant to keep the soil material in suspension. This process can be different from the micellar solubilization, i.e. the surfactant can aggregate around
  the soil particle at a concentration lower than its cmc in the
  washing fluid. For instance, the detergency of SDS aqueous solutions rises before the cmc is reached and remains practically
  constant thereafter[27]. Thus micelle formation can be a competing
  rather than a contributing process.

The soil that can be found on fabrics is generally oily and
the current washing fluids are aqueous surfactant solutions. However the suspending power or "anti-redeposition properties" were
found to be better when microemulsions were used as washing fluids.
The precursor of the modern formulations was a w/o microemulsion
containing pine oil, wood rosin, sodium oleate and water[2]. Upon
addition of further water, this microemulsion inverts to a o/w
microemulsion, if the initial amount of soap is sufficient. Currently
research[28] is made in this field to improve the solubilization efficiency which seems to be the most important role played by microemulsions.

Reverse micelles are also used in dry cleaning processes to
remove polar dirt substances on fabrics. In these formulations, a
small amount of water is incorporated in the dry cleaning fluid,
thus improving the solubility of polar substances.

DYE SOLUBILIZATION

Polar dyes can be solubilized in non-polar solvents with the help of reverse micelles. This is of considerable importance for the technology of dye media. This phenomenon was studied as early as 1940 by McBain who found that solubilization was highly specific[29]. Solubilization is generally accompanied by a significant shift of the fluorescence emission wavelength and large variations in quantum yields of fluorescence. These effects have been attributed to the breakdown of the aggregates that these dyes form in water and to the inhibition of radiationless decay by an increase of the effective viscosity in the micelles. These effects can have interesting potential applications, for instance, in laser dyes and analytical probes. Let us also mention that dyes are widely used for visual cmc measurements and that more refined studies of the fluorescence, its depolarization and lifetime, give interesting information about the micellar structure, e.g. aggregation number and microviscosity[30].

CATALYSIS

The remarkable rate enhancement of certain chemical reactions in reverse micelles is of great potential for industrial application[31]. One of the mechanisms that possibly lead to the rate enhancement is the considerable increase of contact surface between reactants attached to the micelle (at the surface or in the interior). However, the specific increase in reaction rate can be much greater than that expected from the simple partitioning of substrates between micelles and a continuous phase.

In these cases the nature of the microenvironment may influence the passage of a reacting system through its transition states. The microenvironment may indeed resemble the active sites of enzymes, and micelles could be able to bind substrates in fixed configurations and orientations.

Finally, the fluidity of the surfactant layers may have a pronounced influence on diffusion controlled reactions of the reactants in the water pool. This last phenomenon is being studied in detail with a model reaction by B. Robinson and coworkers[33]. By complexing nickel ions with nitrogen ligands (murexide) in w/o microemulsions (AOT – water – heptane) they were able to estimate the number of "sticky" collisions, i.e. the collisions leading to an exchange between the water content of the droplets. From these experiments, it may be postulated that, if the reaction rates are faster than the exchange times (typically 1 µs), droplets are essentially isolated and "frozen" and if the reaction rates are longer, the water phase can be regarded as "continuous" from the reaction kinetics point of view. It must also be pointed out that in some microemulsion systems in which droplets interactions are hard-sphere-like, the droplet interface is very rigid and it appears, as evidenced by electrical conductivity and ultrasonic absorption measurements, that the number of "sticky" collisions is significantly reduced[33]. Although a general correlation between interfacial structure and reaction rates in diffusion controlled reactions has not been established, it is clear that such a correlation will have important general applications.

Let us now turn to specific cases of catalysis by reverse micelles. Water-free micelles of anionic surfactants have interiors which have been likened to molten salts. Metal cations therein have proven to be effective catalysts for a variety of organic reactions[34]. The acidity of the cation dominates the core microenvironment because of the usually weak Lewis basicity of the surfactant anions. Possible reactions include the hydrolysis of esters, where both water and ester are basic and will be attracted by the acid-catalyst and the freed acid will be rejected from the core. With a simpler reaction (unimolecular decomposition of benzylchloroformate) in reverse micelles of metal salts of Aerosol OT, F.M. Fowkes and coworkers showed that the order of effectiveness of the metal ions was the same as that of their Lewis acidities[34]. Other reactions catalyzed by metal ions in reverse micelles are described in the patent literature.

Another important application of reverse micelles is the ionic polymerization in non-polar solvents. A micellar version of the Ziegler catalysis or diene polymerization is described in the patent literature[34,35].

We cannot conclude this section without mentioning the photochemical aspects of catalysis. A typical photochemical reaction can be:

$$A + B + h\upsilon \rightarrow A^- + B^+ \qquad\qquad Equation\ 3$$

Energy storage occurs when the lifetime of the ion pair is long enough. Micelles, vesicles and microemulsions have been investigated recently for this purpose[36,37]. If for instance a cationic reverse micelle is used, B^+ will be repelled after formation from a radiation. Its ejection from the micelle core can increase the ion pair lifetime. On the other hand, solubilization of A and B in the micelles increases the quantum efficiency of the process.

PREPARATION OF SMALL PARTICLES

Micellar systems have been widely used for the elaboration of small particles like catalysts and magnetic colloids. The general method consists of preparing the particles directly in the micellar system with a suitable chemical reaction. However up to now the relationship between the final size of the particle and the initial micellar radius is not clear.

A. Catalysts

Heterogeneous catalysis plays an important role in modern industrial processes. Nearly all industrial chemicals come into contact with a catalyst before they reach the consumer. Metal catalysts are an important class among the usual hetereogeneous catalysts. More dispersion increases their activities, due either to the increased number of active sites and/or to the change in nature of these active sites. Micellar systems have the great advantage over the more clasical methods in that they allow the preparation of small (lower than 100 Å) and monodisperse particles. For instance stable iron and nickel boride particles can be prepared by reducing iron and nickel ions by sodium borohydride in CTAB-n hexanol-water reverse micelles. Particle radii can be varied between 30 and 80 Å and the polydispersity is very small: 2-5 Å. The average size of the particle does not seem to be directly related to the size of the water core and larger micelles may lead to smaller solid particles[38].

Microemulsions have also recently been used to prepare metal particles[28,39]. A metal salt is solubilized in the aqueous regions of polyethylene glycol diethylether-hexadecane-water microemulsions. If the water is acidified, the metal ions are reduced and metal particles grow to a final stable size: 27 ± 3 Å for platinum particles. Rhodium, palladium and irridium particles have also been prepared in this way[28].

B. Photographic emulsions

In photography, as in catalysis, very small and monodisperse particles such as silver chloride are required. An attempt to achieve a better understanding of particle growth has recently been made on oil-external Aerosol OT and non-ionic microemulsions by combining X-ray and light scattering techniques[40]. Two processes were clearly distinguished: crystal formation with a small final diameter (60 Å) and flocculation of these crystals. But the reason for the constant size of the first crystal remains unknown.

C. Magnetic colloids

Magnetic colloids are systems that have both fluid and magnetic properties. As they can easily be confined in a restricted zone by using permanent magnets, they open a broad field for technological applications. Some are under development, others are already in use: e.g. difficult seals and lubrication problems in vacuum pumps, the chemical industry, computer memory protection, separation of minerals on a semi-industrial scale, in the manufacture of loud speakers, by improving thermal contacts and thereby increasing the dissipative power and drug delivery in medicine[41].

Cobalt ferrofluids can be prepared by thermal decomposition of dicobaltoctacarbonyl dissolved in reverse micelles made with a polymer surfactant[42]. Microemulsions can also be used. For instance, the precipitation of iron chlorides in AOT microemulsions by ammonia solutions led to very monodisperse magnetic dispersions. The diameter of the particles is around 1000 Å[43].

D. Microlattices

Emulsion polymerization is a very widely used method for the preparation of latex dispersions. It has however some disadvantages. The latex particles are never smaller than about 100 Å and the polymerization reactions cannot be photochemically induced because of the turbidity of the emulsion. Recently, w/o microemulsions made with copolymers have been used to photopolymerize the acrylamide with azobisisobutyronitrile[44]. For the same microemulsion system different latex particle sizes were obtained (in the range of 200 Å diameter), depending on initiator concentrations. Again the details of the growth mechanism are not entirely understood.

LIQUID-LIQUID EXTRACTION

Liquid extraction is generally used to extract minerals from low-grade ores. The ore is attacked by acidic, basic or salt solutions and one obtains a dilute solution of several substances from which a particular mineral has to be extracted. This is achieved by equilibrating the aqueous solution with an organic phase containing reactants that are specific for the mineral that is to be extracted. It is known that the extraction efficiency increases with the tensioactive character of the extractant and also with the use of alcohol additives. This strongly suggests that micelles might be involved in the extraction process.

A recent study on a model system for nickel extraction in which the extractant is an oil-soluble acid (dialkyldithiophosphoric acid) showed that there is a strong correlation between extraction efficiency and water content of the organic phase. Above a typical water content of about 1M/l the kinetics of the extraction reaction is similar to that in a homogeneous medium[45,46].

CONCLUSION

It appears that the field of technological applications of reverse micelles and oil-continuous microemulsions is very broad. Many of these applications have already been used commercially in past years but other potential ones are still under development. A better knowledge of the structure and dynamic properties of these systems is certainly needed for a better understanding of the mechanisms involved in these applications. In this way it can be expected that their utilization could be rationalized and that the field of applications will be further extended.

ACKNOWLEDGEMENTS

Our work related to oil recovery was supported in part by the DGRST (contracts n° 78-7-2201 and 80 D 849), by the I.F.P. and Rhone Poulenc. I have benefited for the preparation of this paper from helpful discussions with A.M. Cazabat, J. Israelachvili, J. Mitchell, B. Ninham, M.P. Pileni and P. Stenius.

REFERENCES

1. P. Stenius, this volume.
2. L.M. Prince in "Microemulsions" Acad Press (1977).
3. P. Speiser, this volume.
4. A.M. Cazabat, D. Chatenay, P. Guering, D. Langevin, J. Meunier and O. Sorba, this volume.
5. L.E. Scriven, Nature 263 : 123 (1976).
6. Y. Talmon and S. Prager, J. Chem. Phys. 69 : 2984 (1978).
7. P.C. De Gennes and C. Taupin, J. Phys. Chem. 86 : 2294 (1982).
9. Related discussions in this volume.
10. J.J. Taber in "Surface Phenomena in Enhanced Oil Recovery" ed. D.O. Shah, Plenum Press, 1979.
11. J.H. Shulman and J.B. Montagne, Ann. N.Y. Acad. Sci. 92 : 366 (1961).
12. J.E. Puig, E.I. Franses, H.T. Davis, W.G. Miller and L.E. Scriven Soc. Pet. Eng. J., SPE 7055, April 1971.
13. C. Miller, R. Hwan, W. Benton and T. Fort, J. Coll. Int. Sci. 61 : 554 (1977).

14. P.D. Fleming and J.E. Vinatieri, AIChE J. 25 : 493 (1979).
15. P.G. De Gennes, Lectures at the Collège de France, Paris 1982.
16. M.L. Robbins in "Micellization, Solubilization and Microemulsions"
    Vol. 2, Ed. K.L. Mittal, Plenum Press (1977).
17. J.D. Mitchell and B.W. Ninham, Far Trans II 77 : 601 (1981).
18. J.D. Mitchell and B.W. Ninham, to be published.
19. K.S. Chan and D.O. Shah, J. Disp Sci. Techn. 1 : 55 (1980).
20. W.H. Wade, J.C. Morgan, R.S. Schechter, J.K. Jacobson and
    J.L. Salager, Soc. Pet. Eng. J., SPE 6844, October 1977.
21. A. Pouchelon, J. Meunier, D. Langevin, D. Chatenay and
    A.M. Cazabat, Chem. Phys. Lett. 76 : 277 (1980).
    A. Pouchelon, D. Chatenay, J. Meunier and D. Langevin, J. Coll.
    Int. Sci. 82 : 418 (1981).
    A.M. Cazabat, D. Langevin, J. Meunier and A. Pouchelon, Adv.
    Coll. Int. Sci. 16 : 175 (1982).
22. D. Chatenay, D. Langevin, J. Meunier, D. Bourbon, P. Lalanne
    and A.M. Bellocq, J. Disp. Sci. Tech., 3 : 245 (1982).
23. J. Meunier and D. Langevin, J. Phys. Lett. 43 L 185 (1982).
24. S. Friberg, Chem. Techn. 6 No. 2, 124 (1976).
25. A. Weisstuch and K.R. Lange, Materials Protection 10 : 29 (1971).
26. A.W. Adamson, "Physical Chemistry of Interfaces", J. Wiley
    (1976).
27. W.C. Preston, J. Phys. Colloid Chem. 52 : 84 (1948).
28. For instance, research at the Swedish Institute for Surface
    Chemistry, Stockholm.
29. J.W. McBain, R.C. Merril and J.R. Vinograd, J. Amer. Chem. Soc.
    62 : 2280 (1940).
30. See for instance in this volume the papers by E. Geladé and
    F.C. de Schryver.
31. J.H. Fendler and E.J. Fendler "Catalysis in Micellar and
    Macromolecular Systems" Acad. Press (1975).
32. B. Robinson, this volume.
33. R. Zana, J. Lang, O. Sorba, A.M. Cazabat and D. Langevin,
    J. Phys. Lett, in press.
34. F.M. Fowkes, D.Z. Becher, M. Marmo, C. Silebi and C.C. Chao,
    in "Micellization, Solubilization and Microemulsions, Plenum
    Press (1977).
35. J. Boor and F.M. Fowkes, USP 3, 234 : 198 (1966).
36. M. Graetzel in "Micellization, Solubilization and Micro-
    emulsions" Plenum Press (1977).
37. See for instance in this volume the paper by M.P. Pileni and
    J.M. Furois.
38. N. Lufimpadis, J.B. Nagy and E.G. Derouane, Proceedings of
    International Symposium of Surfactants in Solution, Lund (1982)
    J.B. Nagy, A. Gourgue and E.G. Derouane, Stud. Surf. Sci. Catal;
    Elsevier Amsterdam, to appear.

39. J.H. Fendler, this volume.
40. M. Dvolaitzky, R. Anthare, X. Auvray, R. Ober, C. Petipas,
    C. Taupin and C. Williams, CRAS Paris and J. Disp. Sci. Tech.
    in preparation.
41. See for instance the review by A. Martinet in "Aggregation
    Processes in Solution" Ed Wyms Jones Cormally, Elsevier (1982).
42. P.H. Hess and P.H. Parker, J. Appl. Polymer Sci. 10 : 1915
    (1966).
43. K. Kon-no, M. Gobe, K. Kandoni and A. Kitahara, presented at
    the International Conference on Surface and Colloid Science,
    Jerusalem (1981).
44. Y.S. Leong, C. Riess and F. Candau, J. Chim. Phys. 78 : 279
    (1981).
45. J.L. Sabot and D. Bauer, presented at the International Solvent
    Extraction Conference, Toronto (1977).
46. O. Bohm, Thesis, ESPCI Paris (1979).

RECENT APPLICATIONS AND POTENTIALS OF SURFACTANT AGGREGATES IN NON-POLAR SOLVENTS

Janos H. Fendler

Department of Chemistry
Clarkson College of Technology
Potsdam, New York 13676, U.S.A.

Recent applications of surfactant aggregates in apolar solvents - reversed micelles, swollen reversed micelles and microemulsions are critically surveyed. The roles of these systems in reactivity control, formation of catalytically efficient colloid particles, tertiary oil recovery, lubrication, corrosion inhibition, artificial photosynthesis, enzyme mediated synthesis, cryoenzymology, macromolecular conformation, membrane fusion and drug encapsulation are summarized. Emphasis is placed on results obtained in our own laboratories. Unresolved problems and future prospects are highlighted.

INTRODUCTION

Surfactants aggregate in non-polar solvents predominantly by dipole - dipole and ion pair interactions[1-5]. Properties of these aggregates, and indeed the terminology used to describe them[6], depend on the amount of co-solubilized water and on the presence or absence of co-surfactants[2]. Reversed micelles, swollen reversed micelles and water-in-oil (w/o) microemulsions, as these systems became known, have been utilized for a long time in a variety of industrial applications. Much of the early basic research was designated as proprietary information and had to be duplicated. Although the colloid chemistry of some surfactant apolar solvent systems had been examined[7-12] the real impetus came from physical and physical organic chemists in the seventies[2,4,5,13]. Detailed experiments have been carried out using alkyl ammonium carboxylates or sodium

bis(2-ethylhexyl)sulfosuccinate (Aerosol-OT) as surfactant. Attention has been focused on the nature and the type of surfactant association in apolar solvents, on the size of the aggregates formed,
and most importantly, on the properties of the co-solubilized water.
From the practical point of view, surfactant/hydrocarbon/water three
component systems have proven to be most useful. These systems can
be considered to be surfactant entrapped water pools floating in the
hydrocarbon. Increasing the size of the water pools results in
larger aggregates. At some point there is a transition from reversed
micelles to w/o microemulsions. Over-riding importance is attached
to the unique nature of the reversed micelle solubilized water. At
relatively small water to surfactant ratios ($\leq$ 5), all water molecules are tightly bound to the surfactant headgroups at the polar
cores of reversed micelles. These water molecules have high viscosi-
ties, low mobilities, polarities which are similar to hydrocarbons,
and altered pH's. Surfactant bound water in reversed micelles resembles the polar pockets in enzymes and is responsible for many of
the useful applications of these systems[1].

The long time that surfactant aggregates in apolar solvents
have been studied warranted the critical assessment undertaken by
the present workshop. Other presentations surveyed the historical
developments and the current state of our basic understanding of the
field. Recent applications and potential utilizations of organized
surfactant aggregates in apolar solvents will be summarized here.
The extent of the coverage will be inversely proportional to the
available information. Inevitably, a greater emphasis will be
placed on work carried out in our own laboratories.

SURVEY OF APPLICATIONS

Reactivity Control

Alteration of reaction rates, paths and stereochemistry have
been well documented for aqueous and reversed micelles as well as
for microemulsions[1]. The role of organized assemblies is to concentrate the reactants in the microenvironment(s) provided by the
aggregates. Kinetic treatments of reactivities in reversed micelles
and microemulsions need to consider the timescale of substrate reactions in comparison with the timescale of micelle dissociation
and dissolution. The mean lifetime of a surfactant molecule in reversed micelles and microemulsions is in the order of microse-
conds[14,16]. An additional factor is the inter-vesicle exchange of
surfactant solubilized water pools. This event occurs, at least for
Aerosol-OT reversed micelles, on the millisecond timescale[2,16,17].
Reactions occurring in reversed micelles or microemulsions can be

classified into three time domains: ultrafast, fast and slow events. Ultrafast photophysical events occur faster than the fluctuation of the aggregate ($t_{1/2} > 10^{-7}$ sec). On this timescale reversed micelles or microemulsions can be considered to be frozen. Fast chemical reactions (10^{-7} sec $> 10^{-5}$ sec, electron and proton transfer, for example) can go to completion faster than the exchange of water pools among neighboring aggregates. These events are confined, therefore, to defined reversed micelles or microemulsions. Finally, most chemical processes are relatively slow ($t_{1/2} < 10^{-4}$ sec) and their rates are governed by statistical distributions and inter-vesicle exchange of water pools and reactants.

Kinetic treatments derived for aqueous micelles[18-22] are applicable to reversed micelles[23] and vesicles[24,25]. Processes occurring at rates comparable to the dissolution or dissociation of micelles are treated in terms of substrate entry or exit[26,27]. Alternatively, a stochastic approach can be used[28]. Processes occurring at timescales slower than the dissolution of dissociation of aggregates, on the other hand, are treated in terms of regular rate equations familiar to physical organic chemists.

Most treatments consider the overall reaction rate T_{Total} to be the sum of the reaction rates occurring in the bulk aqueous (R_w) and in the pseudo-phase provided by the aggregate (R_M):

$$R_{Total} = R_w + R_M \qquad\qquad Equation\ 1$$

For unimolecular reactions, partitioning of only one substrate needs to be considered. Assuming that the substrate does not perturb the equilibria of the system, unimolecular reactions in the aggregates are described by:

$$S + M \underset{}{\overset{K_S}{\rightleftharpoons}} SM \qquad\qquad Equation\ 2$$

$$\downarrow k_w' \qquad\qquad \downarrow k_M'$$

$$Products \qquad\quad Products$$

where S is the substrate, M is the aggregate, SM is the substrate-aggregate complex, K_S the corresponding dissociation constant, and k_w' and k_M' are the first order rate constants for the two phases. The observed first order rate constant for the reaction, k_{obs} is described by:

$$k_{obs} = \frac{k'_w + k'_M K_s [M]}{1 + K_s [M]}$$ Equation 3

Recognizing the analogy between Equation 3 and the Michaelis–Menten equation for enzyme catalyzed reaction allows the treatment of data by rearranging Equation 3 to[28]:

$$\frac{1}{k'_w + k_{obs}} = \frac{1}{k'_w - k'_M} + \frac{1}{(k'_w - k'_M)K_2 [M]}$$ Equation 4

which is similar to the Lineweaver–Burke equation of enzyme kinetics. Equation 4 well describes the kinetics of both inhibition and catalysis for unimolecular processes. Equation 3 does not adequately describe the kinetics of bimolecular reactions in aggregate systems. Partitioning of both reactants (A and B) between the bulk and the pseudo-phase of the aggregate has to be considered[18,19]:

$$(A + B)_w \xrightarrow{k'_w} Products$$

$$\qquad\qquad\qquad\qquad\qquad\qquad\qquad\qquad\qquad\qquad\text{Equation 5}$$

$$(A + B)_M \xrightarrow{k'_M} Products$$

where the subindices w and M refer to the bulk water and the pseudo-phases of the aggregate. The overall rate, described by Equation 1 is modified to:

$$R_{Total} = k'_M [A]_M [B]_M [M]\bar{V} + k'_w [A]_w [B]_w (1 - M\bar{V})$$ Equation 6

where $\bar{V}$ is the molar volume of the surfactant aggregate and the concentrations of reagents A and B are given by mass balances:

$$[A]_{Total} = [A]_M [M]\bar{V} + [A]_w (1 - [M]\bar{V})$$ Equation 7

$$[B]_{Total} = [B]_M [M]\bar{V} + [B]_w (1 - [M]\bar{V})$$ Equation 8

If the chemical reaction described by Equation 5 does not affect partition equilibria:

$$[A]_w \rightleftharpoons [A]_M$$ Equation 9

and

$$[B]_W \; \rightleftharpoons \; [B]_M \hspace{3cm} \textit{Equation 10}$$

the observed second order rate constant for reactions in the presence of aggregate is given by[18-21]:

$$k_2 = \frac{k_M P_A P_B [M]\bar{V} + k_w(1- [M]\bar{V})}{(1 + (P_A-1)[M]\bar{V})(1 + (P_B-1)[M]\bar{V})} \hspace{2cm} \textit{Equation 11}$$

If both A and B bind strongly to the host ($P_A \gg 1$ and $P_B \gg 1$) then Equation 11 simplifies to[18-21]:

$$k_2 = \frac{k_M K_A K_B [M]/\bar{V} + k_w(1 - [M]\bar{V})}{(1 + K_A[M])(1 + K_B[M])} \hspace{2cm} \textit{Equation 12}$$

where the binding constants are expressed by:

$$K_A = (P_A - 1)\bar{V}$$

$$\hspace{10cm} \textit{Equation 13}$$

$$K_B = (P_B - 1)\bar{V}$$

The experimental data on the dependence of k_2 on [M] can be used to calculate k_M, K_A and K_B. For this purpose Equation 12 is transformed to:

$$\frac{[M]}{k_2 - k_w} = x + y[M]\frac{k_2}{k_2 - k_w} + z[M]^2 \frac{k_2}{k_2 - k_w} \hspace{1.5cm} \textit{Equation 14}$$

where,

$$x = \frac{\bar{V}}{k_M K_A K_B} \hspace{4cm} \textit{Equation 15}$$

$$y = x(K_A + K_B + k_w \cdot \bar{V}) \hspace{3cm} \textit{Equation 16}$$

$$z = x K_A K_B \hspace{5cm} \textit{Equation 17}$$

A plot of the data according to Equation 14 gives a value for the intercept, x. This value allows further analysis in terms of the rearranged equation:

$$\left[\frac{1}{k_2} - \frac{x}{M} \right] \left[1 - \frac{k_w}{k_2} \right] = y + z[M] \qquad\qquad Equation\ 18$$

which, in turn provides numerical values for y and z and hence for k_M, K_A and K_B.

Proper understanding of the kinetic processes is highly rele-vant for the meaningful design of appropriate systems for lubrica-tion and corrosion control.

Formation of Catalytically Efficient Colloidal Particles

Small uniformly sized metal particles have been used as cata-lysts for a long time. The smaller and the more uniform the parti-cles, the more efficient and selective are their catalytic activi-ties. Recently, water-in-oil (w/o) microemulsions have been employed for the in-situ generation of colloidal platinum, palladium, rho-dium, iridium, nickel, iron and gold particles[29-33]. The rationale behind this approach is that the aggregation of reduced metal par-ticles is necessarily limited by the size of the water pool (pre-sumably controllable). Further, advantage can be taken of distribu-ting a known concentration of the metal ions per aggregate and of controlling rates of inter-vesicle exchange of water and growing metal particles. Both cationic (water : hexanol : CTAB = 4.0 or 8.0 or 12.0 or 16.0 or 20.0 : 90.0 or 80.0 or 70.0 or 60.0 or 20.0 : 90.0 or 80.0 or 70.0 or 60.0 or 50.0 : 6.0 or 12.0 or 18.0 or 24.0 ôr 30.0 wt %) and neutral (water : pentaethyleneglycol dodecylether, PEDGE, :n-hexane = 1 : 5 : 44, w/w)[29,33] microemulsions have been used. The metal salts have been reduced by either hydrazine or hydrogen or by irradiation. The obtained colloidal particles have been found to be uniform in the range of 3-5 nm[29]. Significantly, methods have been developed for depositing the colloidal metallic clusters, subsequent to their formation in microemulsions, on solid surfaces[34]. These par-ticles showed superior catalytic behavior[34].

Our own interest centers upon the elucidation of the kinetics and mechanisms of colloidal catalyst formation in microemulsions and in polymeric microemulsions and vesicles. Initially the reduction of $HAuCl_4$ by hydrated electrons, generated by pulse radiolysis, or by nanosecond laser flash photolysis have been examined in water and in water pools entrapped by PEDGE microemulsions[33]. Rate constants have been determined for $Au^{3+} + e^-_{aq} \longrightarrow Au^{2+}$, $2Au^{2+} \longrightarrow Au^{3+} + Au^+$,

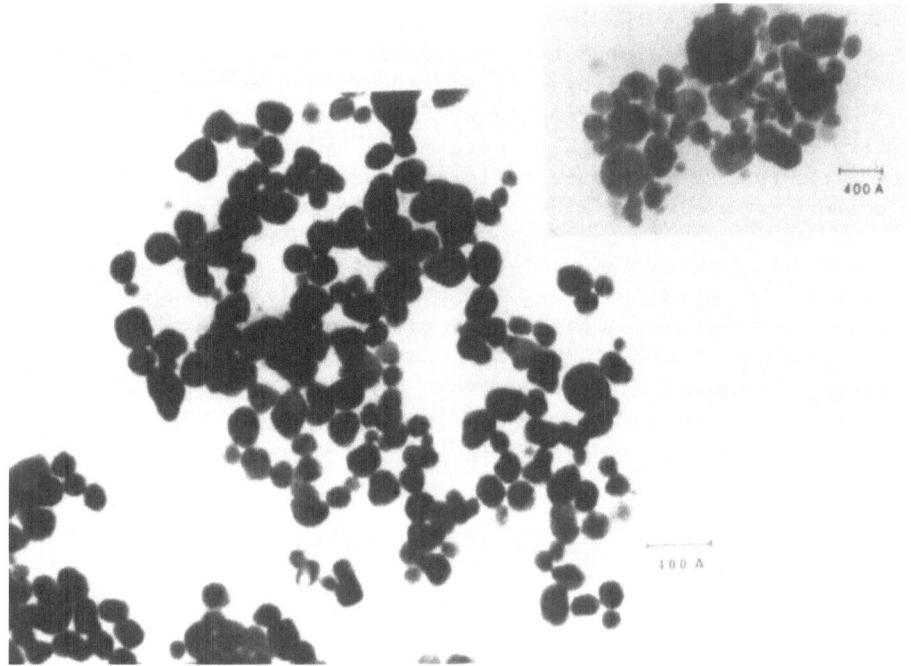

Figure 1. *Electron micrographs of photolytically formed colloidal gold in microemulsions and in water (Insert).*

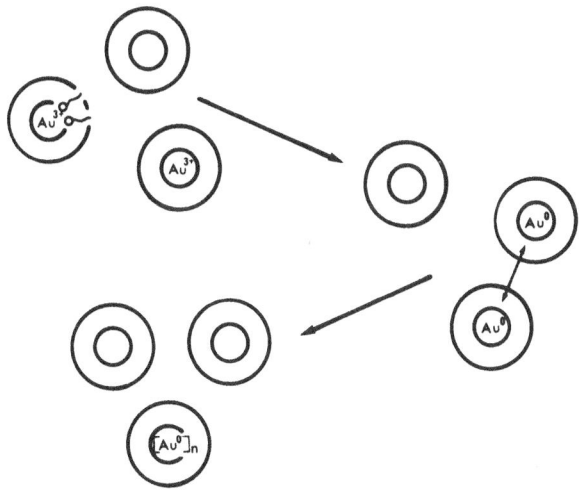

Figure 2. *Schematics of colloidal gold formation in microemulsions.*

$Au^+ + R\cdot \longrightarrow Au^o$. On the longer timescale, formation of colloidal gold, $nAu^o \longrightarrow (Au^o)n$, has been observed. Diameters of empty and of colloidal gold containing microemulsions have been determined by dynamic laser light scattering to be 150 Å and 220 Å, respectively. Morphologies of colloidal gold, determined by electron microscopy, indicate that there is a more efficient formation of smaller and more uniform particles in microemulsions than in water (Figure 1)[33]. Under the experimental conditions, each microemulsion contained approximately eight Au^3 ions which ultimately led to the formation of less than one molecule of Au^{2+} per microemulsion. Under these conditions, Au^{2+} disproportionation and the formation of colloidal $(Au^o)n$ particles occur through the exchange of the contents of neighboring microemulsions (Figure 2). Currently, similar experiments are being carried out in our laboratories on other metal ion reductions in microemulsions and in polymeric surfactant vesicles.

Tertiary Oil Recovery, Lubrication and Corrosion Inhibition

Surfactants have been extensively used in enhancing oil recovery from petroleum reservoirs. Their function is to reduce oil-water interfacial tensions to extremely low values which are needed so that entrapped oils can be transported by capillary forces[35-40]. A large variety of complex surfactant phases, including microemulsions, are responsible for reductions in surface tension. Reversed micelle and/or microemulsion solubilization is central to lubrication and corrosion inhibition. Much of the basic research is currently being carried out in industrial laboratories. A systematic approach is badly needed for the meaningful optimization of the numerous parameters involved.

Artificial Photosynthesis

In natural photosynthesis the captured visible light is transformed to chemical energy which is then utilized for carbon dioxide reduction. The process is described by the "Z-scheme". Briefly, light is absorbed by two pigment systems: photosystem I, PS I (P7000), and photosystem II, PS II (P680). These two systems operate in series; two photons are absorbed for every electron liberated from water. Light induced charge separation in PS II leads to the formation of a strong oxidant, Z^+ (E_o = +0.8V), and a weak reductant, Q^- (E_o + 0.0V). Although the reduction potential of Z^+ is sufficient for water oxidation, evolution of molecular oxygen demands the accumulation of four positive charges. Electron flows from Q^-, via a pool of plastoquinones and other carriers, to a weak oxidant, (E_o = +0.4V), generated along with a strong reductant,

X^- (E_0 = -0.6V) in PS I. This electron flow is coupled to phosphorylation, which converts adenosine diphosphate, ADP, and inorganic
phosphate to adenine triphosphate, ATP. With the aid of ATP, X^-,
reduces carbon dioxide to carbohydrate[41]. The precise, yet not completely understood, arrangements in the thylakoid membrane are responsible for the efficient energy deposition and transmission, for
prevention of charge recombination, and for creating a proton gradient essential for photophosphorylation[42].

In artificial photosynthesis PS I and PS II are substituted
by simple sensitizers, S, and electron relays, R. The thermodynamics
of converting photochemical energy to chemical energy is shown in
Scheme 19. It is important to recognize that the excited state of

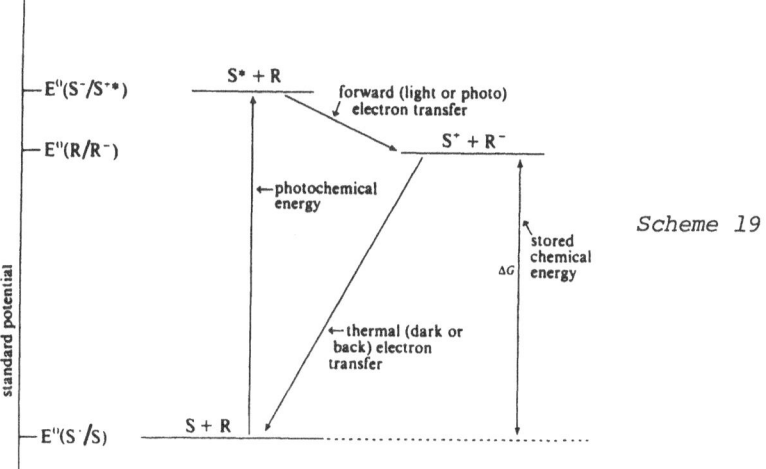

Scheme 19

the sensitizers, S^+, is a better electron donor and, at the same
time, a better electron acceptor than the ground state, S. Absorption of light (hν) can drive therefore, a redox reaction and result
in the storage of energy, ΔG, in S^+ and R^-. In Scheme 19, S* functions as an electron donor. S* may just as well function as an
electron acceptor if R is a suitable donor.

The role of microemulsions is to perform the functions of the
thylakoid membrane. Judicious organization of S and R should bring
about favorable energy deposition and transmission and, importantly,
prevent back electron transfer between R^- and S^+ in Scheme 19.

Ideally (i) the reduced relay (R^-) and the oxidized sensitizer
(S^+) should be thermodynamically capable of generating hydrogen and
ogygen from water (Equations 20 and 21).

$$2R^- + 2H_2O \longrightarrow 2R + 2OH^- + H_2 \qquad\qquad \textit{Equation 20}$$

$$(2e^- \ reduction)$$

$$4S^+ + 2H_2O \longrightarrow 2S + 4H^+ + O_2 \qquad\qquad \textit{Equation 21}$$

$$(4e^- \ oxidation)$$

and (ii) the process should be cyclic. In practice, the multielectron steps demand the use of catalysts:

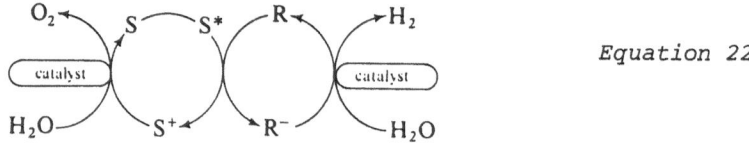

$$\textit{Equation 22}$$

Artificial photosynthesis studies have been further simplified by separating the half cells (in Equation 22) and by limiting the investigations to water photoreduction. Since these half cells are necessarily non-cyclic, their use requires the use of sacrificial electron donors (D) for hydrogen production:

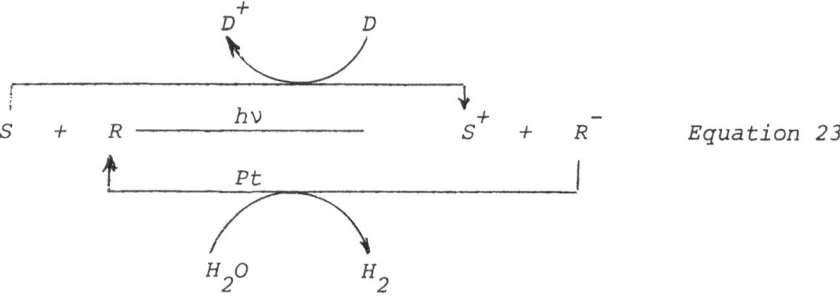

$$\textit{Equation 23}$$

Photosensitized electron transfer and the inhibition of the undesirable back reaction have been realized in several systems[43-46]. For example, photosensitized tris(2,2'-bipyridine)ruthenium cation, (Ru^{2+}), localized in water pools entrapped by dodecyl ammonium propionate (DAP) reversed micelles, transferred an electron to 1,1'-dihexadecal-4,4'-bipyridinium chloride, (HV^{2+}), intercalated in the micelle water interface. When EDTA was used as an electron donor (localized in the water pools), photosensitized reduction of 4-dimethylaminoazobenzene, (dye) was mediated by benzylnicotinamide, BNA^+[43]:

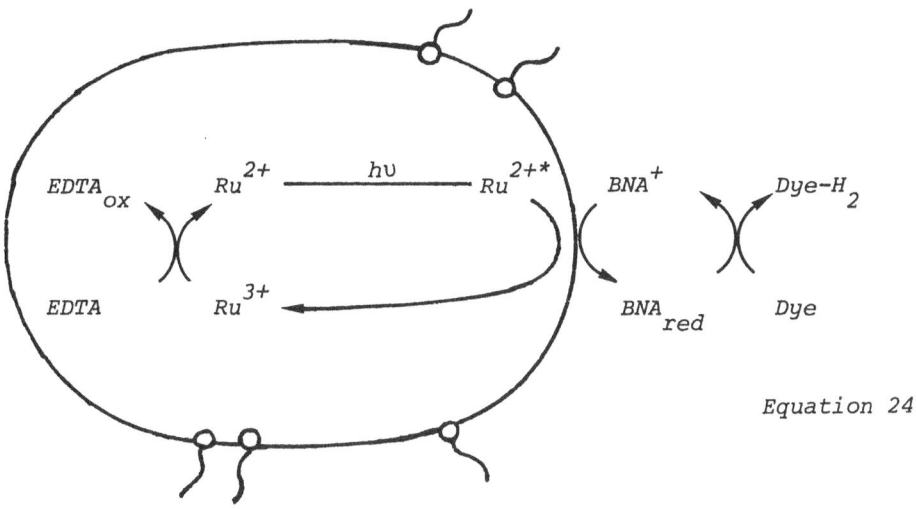

$$Equation\ 24$$

More recently, reversed micelle entrapped hydrogenase was shown to generate hydrogen by Ru^{2+} photosensitized electron transfer from the sacrificial thiophenal donor (distributed in the organic solvent) via the methylviolgen relay[46].

Further improvements in electron transfer hydrogen generation and the coupling of the oxidation and reduction half cells in microemulsions or in related membrane mimetic system[1] are fully expected.

Macromolecules in Reversed Micelles, Cryoenzymology and Enzyme Mediated Synthesis

One of the most significant accomplishments in this area is the incorporation of enzymes into surfactant solubilized reversed micelles[47-54]. Cytochrome-C, chymotrypsin, rhodopsin, ribonuclease, peroxidase, phospholipase, lysozyme, lactate dehydrogenase, pyruvate kinase are the enzymes and Aerosol-OT and phospholipids are the typical surfactants used. The mode of enzyme-reversed micelle interaction is expected to vary from system to system. In general the enzyme activity is retained, although in some cases it is enhanced. Enzyme catalyzed reactions can be, therefore, fruitfully investigated in surfactant entrapped water pools in the hydrocarbon solvent. This certainly is a better model of the real _in vivo_ situation than that provided by normal water solutions. Indeed reversed micelles have been utilized as workable membrane models for investigating polypeptide conformations[55].

Microemulsions also have the potential to act as carriers for substrates and enzymes in multiphase environments[56,57].

Reversed micelles also provide convenient media for studying enzyme mediated processes at subzero temperatures. This in turn considerably slows down reactions and thus allows the leisurely examination of the kinetics of formation and decomposition of short-lived intermediates[58]. Our own interest lies in the elucidation of the intermediates formed in the electron transfer events, relating to photosynthetic pathways. We hope to slow down picosecond events to the nanosecond time domain and observe them by laser flash photolysis[59,60].

Incorporation of enzymes into reversed micelles is directed towards the realization of enzyme mediated syntheses. Formation of hydrocarbon soluble polymers (or biopolymers) from water soluble monomers is the most obvious possibility. The monomer could be continuously fed into the reversed micelle entrapped water pool where the enzyme would mediate the polymerization. Once the polymers formed, they would partition into the organic solvent phase. If appropriate membrane filtration techniques are used, reactants, products and reacting reversed micelles could be separated.

Reversed Micelles and Membrane Fusion

Lipids, proteins and other components of the biological membrane can arrange themselves in a large variety of dynamic structures. Changes in temperature, lipid and electrolyte concentration are manifested in microscopic and/or macroscopic structural alterations. These changes are related to such cellular functions as membrane fusion, exo- and endocytosis. Reversed micelles were suggested to play a role in these porcesses[61,62]. If they do, studies of reversed micelles will gain an additional dimension[63].

Drug Delivery

Target directed efficient drug delivery is the dream of every pharmacologist. The general idea is to entrap the drug in a carrier which would be transported intact to the disease site. Through fusion, endocytosis or another mechanism the drug containing carrier would enter the cell where the drug would be released. Beneficial effects of carrier mediated drug delivery include reduced dosages, allergic and immunological reactions, increased cellular permeability and delayed drug elimination. Liposomes have been extensively explored as potential drug carriers[64-68]. Microemulsions and poly-

merized microemulsions may provide alternative drug carriers[69].
Additional research is clearly required in this highly relevant
area.

UNRESOLVED PROBLEMS

The following is a necessarily incomplete list of basic problems
which merit our attention:

(1) Appropriate theories need to be developed for the rationalization (both in terms of thermodynamics and kinetics) of the formation of reversed micelles and w/o microemulsions. Approaches utilized for aqueous micelles[1,70-79] could profitably be extended to
surfactant aggregates in non-polar solvents. It is hoped that the
obtained information would become part of a unified theory encompas-
sing all conceivable structures formed from surfactants in different
media. Particular attention should be given to the biological role
of the different surfactant structures and to the parameters which
influence their interconversions.

(2) Increasing attention should be focused on determining
structures and properties of surfactant solubilized water pools in
the hydrocarbon solvent. Particular attention should be given to
the highly bound water molecules which hydrate the surfactant headgroups in reversed micelles and microemulsions.

(3) Thorough understanding of electrolyte and hydrogen ion (pH)
concentrations in reversed micelles and microemulsions must be obtained. These "real" concentrations are needed for treatments of
interactions and reactions in the presence of surfactant aggregates
in non-polar solvents.

(4) The kinetics of aggregate formation (and dissociation) and
of the exchange of intervesicular water pools and their contents
should be elucidated as functions of the type and the nature of
aggregates. The obtained data should be compared to those available
for micelles and vesicles.

(5) Structural studies should be extended to systems other than
Aerosol-OT and alkylammonium carboxylate reversed micelles and microemulsions. Systematic investigations of phase diagrams, morphologies
(by steady-state and dynamic light and neutron scattering), and
structures should be carried out on well defined, reproducible
systems.

FUTURE PROJECTS

The wide variety of practical applications will ensure the future utilization of surfactant aggregates in apolar solvents. Our own interest lies in the development of polymerized surfactant aggregates[80]. Vesicles, formed from surfactants having double bonds in their hydrocarbon tails or polar headgroups, could be polymerized either across their bilayers or headgroups[81-85]. These latter polymerizations offer opportunities for selectively "zipping-up" either the outer or the inner surface of vesicles. Polymeric vesicles are considerably more stable than their non-polymeric counterparts. They have extensive shelf lives and remain stable in up to 25% alcohol. Extension of the polymeric vesicle concept to w/o microemulsions should lead to exciting novel chemistry.

ACKNOWLEDGEMENT

Support for this work by the National Science Foundation is gratefully acknowledged.

REFERENCES

1. J.H. Fendler, "Membrane Mimetic Chemistry", John Wiley & Sons, New York, (1982).
2. H.F. Eicke, Top. Curr. Chem. 87 : 85 (1980).
3. A. Kitahara, Advan.Colloid Interface Sci., 12 : 109 (1980).
4. J.H. Fendler, Acc. Chem. Res. 9 : 153 (1976).
5. J.H. Fendler and E.J. Fendler, "Catalysis in Micellar and Macromolecular Systems", Academic Press, New York, 1975.
6. S.E. Friberg, Colloids and Surfaces, 4 : 201 (1982).
7. C.R. Singleterry, J. Am. Oil Chem. Soc. 32 : 446 (1955).
8. N. Pilpel, Chem. Revs. 63 : 221 (1963).
9. R.C. Little and C.R. Singleterry, J. Phys. Chem. 68 : 3453 (1964).
10. R.C. Little, J. Colloid Interface Sci. 21 : 266 (1966).
11. F.M. Fowkes in "Solvent Properties of Surfactant Solutions" (K. Shinoda, ed.), Marcel Dekker, New York (1967), p. 65.
12. P. Becher, in "Nonionic Surfactants" (M.J. Schick, ed.) Marcel Dekker, New York (1967), p. 478.
13. A.S. Kertes and H. Gutman, in "Surface and Colloid Science (E. Matijevic, ed.) Vol. 8 John Wiley, New York (1976), p. 194.
14. M. Almgren, F. Grieser and J.K. Thomas, J. Am. Chem. Soc. 102 : 3188 (1980).
15. K. Tamura and Z. Schelly, J. Am. Chem. Soc. 103 : 1018 (1981).
16. M. Zulauf, H.F. Eicke, J. Phys. Chem. 83 : 408 (1979).

17. B. Robinson, presentation in this workshop.
18. I.V. Berezin, K. Martinek and A.K. Yatsimirski, Russ. Chem. Rev. Eng. Transl. 42 : 787-802 (1973).
19. K. Martinek, A.K. Yatsimirski, A.V. Levashov and I.V. Berezin, in "Micellization, Solubilization and Microemulsions", (K. Mittal, ed.), Plenum Press, New York 489-508 (1977).
20. L.S. Romsted and Ph.D. Thesis, Indiana University, (1975); in Micellization, Solubilization and Microemulsions", (K. Mittal, ed.), Plenum Press, New York 509-530 (1977).
21. C.A. Bunton, Y.S. Hong, L.S. Hong and L.S. Romsted, in "Solution Behavior of Surfactants, Theoretical and Applied Spects", (K. Mittal and E.J. Fenler, ed.), Plenum Press, New York (1982).
22. H. Chaimovich, J.B.S. Bonilha, M.J. Politi and F.H. Quina, J. Phys. Chem. 83 : 1851-1854 (1979); F.H. Quina and H. Chaimovich, J. Phys. Chem. 83 : 1844-1850 (1979); F.H. Quina, M.J. Politi, I.M. Cuccovia, E. Baumgarten, S.M. Martins-Frauchetti and H. Chaimovich, J. Phys. Chem. 84 : 361-365 (1980); H. Chaimovich, R.M. Aleixo, I.M. Cuccovia, D. Zanatte and F.G. Quina, in "Solution Behavior of Surfactants, Theoretical and Applied Aspects", (K. Mittal and E.J. Fendler, ed.), Plenum Press, New York (1982).
23. Y.Y. Lim and J.H. Fendler, J. Am. Chem. Soc. 100 : 7490 (1978).
24. J.H. Fendler and W.L. Hinze, J. Am. Chem. Soc. 103 : 5439 (1981).
25. J.H. Fendler, Pure and Applied Chem. 54 : 1809 (1982).
26. P.P. Infetta and M. Gratzel,J. Chem. Phys. 70 : 179-186 (1979).
27. A. Yetha, M. Aikawa and N.J. Turro, Chem. Phys. 72 : 4358-4367 (1980), 74 : 1098-1109 (1981), 74 : 5627-5635 (1981).
28. First time used for micelle catalyzed reactions by F.M. Meyer and C.E. Portnoy, J. Am. Chem. Soc. 89 : 4698 (1976).
29. M. Boutonnet, C. Anderson and R. Carsson, Acta Chem. Scand. A34 : 639 (1980).
30. M. Boutonnet, J. Kizling, P. Stenius and G. Maire, Colloids and Surfaces (1982) in press.
31. J.B. Nagy, A. Gourgue and E.G. Deroune, Stud. Sci. Catal. (1982). Preparation of Catalysts. Third International Symposium on Scientific Bases for the Preparation of Hetereogeneous Catalysts, Louvan-la Neuve, September 6-9, 1982, Elsevier, Amsterdam, 1982.
32. N. Lufimpadio, J.B. Nagy and E.G. Deroune, Proc. Symposium on Surfactants in Solution, Lund, Sweden, June 27 - July 2, 1982.
33. K. Kurihara, J. Kizling, P. Stenius and J.H. Fendler, J. Am. Chem. Soc., in press.
34. P. Stenius and coworkers, private communication, 1982; Patents pending.
35. U.K. Bansal and D.O. Shah in "Micellization, Solubilization and Microemulsions" (K.L. Mittal, ed.), Plenum Press, New York (1977) p. 87.

36. M.M. Schumacher, "Enhanced Oil Recovery, Secondary and Tertiary Methods", Moyer Data Corp. Park Ride, NJ (1978).

37. D.O. Shah and R.S. Schechter, "Improved Oil Recovery by Surfactant and Polymer Flooding", Academic Press, New York (1977).

38. D.O. Shah, "Surface Phenomena in Enhanced Oil Recovery", Proc. of Stockholm Symposium, August 1979, Plenum Press, New York (1981).

39. W.H. Wade, J.C. Morgan, R.S. Schechter and J.K. Jacobson, Society of Petroleum Engineers of AIME, SPE 6844 (1977).

40. L.E. Scriven, in "Micellization, Solubilization and Microemulsions", (K.L. Mittal, ed.), Plenum Press, New York (1977), p.877.

41. Govinjee, "Bioenergetics of Photosynthesis", Academic Press, New York (1975).

42. P.D. Boyer, B. Chance, L. Ernster, P. Mitchell, E. Racker and E. Slater, Ann. Rev. Biochem. 46 : 957 (1977).

43. I. Willner, W.E. Ford, J.W. Otvos and M. Calvin, Nature, 280 : 823 (1979).

44. I. Willner, C. Laane, J.W. Otvos and M. Calvin, ACS Symposium Series 177 : 71 (1982).

45. J. Kiwi and M. Gratzel, J. Phys. Chem. 84 : 1503 (1980).

46. R. Hilhorst, C. Laane and C. Veeger, Proc. Natl. Acad. Sci., USA 79 : 3927 (1980).

47. P.L. Luisi, F. Henninger, M. Joppich, A. Dossena and G. Casnati, Biochem. Biophys. Res. Commun. 74 : 1384 (1977).

48. P.L. Luisi, F.J. Bonner, A. Pellegrini, P. Wiget and R. Wolf, Helv. Chim. Acta 62 : 740 (1979).

49. F.M. Menger and K. Yamada, J. Am. Chem. Soc. 101 : 6731 (1979).

50. P. Douzou, E. Keh and C. Balny, Proc. Natl. Acad. Sci. USA 67 : 681 (1979).

51. C. Balny and P. Douzou, Biochemie 61 : 445 (1979).

52. S. Barbaric and P.L. Luisi, J. Am. Chem. Soc. 103 : 4239 (1981).

53. K. Martinek, A.N. Semenov and I.A. Berezin, Biochim. Biophys. Acta 658 : 76 (1981).

54. A.V. Levashov, Y.L. Khmelmitsky, N.L. Klyachko, V.Y. Chernyak and K. Martinek, J. Colloid Interface Sci. 88 : 444 (1982).

55. L.M. Gierasch, J.E. Lacy, K.F. Thompson, A.L. Rockwell and P.I. Watnick, Biophysical J. 37 : 275 (1982).

56. P.L. Luisi and R. Wolf, in "Solution Behavior of Surfactants. Theoretical and Applied Aspects" (E.J. Fendler and K.L. Mittal, eds.), Plenum Press, New York (1982).

57. C. Tandre and A. Xenakis, in "Surfactants in Solutions" (B. Lindman and K.L. Mittal, eds.), Plenum Press, New York (1983).

58. P. Douzou, "Cryobiochemistry", Academic Press, New York (1977).

59. A.J.W.G. Visser and J.H. Fendler, J. Phys. Chem. 86 : 947 (1982).

60. J.H. Fendler and coworkers, unpublished results.

61. P.R. Cullis, B. DeKruijff, M.J. Hope, R. Nayar and S.L. Schmid, Canad. J. Biochem. 58 : 1091 (1980).

62. A. Sen, W.P. William, A.P.R. Brain, M.J. Dickens and P.J. Quinn, Nature 293 : 488 (1981).

63. J. Seelig, Presentation at the IV ESF Workshop on Polymer Sciences, "Biological and Technological Relevance of Reverse Micelles and other Amphiphilic Structures in Apolar Media", Rigi-Kaltbad, Switzerland, Sept. 29 - Oct. 2, 1982.

64. H.K. Kimelberg and E.G. Mayhew, CRC Crit. Rev. Toxicol. 6 : 25 (1978).

65. G. Gregoriadis and C. Allison, "Liposomes in Biological Systems", John Wiley, New York (1980).

66. J.H. Fendler and A. Romero, Life Sciences 20 : 1109 (1977).

67. B.E. Ryman and D.A. Tyrell, Essays Biochem. 16 : 49 (1980).

68. D.A. Tyrell, T.D. Heath, C.M. Colley and B.E. Ryman, Biochim. Biophys. Acta 108 : 331 (1980).

69. P. Speiser, Contribution at the IV ESF Workshop on Polymer Sciences, "Biological and Technological Relevance of Reverse Micelles and other Amphiphilic Structures in Apolar Media", Rigi-Kaltbad, Switzerland, Sept. 29 - Oct. 2, 1982.

70. F.M. Menger, Acc. Chem. Res. 12 : 111 (1979).

71. P. Fromherz, Chem. Phys. Lett. 77 : 460 (1980).

72. P. Fromherz, Ber. Bunsengesell. Phys. Chem. 85 : 891 (1981).

73. K.A. Dill and P.J. Flory, Proc. Natl. Acad. Sci. USA 78 : 676 (1981).

74. D.G. Hall, J. Phys. Chem. Soc. Faraday Transactions II 77 : 1973 (1981).

75. S. Vass, in "Microemulsions" (I.D. Robb, ed.), Plenum Press, New York, 1982, p. 173.

76. K. Shinoda, Pure and Appl. Chem. 52 : 1195 (1980).

77. F.H. Stillinger, Variational Model for Micelle Structures, in press.

78. G. Sunnarsson, B. Jonsson and H. Wennerstrom, J. Phys. Chem. 84 : 3114 (1980).

79. G. Kegeles, J. Phys. Chem. 83 : 1728 (1979).

80. J.H. Fendler, in "Surfactants in Solution" (K.L. Mittal, ed.), Plenum Press, New York 1983 (Proceedings of the International Symposium, June 27 - July 2, 1982, Lund, Sweden).

81. S.L. Regen, B. Czech and A. Singh, J. Am. Chem. Soc. 102 : 6638 (1980).

82. L. Gross, H. Ringsdorf and H. Schupp, Angew. Chem. Int. Ed. Eng. 20 : 305 (1981).

83. P. Tundo, D.J. Kippenberger, P.L. Klahn, N.E. Prieto, T.C. Jao and J.H. Fendler, J. Am. Chem. Soc. 104 : 456 (1982).

84. P. Tundo, D.J. Kippenberger, M.J. Politi, P. Klahn and J.H. Fendler, J. Am. Chem. Soc. 104 : 5352 (1982).

85. P. Tundo, K. Kurihara, D.J. Kippenberger, M. Politi and
    J.H. Fendler, Angewandte Chem. Int. Ed. Eng. 21 : 81 (1982).

ENZYMES AND NUCLEIC ACIDS SOLUBILIZED IN HYDROCARBON SOLVENTS WITH THE HELP OF REVERSE MICELLES

P.L. Luisi, P. Meier, V.E. Imre and A. Pande

Technisch-Chemisches Laboratorium
ETH-Zentrum
CH-8092 Zurich/Switzerland

Various aspects of the reverse micellar systems have been discussed in these Proceedings. In general, they can be divided into two main classes: i) structural and thermodynamic studies, and ii) study of reactivity and function. Our contribution belongs to the latter class. In particular, we like to view reverse micelles as microreactors, whose dimensions and physical properties can be altered by changing the water content of the macroscopic system. Of interest to us are questions related to the peculiarities of these microreactors, namely, a) whether and to what extent they confer novel structural and reactivity properties to guest molecules, and b) whether these microreactors can be useful either for biotechnology or as biological models. In a previous review[1] we have discussed various structural and reactivity properties of enzymes solubilized in AOT reverse micelles (AOT = bis(2-ethylhexyl) sodium sulfosuccinate). In this article we would like to review our recent work in the area, emphasizing in particular the characterization and purity of AOT reverse micelles.

THE PROBLEM OF AOT PURITIY

In order to understand the role of these microreactors in various chemical reactions, one needs to know the nature and physical characteristics of the milieu where reactions generally take place (the water pool). Unfortunately, this proves to be somewhat difficult, due to experimental as well as conceptual problems, some of which are elaborated below.

The purity of AOT itself is one problem (see also Ref. 2). For our purpose, we classify the impurities present in AOT in three categories. First, UV absorbing impurities, i.e. those absorbing from 250 to 320 nm, well above the spectral region of AOT. The most obvious effect of such impurities is to obscure the spectral (UV absorption, CD, and fluorescence) characteristics of the guest molecules. Secondly, there are acidic impurities in commercial AOT samples, as well as in samples that are purified according to methods reported in the literature. Finally, a third problem in the chemical purity of AOT is due to the presence of salts, both those present in the commercial AOT preparations and those generated by the neutralization of the acidic impurities. These three classes of impurities are present in different amounts in different commercial preparations and are removed to different extents by various purification procedures.

There is a small but significant UV-absorption extending up to 300 nm even in the samples purified according to literature (Figure 1A). For a 50 mM AOT solution (a typical concentration we use), the absorption at 280 nm is in the range 0.02-0.05 OD. This may appear trivial; however, when running spectra of dilute biopolymer solutions this may be an appreciable percentage of the total absorbance of the guest molecules. Furthermore, such an absorption may correspond to a molar concentration of an aromatic impurity up to 10µM (based on an extinction coefficient of 5000 1 mol^{-1} cm^{-1}) – which is comparable or higher than, say, the concentration of enzymes used for kinetic experiments. Incidentally, these absorption profiles in the 250-320 nm range are pH-independent, except for the absorption at 250 nm which increases markedly with increasing pH.

Of course, there is a compensation for this undesired optical density contribution, since in UV-absorption studies the reference cell contains AOT in the same concentration as in the sample cell. In CD studies the impurities do not give any significant contribution and the base line (obtained with the "empty" AOT micelle solution alone) is subtracted. However, the absorption of AOT solutions below 250 nm becomes almost prohibitive. This is why, for most of our previous CD studies with proteins or nucleic acids in reverse micelles, we have to decrease the AOT concentration down to 10 mM.

We have now tried two new methods of AOT purification. One is essentially a modified version of the Martin and Magid[3] procedure (see Appendix) and the other is by reverse phase HPLC with a µ Bondapak C 18 (Altech Associates Inc.) column (25 cm x 2.5 cm) using methanol-water, 78:22 (v/v) as the elution mixture and a flow

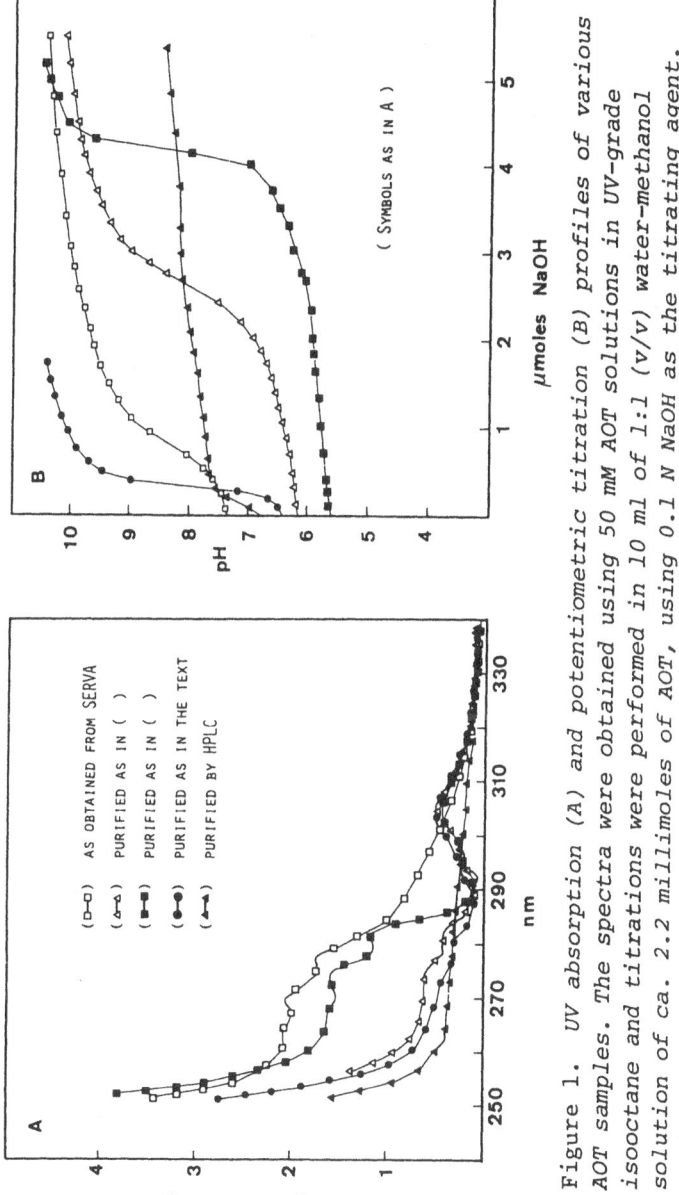

Figure 1. *UV absorption (A) and potentiometric titration (B) profiles of various AOT samples. The spectra were obtained using 50 mM AOT solutions in UV-grade isooctane and titrations were performed in 10 ml of 1:1 (v/v) water-methanol solution of ca. 2.2 millimoles of AOT, using 0.1 N NaOH as the titrating agent. In (A) the extinction coefficient, ε, was calculated as if the absorption were due only to AOT. Thus, an ε of 1 l mole⁻¹ cm⁻¹, for example, refers to an absorbance of 0.05 in a 1 cm cell.*

rate of 6.5 ml per minute (this was carried out in collaboration with Dr. A. Dossena of the University of Parma, Italy). The UV spectra (Figure 1A) and pH titrations (Figure 1B) of the various AOT preparations obtained by the two methods indicated above and two others[3,4] published earlier, are shown in Figure 1. Apparently there are several UV absorbing impurities (Figure 1A); some of these, namely those absorbing around 290 nm, were absent from all of the purified samples while there was a lower concentration of impurities absorbing from 250 to 285 nm in the sample purified according to Wong et al[4]. The Martin and Magid procedure[3] reduced the concentration of these impurities considerably and they were further reduced in the sample purified according to our procedure (Appendix). The sample purified by HPLC shows no UV absorbing impurity but shows a gradual, although slight increase in absorbance with decreasing wavelength, which could be due to light scattering.

The pH-titrations clearly show an increase in the acid content in the samples purified by the published procedures[2,4]. The acid is apparently generated during the purification procedure, most likely due to a heating step. Our modified version of the purification procedure (see Appendix) of Martin and Magid[3] differs from the parent version in that the temperature is always kept at or below room temperature and the sample is treated with basic Alox at the end of the purification. The sample purified by HPLC shows a peculiar titration profile, with minor amounts of contaminant(s) which are easily hydrolyzed above pH 7.5, thereby consuming most of the added base. These contaminant(s) may also be responsible for the scattering observed in the absorption spectrum (see Figure 1A).

What do these titration results signify, from a practical point of view? When 2 μmoles NaOH are used to neutralize 2.2 mmoles of AOT (see Figure 1B), the percentage of acid impurity is around 0.1% (in molarity). Again, this figure exceeds the overall molar concentration of the guest biopolymer and clearly this sets some warning as to the type and concentration of buffer to be used. All this is further elaborated in Ref. 2. It must be kept in mind that in addition to the differences in the nature and amount of the impurities caused by the use of different purification procedures, as shown above, there are significant differences in the various AOT samples obtained from different manufacturers and also from the same manufacturer.

It would be desirable if in the literature to come each AOT preparation would be characterized by the following parameters: the UV absorption spectrum (or at least the absorbance of a standard solution at 280 nm), the pH-titration profile, and possibly a measure of salt content. Knowledge of the last parameter is impor-

tant because the amount of water that can be solubilized in the
micellar system depends on the salt content[3]. This sort of purification and characterization work, seemingly so trivial, has not yet
been systematically carried out. It is probably due to the fact that
it has not been necessary for most of the technically oriented work
carried out with this surfactant or even for the research work usually carried out using high water concentrations. It is mostly when
one studies the chemical and spectroscopic properties of the guest
molecules present in the reverse micellar solutions at low water
concentrations that such purity problems would become critical and
might result in conflicting reports from different laboratories.
The removal of such impurities or a better characterization of AOT
corresponds to the new level of sophistication necessitated by recent developments in the area of reverse micelles.

Several laboratories are now concerned with the problem of removal and characterization of the impurities and it is hoped, also
based on the data that we and Fletcher et al[2] presented in these
Proceedings, that a final and generally acceptable protocol for the
purification and use of AOT solutions will soon be in use. For this,
it would be important that other groups test the two methods we have
proposed here, the one described in the Appendix, and the other based
on HPLC.

Figure 2 illustrates the effects that the different AOT preparations may have on actual enzymatic studies. The figure compares
the activity profiles of ribonuclease (the normalized initial velocity, v/E_0, is used, see also Ref. 5) as a function of pH in water
and in reverse micellar systems with AOT purified in different ways.
The pH in reverse micelles has been taken as the pH of the stock
buffer solutions used to prepare the micelles with the injection
method[1] and of course all the conceptual as well as the practical
difficulties concerning the definition of pH in the isooctane micellar solution apply[6]. We will return to this point later on. This
is however not so important for the point we want to make on the
basis of Figure 2. This is the following: that the pH optimum for
enzyme reactivity is shifted by about three pH units with the use
of an AOT sample with a relatively high acid content, whereas the
use of AOT purified according to the procedure outlined in the
Appendix (and having a negligible content of acid impurities)
brings the pH profile closer to that found with bulk water solutions. This experiment then shows two important features of the reverse micelles seen as microreactors for enzymatic reactions: i) It
is easy to get misleading results if proper care in assessing the
AOT purity is not taken. In this regard, it is timely to point out
that our previous data on α-chymotrypsin[7], where the pH and/or w_0
($w_0 = [H_2O]/[AOT]$ in isooctane) influence on activity was very

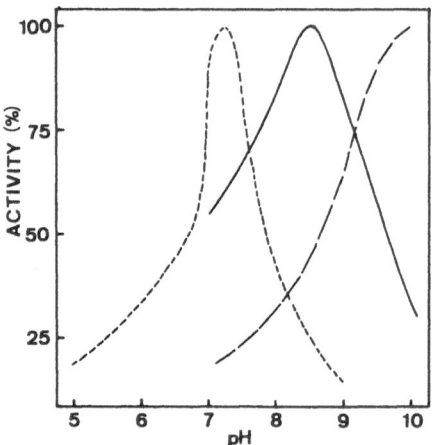

Figure 2. *pH-activity profiles for RNase activity in aqueous solutions (--------)[26]. in AOT-isooctane reverse micelles at w_O = 7.4, using AOT purified as in the Appendix (———); and in micelles with AOT purified as in Reference 4 (— — —). Activity for each profile is calculated as a percentage of the maximum normalized initial velocity observed for that case. Cytidine 2',3'-phosphate was used as a substrate.*

dramatic, and which were obtained with AOT purified according to literature[4] are partly influenced by the particular AOT preparation used. Smaller pH effects are observed with AOT samples purified with the new procedure (P. Lüthi and P.L. Luisi, unpublished results). ii) there is a difference in the activity/pH profile of ribonuclease even with AOT samples practically devoid of acid impurities (Figure 2). Thus, this difference must result from the intrinsic peculiarity of the micellar water pool. Whether this is due to pH or pK changes or to conformational alterations of the enzyme remains to be seen. The problem is also complicated by the fact that the pH profile and the position of the pH optimum appear to depend on w_O[7,8].

PROBLEMS OF CHARACTERIZATION OF THE WATER POOL

The above observation, which implies that the water of the reverse micelle will probably have a different acidity than the stock water solution used to prepare the reverse micellar system impinges on a problem of paramount importance in this area: the characterization of the aqueous milieu of the water pool in which the guest biopolymer is located and in which reactions generally take place.

It has been shown by various physico-chemical techniques that this water behaves differently from normal water[9,10,11,12], especially at low concentrations (e.g. below $w_0 = 10$). Micelles have also been used as a medium for cryospectroscopic work, since the freezing point of this water is apparently significantly lowered[9,13].

In view of this, it is not surprising that the definition and the experimental determination of the pH of the water pool are difficult[11,12]. All the methods are based on the determination of the acid-base equilibrium of an indicator molecule or ion and the implicit assumption is that its pK_a in the reverse micelles is the same as that in the bulk water. Since the water in the reverse micelles is a new solvent, this assumption, at least in principle, is invalidated[6]. The best we can do is to determine an acidity scale based on this assumption. The various methods proposed in the literature differ from each other in the selection of the indicator and in the techniques used to determine the equilibrium. We have used a method based on phosphorus NMR in which ^{31}P chemical shifts of the phosphate buffer are observed[6]. We feel that this is a reliable method because the chemical shifts are a sensitive measure of this equilibrium and since the phosphate buffer ions are small, hydrophilic and anionic, they would remain in the aqueous pool away from the initial interface of the micelle.

With regard to the nature of water in the reverse micelles, attempts are still being made to improve its characterization[14,15]. Generally, this water is considered to be a composite of two different types, the "bound water", i.e. the water lining the interior wall of the AOT micelle and the remaining "free water"[12]. Still finer subdivisions of the water pool have also been proposed. For example, on the basis of their IR data, Gierasch et al[15] further sub-divide the bound water region. Up to a $w_0 = 4$, the water solvates the AOT ion-pair. Further increases in the water concentration, up to a $w_0 = 10$, probably give rise to a hydration shell around the now-separated ions of AOT and still further increase gives rise to the so-called "free water". Although such classifications are largely empirical, they provide a useful conceptual framework. One may then be able to better interpret the data from the studies of chemical or biochemical reactions.

THE LOCATION OF THE GUEST BIOPOLYMERS

As far as the location of the guest biopolymer in this aqueous medium is concerned, based on several experimental observations, we have proposed a "water shell model". Accordingly, the hydrophilic protein or nucleic acid is located in the middle of the water pool

and is separated from the charged micelle internal wall by a layer
of water[1,5,8]. If the guest molecules were located close to the
polyanionic interior surface of the micellar walls, they would experience a situation similar to that of an enzyme bound to a polyanionic support. In this case, the pH experienced by the enzyme is
lower than the pH of the buffer in equilibrium with it and its pH-
activity profile shifts to higher pH values as compared to the profile for the aqueous buffer solution[16]. The situation is simpler
than in the reverse micelles[12], especially since the bound enzyme
is in equilibrium with the bulk aqueous buffer solution, and is thus
more amenable to a simple theoretical treatment which yield the electrostatic potential, ψ, from the difference of pH optimum of the
free and the bound enzyme. Douzou[9] has considered the electrostatic
effect of the polyelectrolyte and other mobile charges inside the
AOT reverse micelle on the pK_a's of the proton function of the guest
molecules. For simplicity, he considers an average value of the
electrostatic potential within the micellar environment (called ψ)
and shows that its effect would be to shift the pK_a towards alkaline
pH values. Furthermore, using similar considerations, he shows that
the positively charged substrate molecules in the AOT reverse micelles would give rise to a lower K_m and vice versa. It must be pointed
out that the use of an average electrostatic potential for the micellar interior becomes questionable when one considers the above
mentioned difficulties in defining and measuring the pH of the water
pool. For example. there are differences of up to four pH units in
the pK_a of the indicator p-nitrophenol, depending on where it is
located in the micellar phase[12]. In conclusion, it appears that the
description of the interior of the reverse micelles in terms of
electrostatic potential is still a long way from being complete, but
it is certainly a lead to be followed.

STRUCTURAL EFFECTS

     What is the effect of the solubilizing water pool on the struc-
ture of the guest biopolymer? We have recently documented that the
solubilization of the aqueous solution of nucleic acids in reverse
micelles is attended by spectroscopic changes[17] which are diagnostic
of structural alterations. In particular, with low molecular weight
DNA there is a decrease of absorbance in the 260 nm region which is
attended by an increase in molar ellipticity. This effect is more
pronounced when the water concentration is low and has been as-
cribed[18] to an increase in the extent of base stacking, which in
turn may be connected to an increased extent of hydrogen bonding.
The same picture emerges for proteins[7,8]. The fact that the effect
is more pronounced for RNA than for low molecular weight DNA, which
is more rigid and more hydrogen bonded (due to the double helix),

also supports the view that the guest biopolymers undergo an increase in conformational rigidity. This can be attributed to the particular nature of the microenvironment, namely, the nature of the water in the reverse micelles. The fact that the spectroscopic conformational effects tend to vanish by increasing the water concentration (i.e. when the water in the reverse micelles becomes similar to bulk water) is another indication of the validity of such a view.

In the case of plasmids as well as the high molecule weight DNA (under particular micellar conditions) we have found evidence for the so-called ψ-spectrum. This is a CD-spectrum corresponding to a condensed, or highly packed form of DNA[19]. Such a form of DNA may be present in the head of certain phages, in chromatin and generally in those biological small compartments where a macromolecule is obliged, due to space restrictions, to be compressed in a supercoiled form. In this sense the plasmid or high molecular weight DNA contained in reverse micelles can be seen as suitable models for a head virus (see Figure 3).

Transfer RNA can also be readily solubilized in AOT/isooctane reverse micelles. We have chosen the well characterized phenylalanine transfer RNA from yeast, with 77 nucleotides and a molecular weight of ca. 22,500. Figure 4 shows the UV (a) and CD (b) spectra of tRNA-Phe solubilized in reverse micelles, as a function of the water content. Notice that in Figure 4a the absorbance is maximum for water solution, lower for reverse micelles with a high water content (w_O = 22.2) and lowest for the micelles with a low water content (w_O = 5.9). On the other hand (in Figure 4b), the CD spectra of the same solutions show the opposite trend: the ellipticity is the lowest for water, higher at w_O = 22.2 and the highest at w_O = 5.9.

We are now conducting studies of the codon – anticodon inter-actions in reverse micelles in order to assess whether binding in reverse micelles is stronger than in bulk water. If hydrogen bon-ding in the nucleic acids is enhanced in the reverse micelles, it is quite plausible that interactions, for example between poly U and tRNA-Phe, would also increase in this environment.

It will be highly desirable to eventually obtain approximate but reliable values of the physical properties of the aqueous core of the reverse micelles. In particular, knowledge of the dielectric constant, microviscosity, the structure of water and its acidity would greatly help in our understanding chemical reactivity.

ENZYMES IN REVERSE MICELLES

We have recently reviewed the reactivity of enzymes in reverse micelles[1] and it is not necessary to repeat the matter here. Only a couple of points should be mentioned here. It can be considered that it is now well ascertained that the maximal activity for enzymes in reverse micelles is not found (as one might have expected) at the largest possible water concentrations, but rather at w_o values well below 20[7,8,20,21,22]. This is so, despite significant conformational changes experienced by some enzymes under such conditions[7,8]. It is still too early to offer an explanation for this.

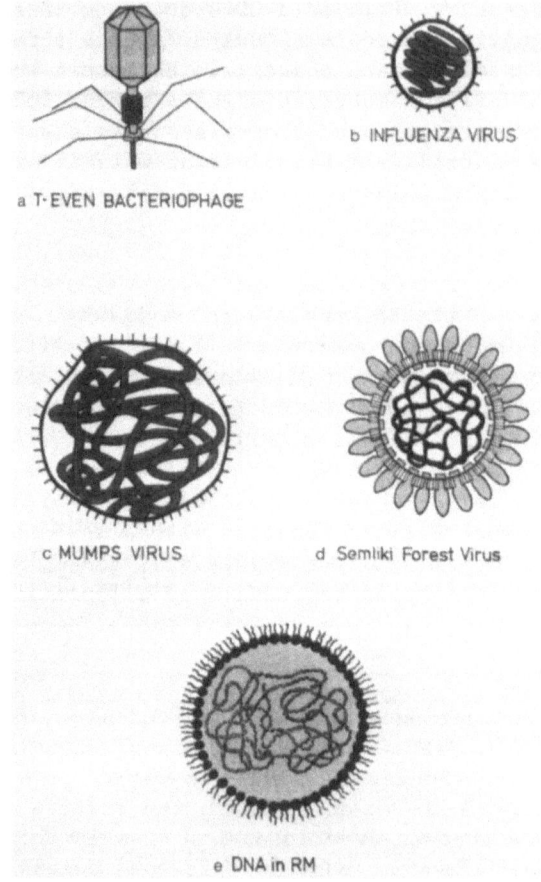

Figure 3. *A proposal for the structure of a DNA containing reverse micelle (e) and its comparison with known structures adopted from References 31 and 32 of bacteriophage T-Even and different viruses (a,b,c,d).*
Note: The scale used for (e) is approximately one fourth that for the rest.

The particular structure and properties of water in the reverse micelles may be one important factor; it is also possible that at large
w_0-values the thermodynamic stability of the enzyme-containing reverse micelles is decreased, with a consequent denaturation of the
enzyme. Investigations of phase-diagrams of reverse micellar systems
with and without enzymes are in progress in our group to address
this problem.

The rate constant k_{cat} determined for reverse micelles is generally close to that observed for aqueous solutions. However, a
slight increase (about 3 to 4 fold) in the k_{cat} values for some enzymes in the reverse micelles has been observed[7],[20],[23],[24]. A larger
increase (about 20 fold) has also recently been reported for peroxidase[25]. Some caution, however, is needed in the precise determination of this rate enhancement, as it is very easy to make misinter-
pretations. First of all, one should compare saturated rates and not
simply initial velocities; secondly, one should compare k_{cat} for
water and for reverse micelles at their respective pH optima, not
simply at one given nominal pH. In fact, as evidenced from Figure 2,

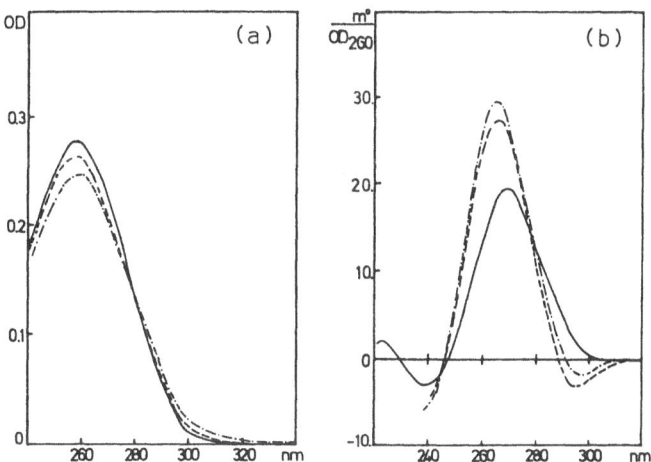

Figure 4. *Absorption (a) and circular dichroism properties (b)*
(1 cm light path) of tRNA-Phe (mol. weight 22,500 Daltons) in water
and in hydrocarbon micellar solutions. (——) tRNA-Phe in aqueous
solution borate buffer pH=9.0; (---) in isooctane/AOT solution at
w_O=22.2 (prepared from a borate buffer stock solution pH=9.0);
(-·-) same as (---) but w_o=5.9. These solutions were prepared by
the addition of equal volumes (8μl) of the stock solution using
the injection procedure[1] to either 3 ml of water or of hydrocarbon
micellar solution.

there is usually a shift of the pH-profile for the enzymatic reaction in reverse micelles which may bring about as an artifact a high increase of k_{cat} in a certain pH-range. The study by Menger and Yamada[23] gives a clear illustration of this aspect.

There is, of course, no reason to ignore the higher k_{cat} values obtained under carefully controlled conditions. In fact, one must look for the novel effects, however small, of these "microreactors" in the case of each enzyme tested. It is possible that a careful study of such effects may shed some light on the physical determinants of enzymatic rates and armed with such information, we may be able to further modulate such rate enhancements.

Concerning the enzymes reactivities and their kinetic parameters, we are still facing the question of the valid K_m in the micellar phase. As we pointed out in one of our earlier publications[5], for compounds which are only soluble in water, one can define two different sets of concentrations in the reverse micellar system, according to whether one views the "overall" volume of the system (hydrocarbon and water), or the aqueous pool alone. Since K_m has the dimensions of concentration, its magnitude varies, depending on whether the overall or water pool concentration is used. For example, with the case of α-chymotrypsin, K_m, expressed as $K_{m,overall}$ $[(K_m)_{ov}]$ is 0.5 mM while $K_{m,water\ pool}$ $[(K_m)_{wp}]$ is 50 mM[7], when the reverse micellar solution contains 1% water. K_m for aqueous solution $[(K_m)_{aq}]$ however, is 0.6 mM[7]. Since K_m is a good measure of the dissociation constant of the enzyme-substrate (ES) complex[27], $(K_m)_{ov}$ would indicate that the stability of the ES complex remains unaltered in the micelles while $(K_m)_{wp}$ would show that the complex is considerably (about 85 fold) destabilized as compared to the aqueous solution. The question is, which one of these K_m's represents the true situation.

In our earlier publications[1,7,21] we argued that if both enzyme and substrate are water soluble, $(K_m)_{wp}$ and not $(K_m)_{ov}$ would be the physically relevant quantity. This agrees with the conclusions of Fletcher et al[28]. Their calculations show that only the intramicellar effects, i.e. those arising from an alteration of the reaction medium within the micelle, would be revealed in $(K_m)_{wp}$ and not the intermicellar effect. However, their calculations presume a low occupancy of micelles by reagent molecules, not a valid assumption when the concentrations of substrate and micelles used in the enzyme reactions are similar, as is indicated in the beginning of this section. Curiously enough, the value of $(K_m)_{ov}$ determined for almost all enzymes we have studied so far, namely α-chymotrypsin[7], lysozyme[8], liver alcohol dehydrogenase[21] and ribonuclease[24], is closer to $(K_m)_{aq}$ and is generally more than an order of magnitude smaller than $(K_m)_{wp}$. Although this similarity in the magnitudes of $(K_m)_{aq}$ and $(K_m)_{ov}$ may be a coincidence, we still would like to keep an open mind about the matter, and we are looking for ways to directly evidence which K_m is the physically relevant one.

Finally, we would like to make a last point concerning the properties of the reverse micelles seen as reactors for enzymatic reactions. This point has never been elaborated, and it is very illustrative of the peculiar nature of enzyme-containing reverse micelles. This has to do with the observation that enzymes hosted in reverse micelles can accept and catalyze not only water soluble substrates (see α-chymotrypsin, lysozyme, ribonuclease, etc.) but also water insoluble (or sparingly water soluble) ones such as the steroids of Laane and coworkers[29], or linoleic acid in the case of lipooxygenase[30] or hexanal or steroids in the case of horse liver alcohol dehydrogenase[21,30]. This behavior is surprising. It probably can be explained on the basis of the peculiar properties of the solvent in the water pool. This solvent, although it is basically aqueous, has a lower dielectric constant[11] and has other features which make it compatible also with hydrophobic compounds. This is very important for biotechnological applications, as the expectation of using enzyme-containing reverse micelles for the catalytic transformation of water insoluble compounds relies on this property.

ACKNOWLEDGEMENTS

We would like to thank Dr. Vincenzo Rizzo for help in formulating the AOT purification procedure given in the Appendix and for the pH titrations of AOT, and to Dr. Arnaldo Dossena for the purification of AOT using HPLC.

APPENDIX

AOT Purification Procedure (A modified version of the procedure by Martin and Magid[3])

AOT (80 g, from SERVA) was dissolved in 800 ml benzene and 35 ml of water was added to it. After a few hours of standing, the supernatant was decanted and the flocculant was discarded. The procedure was repeated twice with the decanted benzene solution by adding first 25 ml and then 20 ml of water. The solvent was removed from the final solution with a rotary evaporator which was always used without heating the solution. The residue was redissolved in 250 ml methanol and mixed with activated charcoal (10 g) for about 6 hrs. It was then filtered through a Millipore filter (0.45 μ).

The solvent was reduced to about half the volume using a rotary
evaporator and water was added to make the volume up to 500 ml.
This solution was washed four times with 100 ml of isooctane each
time. The aqueous phase was concentrated with a rotary evaporator
(in order to remove methanol) and then freeze dried. The product
was redissolved in methanol and shaken for a few minutes with 5 g
of basic Alox (Woelme) and filtered. The solution was kept at -20°C.
After 48 hrs, the solution was filtered through a Millipore filter
(0.45 µ) and dried in a rotary evaporator and finally kept in vacuo
for 2 days. The yield was about 50%. About 0.5 mole of water per
mole of AOT remained in this AOT preparation as determined by NMR.
The product was stored in a dessicator over P_2O_5.

REFERENCES

1. P.L. Luisi and R. Wolf, in "Solution Behavior of Surfactants,
   Vol. 2, K.L. Mittal and E.J. Fendler, Eds., Plenum Publishing
   Corp., New York (1982).
2. P.D.I. Fletcher, N.F. Perrins, B.H. Robinson and C. Toprakcioglu,
   in these Proceedings.
3. C.A. Martin and L.J. Magid, J. Phys. Chem., 85 : 3938 (1981).
4. M. Wong, J.K. Thomas and M. Grätzel, J. Am. Chem. Soc., 98 : 2391
   (1976).
5. F.J. Bonner, R. Wold and P.L. Luisi, J. Solid-Phase Biochem.,
   5 : 255 (1980).
6. R.E. Smith and P.L. Luisi, Helv. Chimica Acta, 63 : 3202 (1980).
7. S. Barbaric and P.L. Luisi, J. Am. Chem. Soc., 103 : 4239 (1981).
8. C. Grandi, R.E. Smith and P.L. Luisi, J. Biol. Chem. 256 : 837
   (1981).
9. P. Douzou, Adv. in Enzymology, 51 : 1 (1980).
10. F. Nome, S.A. Chang and J.H. Fendler, J. Colloid Interface Sci.,
    56 : 146 (1976).
11. J.H. Fendler, "Membrane Mimetic Chemistry", Chapter 3, Wiley-
    Interscience, New York (1982).
12. O.A. El Seoud, in these Proceedings.
13. C. Balny and P. Douzou, Biochemie, 61 : 445 (1979).
14. H.-F. Eicke and P. Kvita, in these Proceedings.
15. L.M. Gierasch and K.F. Thompson, in these Proceedings.
16. L. Goldstein, Y. Levin and E. Katchalski, Biochemistry, 3 : 1913
    (1964).
17. V.E. Imre and P.L. Luisi, Biochem. Biophys. Res. Commun.,
    107 : 538 (1982).
18. J. Brahms, A.M. Michelson and K.E. Van Holde, J. Mol. Biol.,
    15 : 467 (1966).
19. L.S. Lerman, in "Physico Chemical Properties of Nucleic Acids",
    J. Duchesne, Editor, Vol. 3, 59-76, Academic Press, London,
    New York (1973).

20. B.H. Robinson and P.D.I. Fletcher, personal communication.
21. P. Meier and P.L. Luisi, J. Solid-Phase Biochem., 5 : 269 (1980).
22. K. Martinek, A.V. Levashov, N.L. Klyachko, V.I. Pantin and
    I.V. Berezin, Biochem. Biophys. Acta, 657 : 277 (1981).
23. F.M. Menger and K. Yamada, J. Am. Chem. Soc., 101 : 6731 (1979).
24. A. Pande, H. Gablinger and P.L. Luisi ,unpublished results.
25. K. Martinek, A.V. Levashov, Y.L. Khmelnitsky, N.L. Klyachko and
    I.V. Berezin, Science, 218 : 889 (1982).
26. D. Findlay, A.P. Mathias and B.R. Rabin, Biochem. J., 85 : 139
    (1962).
27. A. Fersht, "Enzyme Structure and Mechanism", p. 90, W.H. Freeman
    and Company, California (1977).
28. P.D.I. Fletcher, A.M. Howe, B.H. Robinson and D.C. Steytler, in
    these Proceedings.
29. R. Hillhorst, C. Laane and C. Veeger, manuscript in preparation.
30. P. Meier, Dissertation (ETH Nr. 7222) (1983).
31. K. Simons, H. Garoff and A. Helenius, Sci. Am., 246 (2) : 46
    (1982).
32. R.W. Horne, Sci. Am., 280 (1) : 48 (1963).

HARDENED REVERSE MICELLES AS DRUG DELIVERY SYSTEMS

P. Speiser

School of Pharmacy ETH
Clausiusstr. 25
CH-8092 Zürich, Switzerland

INTRODUCTION

Micelles and microemulsions are elegant systems for drug delivery through solubilization of lipophilic drugs and transdermal or
parenteral application of drugs. The disadvantage of such systems
is mainly the lack of shelf stability and poor resistance to
heat which is necessary for antimicrobial treatment. The main reason
for this instability is the loose and mobile liquid nature of the
micelle membrane and consequently the high degree of molecular motions, and the varying permeability for drug molecules with changing temperatures. One way to eliminate this instability could be to
harden the micelle itself or its membrane, so as to get a vesicle
with a solid amorphous wall.

The hardening of swollen micelles or microemulsions in optically isotropic aqueous phases has been known for quite a long
time[1]. In an aqueous micellar system the amphiphilic micelle con-
taines styrole as the monomer which is polymerized as the inner
phase. Successively, new monomers join the micelles from the outer
phase. During polymerization, the lipophilic microdroplets grow like
avalanches, forming finally a coarse, non-isotropic lipophilic polymer system in the inner phase.

HARDENED REVERSE MICELLES

Contrary to this aqueous emulsion-polymerization, the reverse
system is hardened differently. Apart from the various hardening
methods, such as polycondensation and polymerization with various

techniques like ionic polymerization etc.[2], usually radical polymerization is preferred and catalysts are used which initiate radical formation at the vinyl double bound with consequent chain
growth[3]. In fact, the radical polymerization occurs at the interface between the inner aqueous phase and the outer non-aqueous
apolar phase, so that the interface is strengthened.

This is demonstrated in Figure 1 with the example of a copolymerization between acrylamide in the inner polar and N,N'-methy-
lenebisacrylamide in the outer non-polar phase. The partition takes
place in or at the membrane by lateral diffusion and space filling.
Therefore, in this case, the radical copolymerization is done directly at the interface[4], which solidifies and forms a solid skin
or wall around each polar microphase.

In order to transfer the radical interface polymerization into
a reverse micellar system (Figure 2), the following operations have
to be performed:

- The incorporation of polar drugs in the aqueous core of micelles
  formed in an apolar outer phase is called primary solubilization.
  This w/o-system is obtained with the aid of classical surfactants.
  In other words, an aqueous drug system is enriched in a polar
  microphase[5]. The drug and water content is increased to a maximum,
  but no percolation should take place. The outer phase usually is
  an organic apolar solvent like hexane, heptane, etc.

- This isotropic reverse micellar system with a drug in the inner
  compartment is secondarily stabilized by tensioactive monomers
  (reactive surfactants) which are preferentially localized at the
  interface between micelle interior and micelle exterior (space
  filling). This secondary solubilization and orientation of tensioactive monomers in the interface is very important for a good
  product.

- In the third step, the hardening of the interface by polycondensation or polymerization techniques can take place by means of
  chemical or physical methods[6]. Usually the initiation of the net-
  work-forming or crosslinking material is achieved with catalysts,
  physiologically inert additives, by light, by ultraviolet light
  or X-rays, but preferably by γ-rays (in the case of methylmeta-
  crylate). In some cases (e.g. cyanoacrylates)[7] the presence of
  water in the inner micelle compartment is sufficient to initiate
  the crosslinking in the interface.

- In the fourth step (purification), the outer polar organic phase
  is replaced by water with the aid of precipitation, ultrafiltra-

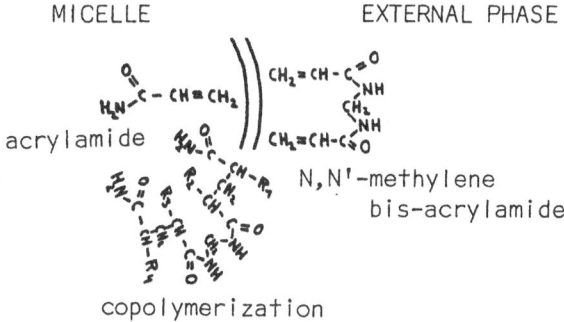

Figure 1. *Secondary solubilization and interface-copolymerization with the example of acryl-amides.*

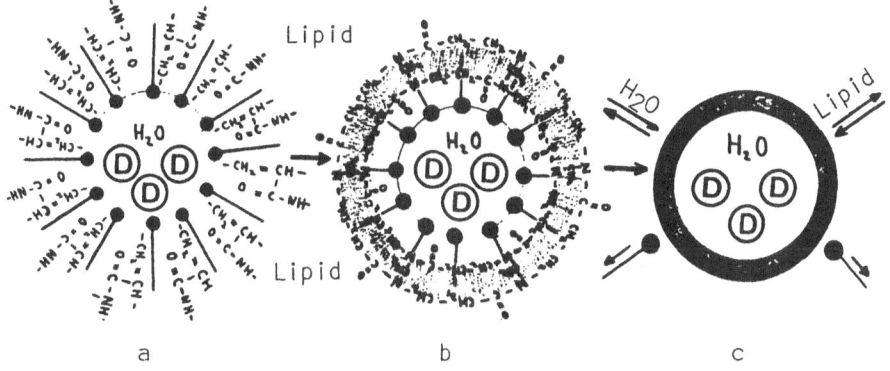

Figure 2. *Schematic representation of the procedure for hardening of reverse micelles; a) micelle formation primary solubilization, with the drug in the inner aqueous phase, the surfactant as the interface, and surrounded by the external organic lipid phase; b) hardening by polymerization and cross-linking of the interface surfactant; c) removal of the lipid external phase and replacement with water.*

tion, ultracentrifugation or ultradialysis methods. During this purification procedure the monomers (surfactants) which are generally toxic and other auxiliary materials are completely washed out.

– Finally, an ultrafine colloidal aqueous suspension system, water in water, is obtained which is suitable for pharmaceutical purposes. The drug encapsulated in the inner system, the former reverse micelle, serves as a reservoir for carrying the maintenance drug dose, and the outer phase can incorporate additional drug molecules for the initial dose of a drug delivery system[7].

- By this procedure a colloidal drug suspension system results from solidified reverse micelles. The drug in this micellar system can be built inside of the solidified micelle, or pallisade-like, in the solidified micelle wall, or peripherally, at the wall surface by adsorption processes[5].

PROPERTIES OF THE HARDENED MICELLE

As chemicals to be used as a reservoir or capsule wall material to carry the drug, the following reactive monomers can be used: styrols, acrylates, acrylamides, methylmetacrylates, and alkylcyanoacrylates. Many of these corresponding polymers have been used for a long time in surgery as suture-materials, tissue adhesives, bone cement, etc. Actually the use of alkylcyanoacrylate is preferred, because no additives and no energy are needed to start the polymerization[8].

PHYSICAL PROPERTIES

The usual particle size of hardened swollen micelles or microphases lies between 30 and 350 nanometers (nm). The particle size distribution of most samples shows a narrow distribution in the range of 200 to 270 nm[7].

The shape is spherical, the specific surface is around $50m^2$ per gram, the surface (texture) is smooth, amorphous and lipophilic. The wall thickness of these so-called drug-reservoir nanocapsules is between 15 and 60 nm, containing pores between 3 and 6 nm in diameter[10].

The scanning electron microscope pictures at a magnification of 4000 always show spherical hardened micelles and agglomerates. Until now no shapes other than spherical micelles have been found.

BIOLOGICAL PROPERTIES

Distribution. Whole body autoradiographies with ^{14}C-radioactively labeled solidified micelles of PMMA (Polymethylmethacrylate carriers) show the following distribution in rats: Thirty minutes after intravenous application, 22% of the micelles are accumulated in the lung and 60% in the liver. Six hours later a complete change takes place and only 14% remain in the lung, 68% in the liver and the rest is distributed in the spleen, enternal epithelium and RHS-system. After 7 days most of these hardened micelles have reached the reticuloendothelial system (RES), especially the vertebral column and bone marrow[11].

Intramuscular application produces quite a different effect: the hardened micelles remain for over 70 days at the site of application, probably due to adsorption. These hardened micelles are therefore suitable as carriers for very long delayed drug delivery.

Elimination. The residence time of such ultrafine drug reservoirs or drug carriers in the body can be measured by following the excretion rate. Around 5.5% to 10% of the hardened micelles are excreted per week in the urine, feces and respiration air (rat)[11]. In general, the half-life of such a hardened micellar system depends on the particle size, molecular weight of the polymers and the animal.

Toxicity. The acute and subacute toxicity of polycyanoacrylate is extremely low for mice and the LD 50 lies between 200 and 230 mg per kg mouse. That means the toxicity is lower than that of the polymerization media for the tensioactive material (334 mg/kilo)[12].

Metabolism. Many polyacrylates are well metabolized. So if polymethylmetacrylate is used as micelle wall material, it will be metabolized partly by hydrolysis and corrosion, but mostly by oxidation. The oxidative pathway can go in two directions:

- β-oxidation, coupled with coenzyme A, which leads directly to succinyl coenzyme A[13,14];

- a simultaneous α- and β-oxidation which leads to pyruvate and to oxal acetate[15].

These metabolic products belong to the citric acid cycle and the hardening materials can therefore be regarded as physiological. The fact that no other metabolic products have been found in the urine until now speaks for this hypothesis.

PHARMACEUTICAL USE OF HARDENED REVERSE MICELLE SYSTEMS

Such colloidal drug carrier systems are currently being studied for the following purposes:

- Analytical optical measurements of pH, pO_2 and pCO_2, in body tissues;

- sustained drug delivery systems;

- immunological preparations to improve antigen-antibody response (adjuvant);

– transcellular and intracellular drug transport systems[7].

Sustained Release. Hardened micelles containing dihydroergota-
mine-tartrate (DHET) in vitro give sustained drug release, and the
rate can be regulated over a wide range. The half-life drug release
can be varied between 7.5 and 48 hours. Such a drug delivery system
can be used for oral sustained release or depot formulations[16].

Analytical Spectroscopic Measurements on Body Tissues. The incorporation of fluorescent indicators such as pyrene-butyric acid
β-methyl umbelliferon or acridine orange can be used to analyze degradation processes in the organism, especially with the measurement of oxygen, CO_2, acidity and alkalinity in plasma, tissues, and
other organs, such as the brain. Calibration values for pH and pO_2
show that it is possible to get reliable and reproducible results
in the body[17].

Immunological Assays. The antigen-antibody response can be improved by incorporating antigens into hardened micelles. Due to
this inclusion, the protection of the antigen from degradation is
high and the antibody formation is increased.

Various mechanisms of such an antibody increase can be proposed, for example:

– the protection of the antigen from a rapid metabolism or degradation;

– a slow antigen release with continuous antibody stimulation of
  immuno-competent cells;

– the stimulation of phagocytosis;

– a general stimulation of the RES-system;

– the stimulation of the lymphocyte proliferation, which facili-
  tates the penetration and the dissemination in the immunosystem,
  lymph nodes and spleen[10].

The adjuvant effect of immuno-gamma-globulin G, (in vesicles
of hardened micelles) is an example of improved antibody formation.
After subcutaneous application of micellized immuno-gamma-globulin G,
the immunization in guineag pigs is much better than with the normally employed adjuvants (such as the adsorption immuno-gamma-globulin G on aluminum oxide or the untreated immuno-gamma-globulin G)[4].

Intracellular Medication. The use of hardened micelles for the
intra- or transcellular transport of drugs to specific cells in organs is a recent approach used to specifically target cells with

drugs. For this purpose the drug in the hardened micelle as a carrier is transported through the cell wall with the aid of endocytosis processes[8]. This target is shown with an antitumor drug, Actinomycin D. With the control (no drug) or with the free Actinomycin D, there is no reduction in tumor growth after intravenous injection of the drug, but with the drug incorporated in hardened micelles with polycyanoacrylate there is a significant decrease of tumor growth in mice[12].

Transintestinal transport of micellized drugs is also possible. In some cases drugs in hardened micelles are transported through the intestinal epithelium by so-called micropinocytoses or cytopempsis, a migration of solid particles from the intestines through the intestinal epithelium to the capillary blood or lymphatic system. By this cytopempsis it is possible to transport drugs in the undissolved ultrafine state to the body systemically. Polymethylmetacrylate vesicles with a molecular weight around 35,000 are able to cross the intestinal barrier and to exert systemic activity. Between 10% and 15% is absorbed in the intestine of rats. This enteral uptake does not take place with relatively hydrphilic polyalkycyanoacrylate vesicles with a low molecular weight of around 5,000. For this endocytosis, a certain lipophilicity, partition, wettability and a relatively high molecular weight must be maintained. Otherwise, no enteral passage takes place[16].

FINAL REMARKS

Hardened reverse micelles, loaded with drugs and suspended as an ultrafine aqueous system can be applied parenterally as well as orally. Some examples have been given to show the practical applicability of such systems in medicine, biology and pharmacology. In spite of the favorable results, it cannot yet be decided, if hardened micellar systems can be used as a drug delivery system on a wide scale, because there are many problems with side-reactions, chronic toxicity, teratogenicity, specific drug targeting etc. which still must be solved.

REFERENCES

1. H. Fikentschen, H. Gerrens and H. Schuller, Angew. Chem. 72 : 856 (1960).
2. H.R. Allcock, Angew. Chem. 89 : 153 (1977).
3. T.M.S. Chang, Science 146 : 524 (1964); Canad. J. Physiol. Pharmacol. 44 : 115 (1966); 47 : 1043 (1969).
4. G. Birrenbach and P. Speiser, J. Pharm. Sci. 65 : 1763 (1976).
5. A.E. Alexander and P. Johnson, Colloid Science, Vol. II, Clarendon Press Oxford, p. 686 (1949).

6. P. Speiser, Progr. Colloid Polymer Sci. 59 : 48 (1976).
   P. Speiser and G. Birrenbach, U.S. Patent 4 021 364 (1977).
7. J.T. Dingle, P.J. Jacques and I.H. Shaw, Lysosomes in Appl.
   Biology and Therapeutics 6, North Holland Publ. Comp., chapter
   23, page 653 (1979).
8. P. Couvreur, P. Tulkens, M. Roland, A. Trouet and P. Speiser,
   FEBS Lett. 84 : 323 (1977).
9. H. Kopf, Thesis ETH-Zürich (1975); H. Kopf, R.K. Joshi,
   M. Soliva and P. Speiser, Pharm. Ind. 38 : 281 (1976); 39 : 993
   (1977).
10. G. Birrenbach, Thesis ETH-Zürich (1973).
11. J. Kreuter, R. Mauler, H. Gruschkau and P. Speiser, Exp. Cell.
    Biol. 44 : 12 (1976); J. Kreuter and P. Speiser, Infect. Immun.
    13 : 204 (1976); J. Pharm. Sci. 65 : 1624 (1976).
12. P. Couvreur et al. in press (1982).
13. W.W. Nowinsky, Fundamental Aspects of Normal and Malignant
    Growth, Elsevier, Amsterdam (1958).
14. D.E. Nicholson, Metabolic Pathways, Koch-Light Lab. Colnbrook,
    Bucks, England (1968).
15. M. Pantůček, Talanta 14 : 643 (1967).
16. Proceedings of the Meran Symposium 1982, Schriftenreihe
    Bundesapothekerkammer, Gelbe Reihe, Band X (1982).
17. D.W. Lübbers, N. Opitz, P. Speiser and H. Bisson, Z. Naturforsch.
    32c : 133 (1977).

SUMMARIZING POINTS

Per Stenius

Institute for Surface Chemistry
Box 5607
114 86 Stockholm, Sweden

It appears to be generally accepted that when talking about
reversed systems we are concerned with thermodynamically stable
systems and hence knowledge about phase equilibria is of great importance. This lays down a solid basis for comparative studies,
but one is confused by the fact that different research groups work
in different systems to an extent which makes comparison of results
quite difficult. An exception appears to be Aerosol OT which has
been used as a model system by many people. However, for reasons
given below, I think that more work should be done with other
systems. At the recent meeting in Lund a number of people from
France, Germany, Switzerland, Sweden und the UK decided to investigate the system Sodium/Octyl/Benzene Sulfonate/water/butanol/hexa-
decane by different methods; we hope that this effort will result
in better possibilities to compare results from different laboratories.

Aggregate size and structure can be determined with reasonable
reliability at relatively low concentrations of water. At higher
concentrations, however, interpretations of the results from different scattering and spectroscopic methods are not unambiguous.
Generally, we still need to conduct an open-minded discussion about
the relative reliability and sensibility of different methods and
how comparable the results obtained from different laboratories
really are. In particular, it would be important to have a clear
picture of the time scales in which different experimental methods
probe the systems. Another case in point is the question of micellar size and interactions in non-ionic systems close to the cloud
point or phase inversion temperature.

Only after a discussion about the experimental method can we begin to propose more definite models of the structure of concentrated reversed systems. The present impasse is the reason why we do not have as yet a good definition of a microemulsion, or of the other types of aggregates in oils, and so far, we have to resort to more or less operational descriptions.

Another very important point is the role of the polar solvent in reversed systems. There are strong indications suggesting that one condition for the formation of highly concentrated systems is the occurrence of a highly flexible or mobile interface between oil and water domains. This flexibility will be determined by the size and amount of the surfactant relative to that of the co-surfactant and by the interaction of these with the solvent. The incorporation of molecules from the solvent into the interfacial structures may be of importance; but the influence of the solvent on the chemical potential of the amphiphiles must also be equally important. The understanding of this point is crucial for the technical utilization of microemulsions. This is why one is perhaps oversimplifying the problem by sticking too strongly to Aerosol OT.

Two additional points should be mentioned.

a)   We need to pay more attention to the use of surfactants based on natural products in microemulsions (in particular lipids).

b)   Closely related to this, we need to pay more attention to the purity of the surfactants we use in microemulsions.

There is no limit to technological or pharmaceutical possibilities of reversed micelles. The results already obtained in the attempts to incorporate enzymes and proteins and the photochemical reactions in microemulsions are exciting. This meeting certainly very successfully has stressed the importance of reversed systems in this respect. The biological relevance appears to be less obvious. and we clearly still have to give definite proof for the existence of reversed structures in biological membranes. Otherwise, what we have seen of the potential application certainly makes research on reversed systems, whether fundamental or applied, very exciting indeed.

AUTHOR INDEX

SUBJECT INDEX